Studies in Computational Intelligence, Volume 77

Editor-in-chief
Prof. Janusz Kacprzyk
Systems Research Institute
Polish Academy of Sciences
ul. Newelska 6
01-447 Warsaw
Poland
E-mail: kacprzyk@ibspan.waw.pl

Further volumes of this series
can be found on our homepage:
springer.com

Vol. 57. Nadia Nedjah, Ajith Abraham and Luiza de Macedo Mourella (Eds.)
Computational Intelligence in Information Assurance and Security, 2007
ISBN 978-3-540-71077-6

Vol. 58. Jeng-Shyang Pan, Hsiang-Cheh Huang, Lakhmi C. Jain and Wai-Chi Fang (Eds.)
Intelligent Multimedia Data Hiding, 2007
ISBN 978-3-540-71168-1

Vol. 59. Andrzej P. Wierzbicki and Yoshiteru Nakamori (Eds.)
Creative Environments, 2007
ISBN 978-3-540-71466-8

Vol. 60. Vladimir G. Ivancevic and Tijana T. Ivancevic
Computational Mind: A Complex Dynamics Perspective, 2007
ISBN 978-3-540-71465-1

Vol. 61. Jacques Teller, John R. Lee and Catherine Roussey (Eds.)
Ontologies for Urban Development, 2007
ISBN 978-3-540-71975-5

Vol. 62. Lakhmi C. Jain, Raymond A. Tedman and Debra K. Tedman (Eds.)
Evolution of Teaching and Learning Paradigms in Intelligent Environment, 2007
ISBN 978-3-540-71973-1

Vol. 63. Wlodzislaw Duch and Jacek Mańdziuk (Eds.)
Challenges for Computational Intelligence, 2007
ISBN 978-3-540-71983-0

Vol. 64. Lorenzo Magnani and Ping Li (Eds.)
Model-Based Reasoning in Science, Technology, and Medicine, 2007
ISBN 978-3-540-71985-4

Vol. 65. S. Vaidya, L.C. Jain and H. Yoshida (Eds.)
Advanced Computational Intelligence Paradigms in Healthcare-2, 2007
ISBN 978-3-540-72374-5

Vol. 66. Lakhmi C. Jain, Vasile Palade and Dipti Srinivasan (Eds.)
Advances in Evolutionary Computing for System Design, 2007
ISBN 978-3-540-72376-9

Vol. 67. Vassilis G. Kaburlasos and Gerhard X. Ritter (Eds.)
Computational Intelligence Based on Lattice Theory, 2007
ISBN 978-3-540-72686-9

Vol. 68. Cipriano Galindo, Juan-Antonio Fernández-Madrigal and Javier Gonzalez
A Multi-Hierarchical Symbolic Model of the Environment for Improving Mobile Robot Operation, 2007
ISBN 978-3-540-72688-3

Vol. 69. Falko Dressler and Iacopo Carreras (Eds.)
Advances in Biologically Inspired Information Systems: Models, Methods, and Tools, 2007
ISBN 978-3-540-72692-0

Vol. 70. Javaan Singh Chahl, Lakhmi C. Jain, Akiko Mizutani and Mika Sato-Ilic (Eds.)
Innovations in Intelligent Machines-1, 2007
ISBN 978-3-540-72695-1

Vol. 71. Norio Baba, Lakhmi C. Jain and Hisashi Handa (Eds.)
Advanced Intelligent Paradigms in Computer Games, 2007
ISBN 978-3-540-72704-0

Vol. 72. Raymond S.T. Lee and Vincenzo Loia (Eds.)
Computation Intelligence for Agent-based Systems, 2007
ISBN 978-3-540-73175-7

Vol. 73. Petra Perner (Ed.)
Case-Based Reasoning on Images and Signals, 2008
ISBN 978-3-540-73178-8

Vol. 74. Robert Schaefer
Foundation of Global Genetic Optimization, 2007
ISBN 978-3-540-73191-7

Vol. 75. Crina Grosan, Ajith Abraham and Hisao Ishibuchi (Eds.)
Hybrid Evolutionary Algorithms, 2007
ISBN 978-3-540-73296-9

Vol. 76. Subhas Chandra Mukhopadhyay and Gourab Sen Gupta (Eds.)
Autonomous Robots and Agents, 2007
ISBN 978-3-540-73423-9

Vol. 77. Barbara Hammer and Pascal Hitzler (Eds.)
Perspectives of Neural-Symbolic Integration, 2007
ISBN 978-3-540-73953-1

Barbara Hammer and Pascal Hitzler (Eds.)

Perspectives of Neural-Symbolic Integration

Printing: Krips bv, Meppel
Binding: Stürtz, Würzburg

Barbara Hammer
Pascal Hitzler
(Eds.)

Perspectives of Neural-Symbolic Integration

With 81 Figures and 26 Tables

 Springer

Prof. Dr. Barbara Hammer
Institute of Computer Science
Clausthal University of Technology
Julius Albert Straβe 4
38678 Clausthal-Zellerfeld
Germany
E-mail: hammer@in.tu-clausthal.de

PD Dr. Pascal Hitzler
Institute AIFB
University of Karlsruhe
76128 Karlsruhe
Germany
E-mail: hitzler@aifb.uni-karlsruhe.de

Library of Congress Control Number: 2007932795

ISSN print edition: 1860-949X
ISSN electronic edition: 1860-9503
ISBN 978-3-540-73953-1 Springer Berlin Heidelberg New York

This work is subject to copyright. All rights are reserved, whether the whole or part of the material is concerned, specifically the rights of translation, reprinting, reuse of illustrations, recitation, broadcasting, reproduction on microfilm or in any other way, and storage in data banks. Duplication of this publication or parts thereof is permitted only under the provisions of the German Copyright Law of September 9, 1965, in its current version, and permission for use must always be obtained from Springer-Verlag. Violations are liable to prosecution under the German Copyright Law.

Springer is a part of Springer Science+Business Media
springer.com
© Springer-Verlag Berlin Heidelberg 2007

The use of general descriptive names, registered names, trademarks, etc. in this publication does not imply, even in the absence of a specific statement, that such names are exempt from the relevant protective laws and regulations and therefore free for general use.

Cover design: deblik, Berlin
Typesetting by the SPi using a Springer LaTeX macro package
Printed on acid-free paper   SPIN: 11744269   89/SPi   5 4 3 2 1 0

... für Anne, Manfred und Michel ...

Contents

Preface IX

Part I Structured Data and Neural Networks

Introduction: Structured Data and Neural Networks 3

1 Kernels for Strings and Graphs
Craig Saunders and Anthony Demco 7

**2 Comparing Sequence Classification Algorithms
for Protein Subcellular Localization**
Fabrizio Costa, Sauro Menchetti, and Paolo Frasconi 23

**3 Mining Structure-Activity Relations in Biological Neural
Networks using *NeuronRank***
Tayfun Gürel, Luc De Raedt, and Stefan Rotter 49

**4 Adaptive Contextual Processing of Structured Data
by Recursive Neural Networks: A Survey of Computational
Properties**
Barbara Hammer, Alessio Micheli, and Alessandro Sperduti 67

**5 Markovian Bias of Neural-based Architectures
With Feedback Connections**
Peter Tiňo, Barbara Hammer, and Mikael Bodén 95

**6 Time Series Prediction with the Self-Organizing Map:
A Review**
Guilherme A. Barreto 135

7 A Dual Interaction Perspective for Robot Cognition: Grasping as a "Rosetta Stone"
Helge Ritter, Robert Haschke, and Jochen J. Steil 159

Part II Logic and Neural Networks

Introduction: Logic and Neural Networks 181

8 SHRUTI: A Neurally Motivated Architecture for Rapid, Scalable Inference
Lokendra Shastri 183

9 The Core Method: Connectionist Model Generation for First-Order Logic Programs
Sebastian Bader, Pascal Hitzler, Steffen Hölldobler, and Andreas Witzel 205

10 Learning Models of Predicate Logical Theories with Neural Networks Based on Topos Theory
Helmar Gust, Kai-Uwe Kühnberger, and Peter Geibel 233

11 Advances in Neural-Symbolic Learning Systems: Modal and Temporal Reasoning
Artur S. d'Avila Garcez 265

12 Connectionist Representation of Multi-Valued Logic Programs
Ekaterina Komendantskaya, Máire Lane and Anthony Karel Seda 283

Index 315

Preface

The human brain possesses the remarkable capability of understanding, interpreting, and producing human language, thereby relying mostly on the left hemisphere. The ability to acquire language is innate as can be seen from disorders such as specific language impairment (SLI), which manifests itself in a missing sense for grammaticality. Language exhibits strong compositionality and structure. Hence biological neural networks are naturally connected to processing and generation of high-level symbolic structures.

Unlike their biological counterparts, artificial neural networks and logic do not form such a close liason. Symbolic inference mechanisms and statistical machine learning constitute two major and very different paradigms in artificial intelligence which both have their strengths and weaknesses: Statistical methods offer flexible and highly effective tools which are ideally suited for possibly corrupted or noisy data, high uncertainty and missing information as occur in everyday life such as sensor streams in robotics, measurements in medicine such as EEG and EKG, financial and market indices, *etc*. The models, however, are often reduced to black box mechanisms which complicate the integration of prior high level knowledge or human inspection, and they lack the ability to cope with a rich structure of objects, classes, and relations. Symbolic mechanisms, on the other hand, are perfectly applicative for intuitive human-machine interaction, the integration of complex prior knowledge, and well founded recursive inference. Their capability of dealing with uncertainty and noise and their efficiency when addressing corrupted large scale real-world data sets, however, is limited. Thus, the inherent strengths and weaknesses of these two methods ideally complement each other.

Neuro-symbolic integration centers at the border of these two paradigms and tries to combine the strengths of the two directions while getting rid of their weaknesses eventually aiming at artificial systems which could be competitive to human capacities of data processing and inference. Different degrees of neuro-symbolic integration exist: (1) Researchers incorporate aspects of symbolic structures into statistical learners or they enrich structural reasoning by statistical aspects to extend the applicability of the respective paradigm. As an example, logical inference mechanisms can be enlarged by statistical reasoning mainly relying on Bayesian statistics. The resulting systems are able to solve complex real-world problems, such as impressively demonstrated in recent advances of statistical-relational learning. (2) Researchers try to exactly map inference mechanisms of one paradigm towards the other such that a direct relation can be established and the paradigm which is ideally suited for the task at hand can be chosen without any limitations on the setting. Recent results on the integration of logic programs into neural networks constitute a very interesting example of this 'core method'.

This book focuses on extensions of neural methodology towards symbolic integration. According to the possible degree of integration, it is split into two parts: 'loose' coupling of neural paradigms and symbolic mechanisms by means of extensions of neural networks to deal with complex structures, and 'strong' coupling of neural and logical paradigms by means of establishing direct equivalences of neural network models and symbolic mechanisms. More detailed introductions to the chapters contained in these two parts are given later, on pages 3 and 181, respectively.

A selection of the most prominent researchers in the area has contributed to this volume. Most of the chapters contain overview articles on important scientific contributions by the authors to the field, and combined they deliver a state-of-the-art overview of the main aspects of neuro-symbolic integration. As such, the book is suitable as a textbook for advanced courses and students, as well as an introduction to the field for the interested researcher.

We thank all contributors, not only for their superb chapters, but also because the production of this book was a very smooth process so that it was a pleasure to work together on its completion. We thank the editor-in-chief of this book series, Janusz Kacprzyk, for suggesting to us to edit this volume, and we thank Thomas Ditzinger from Springer for a very constructive cooperation. Finally, we thank our families for bearing with our ever-increasing workload.

Barbara Hammer & Pascal Hitzler
Clausthal & Karlsruhe
June 2007

List of Contributors

Sebastian Bader
International Center for Computational Logic
Technische Universität Dresden
01062 Dresden, Germany
Sebastian.Bader@inf.tu-dresden.de

Guilherme A. Barreto
Department of Teleinformatics Engineering
Federal University of Ceará
A v. Mister Hull, S/N - C.P. 6005, CEP 60455-760, Center of Technology
Campus of Pici, Fortaleza, Ceará, Brazil
guilherme@deti.ufc.br

Mikael Bodén
School of Information Technology and Electrical Engineering
University of Queensland, Australia
mikael@itee.uq.edu.au

Fabrizio Costa
Machine Learning and Neural Networks Group
Dipartimento di Sistemi e Informatica
Università degli Studi di Firenze
Via Santa Marta 3
50139 Firenze, Italy
costa@dsi.unifi.it

Artur S. d'Avila Garcez
Department of Computing
School of Informatics
City University London
London EC1V 0HB, UK
aag@soi.city.ac.uk

Anthony Demco
ISIS Group, School of Electronics and Computer Science
University of Southampton
Southampton, SO17 1BJ, UK
aad04r@ecs.soton.ac.uk

Luc De Raedt
Departement Computerwetenschappen
Katholieke Universiteit Leuven
Celestijnenlaan 200 A
3001 Heverlee, Belgium
Luc.DeRaedt@cs.kuleuven.be

Paolo Frasconi
Machine Learning and Neural Networks Group
Dipartimento di Sistemi e Informatica
Università degli Studi di Firenze
Via Santa Marta 3, 50139 Firenze, Italy
p-f@dsi.unifi.it

Peter Geibel
Institute of Cognitive Science
University of Osnabrück
Albrechtstraße 28
D-49076 Osnabrück, Germany
pgeibel@uos.de

Tayfun Gürel
Bernstein Center for Computational
Neuroscience and Institute of
Computer Science
Albert Ludwigs Universität Freiburg
Hansastraße 9a
79104 Freiburg, Germany
guerel@informatik.uni-freiburg.de

Helmar Gust
Institute of Cognitive Science
University of Osnabrück
Albrechtstraße 28
D-49076 Osnabrück, Germany
hgust@uos.de

Barbara Hammer
Institute of Informatics
Universität Clausthal
Julius-Albert-Straße 4
38678 Clausthal-Zellerfeld, Germany
hammer@in.tu-clausthal.de

Robert Haschke
Neuroinformatics Group
Bielefeld University
P.O.-Box 100131
33501 Bielefeld, Germany
rhaschke@techfak.uni-bielefeld.de

Pascal Hitzler
Institute AIFB
University of Karlsruhe
76128 Karlsruhe, Germany
pascal@pascal-hitzler.de

Steffen Hölldobler
International Center for Computational Logic
Technische Universität Dresden
01062 Dresden, Germany
sh@iccl.tu-dresden.de

Ekaterina Komendantskaya
Department of Mathematics
National University of Ireland,
University College Cork
Cork, Ireland
komendantskaya@gmail.com

Kai-Uwe Kühnberger
Institute of Cognitive Science
University of Osnabrück
Albrechtstraße 28
D-49076 Osnabrück, Germany
kkuehnbe@uos.de

Máire Lane
Department of Mathematics
National University of Ireland,
University College Cork
Cork, Ireland
maireln@bcri.ucc.ie

Sauro Menchetti
Machine Learning and Neural
Networks Group
Dipartimento di Sistemi e Informatica
Università degli Studi di Firenze
Via Santa Marta 3
50139 Firenze, Italy
menco@inwind.it

Alessio Micheli
Dipartimento di Informatica,
Università di Pisa
Largo Bruno Pontevecoro 3
56127 Pisa, Italy
micheli@di.unipi.it

Helge Ritter
Neuroinformatics Group
Bielefeld University
P.O.-Box 100131
33501 Bielefeld, Germany
helge@techfak.uni-bielefeld.de

Stefan Rotter
Bernstein Center for Computational
Neuroscience and Institute for
Frontier Areas of Psychology and
Mental Health
Wilhelmstraße 3a
79098 Freiburg, Germany
stefan.rotter@biologie.uni-freiburg.de

Craig Saunders
ISIS Group, School of Electronics
and Computer Science
University of Southampton
Southampton, SO17 1BJ, UK
cjs@ecs.soton.ac.uk

Anthony Karel Seda
Department of Mathematics
National University of Ireland,
University College Cork
Cork, Ireland
a.seda@ucc.ie

Lokendra Shastri
International Computer Science
Institute
1947 Center Street, Suite 600
Berkeley, CA 94704, USA
shastri@icsi.berkeley.edu

Alessandro Sperduti
Dipartimento di Matematica Pura es
Applicata
Università di Padova
Via Trieste 63
35121 Padova, Italy
sperduti@math.unipd.it

Jochen J. Steil
Neuroinformatics Group
Bielefeld University
P.O.-Box 100131
33501 Bielefeld, Germany
jsteil@techfak.uni-bielefeld.de

Peter Tiño
School of Computer Science
University of Birmingham
Edgbaston, Birmingham B15 2TT,
UK
p.tino@cs.bham.ac.uk

Andreas Witzel
Institute for Logic, Language and
Computation
Universiteit van Amsterdam
1018 TV Amsterdam, The Netherlands
awitzel@illc.uva.nl

Part I

Structured Data and Neural Networks

Introduction: Structured Data and Neural Networks

The first part of this book addresses approaches which extend standard neural technics towards structural information and which combine statistical methods with structures to tackle complex application scenarios e.g. from bioinformatics or robotics. In the 90s, researchers still claimed that there exist principled limitations of neural methods to represent structures [1]. This criticism was met by a variety of arguments [2] and counterexamples [3, 4]. Today, large scale applications of neural methods in structural domains exist including problems of natural language processing and text analysis, several problems in bioinformatics such as remote homology detection, protein structure prediction, or phylogenetic inference, and chemical challenges such as quantitative structure-activity-relationship prediction. The methodologies can mainly be distinguished into two approaches: one can consider complex structures as a whole, providing an interface which maps structures to entities neural methods can deal with such as distances or kernel values. Alternatively, one can integrate the structure processing into the neural method and recursively integrate basic structural constituents using neural methods within the context provided by the structural relationships. Two prime examples of these ways of structure processing are kernels for structures [5] on the one side and recursive neural networks [6] on the other side. These two approaches constitute state-of-the-art technics which extend neural methods to structured inputs for automatic regression or classification problems. Interestingly, the approaches also open the way towards unsupervised learning scenarios and structured output data, as will be discussed in this book.

In this first part of the book, the focus will first lie on neural structure processing by means of the design of kernels or structural features which can directly be used as an interface between structural information and statistical machine learners. The article 'Kernels for Strings and Graphs' reviews recent developments on kernels for structures with a particular emphasis on graph kernels. The article 'Comparing Sequence Classification Algorithms for Protein Subcellular Localization' presents large scale experiments to compare structure kernels in combination with SVMs and alternative machine learning techniques such as k-nearest neighbor methods with distances specifically tailored to the data at hand. The area of application stems from an important problem in bioinformatics, the localization of a specific functional region of proteins. A different approach is taken in the contribution 'Mining Structure-Activity Relations in Biological Neural Networks using *NeuronRank*'. The task is a prediction of average firing rates of biological neural networks based on machine learning methods. Thereby, structural features are extracted from the networks using techniques which propagate activation through the network similar to the popular PageRank algorithm. These three contributions all center around methods which extract distance values or features from structural data which can then be plugged into a machine learning technique.

The next three contributions take a different perspective and discuss methods which tackle structures by a recursive processing of their basic constituents within the context posed by the structural relations. In all cases, the neural networks dynamics mirrors the input structure as closely as possible. The contribution 'Adaptive Contextual Processing of Structured Data by Recursive Neural Networks: A Survey of Structured Data of Computational Properties' considers extensions of cascade correlation networks to structural prediction and transduction. Thereby, the approximation capabilities of these methods with respect to recursive structures including sequences, trees, and acyclic graphs (together with a mild structural condition) are investigated in depth. Both articles 'Markovian Bias of Neural-based Architectures with Feedback Connections' and 'Time Series Prediction with the Self-Organizing Map: A Review' focus on simple structures, time series, which can be processed by well-establishes recurrent neural networks. Thereby, the latter article presents a state-of-the-art comparison and overview of methodologies to tackle sequences in an unsupervised framework. This issue constitutes a recent and nontrivial area of research with high practical applicability e.g. for time series and sequence mining and visualization. The former article investigates implicit symbolic restrictions posed on recurrent systems by the specific type of training and initialization. It establishes a rigorous equivalence of symbolic Markov models and specific supervised as well as unsupervised recurrent neural architectures.

The final contribution in this first part entitled 'A Dual Interaction Perspective for Robot Cognition: Grasping as a "Rosetta Stone"' deals with an application scenario at the cutting edge of research again, autonomous grasping in the field of humanoid robotics. In this context, low-level signals (realized by means of classical subsymbolic feedback-control) and high level coordination (captured e.g. by symbolic augmented finite state automata) have to be integrated to arrive at reliable autonomous behavior. Besides the integration of information by means of machine learning methods, this article also takes a glimpse at reliable software engineering of these constituents.

Altogether, these contributions constitute a representative albeit subjective overview of recent technologies and applications in neural methods and machine learning for structured data.

References

1. J. Fodor and Z. Pylyshyn. Connectionism and cognitive architecture: A critical analysis. *Cognition*, 28 : 3–71, 1988.
2. T. Van Gelder. Compositionality: a connectionist variation on a classical theme. *Cognitive Science*, 14 : 355–384, 1990.
3. J. Elman. Finding structure in time. *Cognitive Science*, 14 : 179–211, 1990.
4. B. Hammer. Compositionality in Neural Systems. In: M. Arbib (ed.), *Handbook of Brain Theory and Neural Networks*, 2nd edition, pp. 244–248, 2002.
5. T. Gärtner. A Survey of Kernels for Structured Data. *SIGKDD explorations*, 5(1) : 49–58, 2003.
6. P. Frasconi, M. Gori, and A. Sperduti. A general framework for adaptive processing of data structures. *IEEE Transactions on Neural Networks*, 9(5) : 768–786, 1998.

1

Kernels for Strings and Graphs

Craig Saunders and Anthony Demco

ISIS Group, School of Electronics and Computer Science, University of Southampton, Southampton, SO17 1BJ, UK {cjs,aad04r}@ecs.soton.ac.uk

Summary. Over recent years, kernel methods have become a powerful and popular collection of tools that machine learning researchers can add to the box of tricks at their disposal. One desirable aspect of these methods is the 'plug and play' nature of the kernel: if one can construct a kernel function over the data they wish to analyse, then it is straightforward to employ Support Vector Machines, Kernel PCA, Dual Ridge Regression, or any other of the myriad of techniques available.

In order for these techniques to work on symbolic or discrete data, it is therefore sufficient to define a valid kernel for the type of data in question. The kernel approach therefore gives one perspective on learning with symbolic data that is very flexible and widely applicable. Researchers have shown how such kernels can be defined from different perspectives, and this has lead to the successful creation of a number of structure kernels, principally those that operate on strings, trees and graphs. In this chapter we shall provide a brief review of kernels for structures and concentrate on the domains of strings and graphs. The former of which was one of the first structures to be tackled and is still widely used for biologically-related applications; the latter is somewhat newer and here we present the state-of-the-art and some novel additions to the graph kernel family.

1.1 Kernel functions

We shall not go into details regarding any specific kernel-based algorithm here, however for readers not so familiar with such methods we provide a brief description of kernel functions themselves and start by giving a trivial example of a kernel over a discrete structure.

A kernel function in this setting is simply a function which corresponds to an inner product between two data points, often after some non-linear function has been applied. That is one can define a kernel function k between two data points $\mathbf{x}$ and $\mathbf{z}$ as:

$$\kappa(\mathbf{x}, \mathbf{z}) = \phi(\mathbf{x}) \cdot \phi(\mathbf{z}) \tag{1.1}$$

where $\mathbf{x}, \mathbf{z} \in \mathcal{X}$ and $\phi : \mathcal{X} \mapsto \mathcal{F}$. The space $\mathcal{X}$ is referred to as the input space and $\mathcal{F}$ is called the feature space. Traditionally the input space is the

d-dimensional vector space $\mathcal{X} = \mathbb{R}^d$. For structure based kernels however, the input space is usually the set of all possible structures in the domain (e.g. the set of all finite-length sequences constructed from some alphabet $\mathcal{A}$).

Kernel methods share a common trait in that they only depend on inner products between data points; it is never necessary to represent or compute a vector in feature space (e.g. $\phi(\mathbf{x})$) explicitly. This is the power behind the 'kernel trick' – many algorithms can be converted to their dual version where one learns parameter weights for each training example, rather than for each feature dimension. This allows the use of highly non-linear kernels to be used efficiently. For example, one popular kernel is the Gaussian kernel

$$\kappa(\mathbf{x}, \mathbf{z}) = \exp^{\frac{||\mathbf{x}-\mathbf{z}||^2}{2\sigma^2}}$$

for which vectors in the feature space can be shown to be infinite-dimensional. For a full discussion of kernels and an introduction of the techniques available the reader is referred to [1, 2, 3]. Often, the kernel is normalised in the following way

$$\kappa_N(\mathbf{x}, \mathbf{z}) = \frac{\kappa(\mathbf{x}, \mathbf{z})}{\sqrt{\kappa(\mathbf{x}, \mathbf{x})\kappa(\mathbf{z}, \mathbf{z})}}$$

Obviously this is not necessary for the gaussian kernel, but for other kernels it is sometimes beneficial to the algorithm being used to ensure that the data lie on a unit hypersphere. For many structure kernels normalisation aids to take into account that structures are not necessarily the same size.

1.2 Kernels for Strings

Strings formed of sequences from some alphabet was the first domain to be tackled by structure kernels. Haussler [4] introduced a general framework for convolution kernels, which allows kernels for structures to be built by successive comparison of smaller sub-parts of a structure. String kernels were highlighted (c.f. section 4.4 in [4]) as a special case. String kernels for biological sequences were independently proposed by Watkins [5] which was based on pair hidden Markov models, and used a dynamic programming technique to compute the kernel.

1.2.1 A trivial symbolic kernel

Before discussing structure kernels in detail, let us give the simplest definition of a kernel on sequences as an example: the 1-gram kernel. Assume our data points $\mathbf{x_i}$ are sequences of symbols a_j from some alphabet $\mathcal{A}$. In order to construct a valid kernel we can simply count the number of times each symbol occurs in each sequence, i.e. $\phi_{a_j}(\mathbf{x_i}) = \#\{\mathbf{a_j} \in \mathbf{x_i}\}$[1]. The dimension of the

---

[1] with a slight abuse of notation we use $\#\{a_j \in \mathbf{x_i}\}$ to indicate the number of times the symbol a_j occurs in the sequence $\mathbf{x_i}$

feature space in this case is therefore $|\mathcal{A}|$; i.e. the number of symbols in the alphabet. The kernel between two data points can then be computed easily by using (1.1) above. Although it is a trivial example, this indicates how symbolic information can be used with kernels. Indeed extending this idea to count not just the occurance of single symbols, but to measure the frequency of bi-grams, tri-grams and so on leads to the n-gram kernel. Considering contiguous sequences of length n already starts to incorporate information regarding the ordering of symbols in the input sequences (which is often ignored in many feature representation, e.g. when using TF-IDF features for text processing). Indeed n-gram kernels (also referred to as spectrum kernels) have been shown to achieve good performance in text categorisation [6] and protein homology tasks [7]. For larger choices of n efficient computation is not straightforward, however many efficient approaches have been developed, including a recent approach using suffix arrays [8] which counts all matching subsequences up to length n, with linear computation time and very small memory cost.

1.2.2 Non-contiguous strings

A more interesting use of kernels over strings would be one that allows for insertions and deletions in the sequence: i.e. one that compares non-contiguous sequences. This is exactly what the kernel suggested by Watkins computes, and it has an obvious biological motivation. The key idea behind the gap-weighted subsequences kernel is to compare strings by means of the subsequences they contain – the more subsequences in common, the more similar they are – rather than only considering contiguous n-grams, the degree of contiguity of the subsequence in the input string s determines how much it will contribute to the comparison.

In order to deal with non-contiguous substrings, it is necessary to introduce a decay factor $\lambda \in (0, 1)$ that can be used to weight the presence of a certain feature in a string. For an index sequence $\mathbf{i} = (i_1, \ldots, i_k)$ identifying the occurrence of a subsequence $u = s(\mathbf{i})$ in a string s, we use $l(\mathbf{i}) = i_k - i_1 + 1$ to denote the length of the string in s. In the 'gappy kernel', we weight the occurrence of u with the exponentially decaying weight $\lambda^{l(\mathbf{i})}$.

Definition 1 (Gap-weighted subsequences kernel). *The feature space associated with the gap-weighted subsequences kernel of length p is indexed by $I = \Sigma^p$ (i.e. subsequences of length p from some alphabet Σ), with the embedding given by*

$$\phi_u^p(s) = \sum_{\mathbf{i}:\, u=s(\mathbf{i})} \lambda^{l(\mathbf{i})}, u \in \Sigma^p. \tag{1.2}$$

The associated kernel is defined as

$$\kappa_p(s,t) = \langle \phi^p(s), \phi^p(t) \rangle = \sum_{u \in \Sigma^p} \phi_u^p(s)\phi_u^p(t). \tag{1.3}$$

Example 1. Consider the simple strings "cat", "car", "bat", and "bar". Fixing $p = 2$, the words are mapped as follows:

ϕ	ca	ct	at	ba	bt	cr	ar	br
cat	λ^2	λ^3	λ^2	0	0	0	0	0
car	λ^2	0	0	0	0	λ^3	λ^2	0
bat	0	0	λ^2	λ^2	λ^3	0	0	0
bar	0	0	0	λ^2	0	0	λ^2	λ^3

(1.4)

So the unnormalised kernel between "cat" and "car" is $k(\text{"cat"},\text{"car"}) = \lambda^4$, while the normalised version is obtained using

$$\kappa(\text{"cat"},\text{"cat"}) = \kappa(\text{"car"},\text{"car"}) = 2\lambda^4 + \lambda^6 \tag{1.5}$$

as $\hat{\kappa}(\text{"cat"},\text{"car"}) = \lambda^4/(2\lambda^4 + \lambda^6) = (2 + \lambda^2)^{-1}$.

A dynamic programming approach to efficiently compute the kernel was given in [5], however this process has a complexity of $O(k|s||t|)$ where $|s|$ and $|t|$ are the lengths of the two strings being compared. This can be time consuming for larger strings, therefore many approximation methods have been suggested. One method is to compute the string kernel between a string and a set of short 'reference' strings of length k [9] to create an explicit vector representation which can then be used with any learning algorithm. Leslie and Kuang introduce a fast approximation to the string kernel [10] which is based on the observation that sequences with many gaps do not contribute greatly to the kernel sum. By restricting the kernel to only consider at most g gaps the computational complexity is reduced to $O(c(g,k)(|s| + |t|))$ where $c(g,k)$ is a constant that depends on g and k.

1.2.3 Other kernels for strings

Many other kernels for strings have been developed, primarily for application in biologically-related tasks. The feature space for all string kernels is indexed by substrings taken from the alphabet. These may be all of the same length (as in n-gram kernels), of different lengths (e.g. all possible substrings up to length n) with the features themselves being straightforward occurance counts (as with n-grams) or a combination of weighted scores (c.f. the non-contiguous kernel). Two very successful kernels that have been applied to protein homology tasks and other related tasks are the mismatch kernel [11] and the profile kernel [12]. Mismatch kernels result in the same feature vector as n-gram kernels, however counts for each feature are incremented for both exact matches, and matches where at most m mismatches occur. Instead of employing a naive counting system, profile kernels take as input a profile of an input sequence, which in the context of the application is a probability distribution over the likelihood of a particular amino acid occurring at each position in the sequence. Both of these kernels are computable in linear time, and therefore are very attractive computationally. One further advantage they have is that the resulting kernel matrix is not so diagonally dominant. For structure kernels it is often the case that the degree to which two non-identical structures match is very low, which results in a kernel matrix that is very diagonally dominant. This is known not to be a good situation for learning. Therefore by allowing mismatches (either explicitly or by considering probability profiles), this effect is reduced and often results in a better classifier.

1.2.4 General frameworks for structure kernels

In this chapter we are presenting kernels which are specifically tailored for certain structures, however there exists several mechanisms for designing kernels over a wider range of domains. Lack of space prevents us from discussing these systems in detail, however this section seems an appropriate time to give the reader a very brief overview of the main systems and pointers to further literature.

As mentioned above, Haussler presented the general framework of convolution kernels, of which string kernels are a special case (c.f. section 4.3 in [4]). Convolution kernels essentially compare two structured objects by iteratively comparing their "parts". Haussler demonstrates several properties of convolution kernels, shows how many different types of structure can be considered and gives different methods for generating such kernels (e.g. from pair HMMs, random fields, generalized regular expressions, etc.).

More recently, rational kernels have been proposed [13] which proposed a general framework for building kernels from weighted state transducers. Many existing kernels can easily be placed within the rational kernel framework (convolution kernels included), so this method may provide an avenue for defining structured kernels for other domains.

It is also possible to construct kernels from general generative models (such as HMMs). There are two main approaches to this. The first approach, known as P-kernels [1] uses the probabilistic parameters of the model (for example emission and transition probabilities of an HMM) directly as features. An alternative approach is to take the partial log-derivative of each parameter setting to obtain a Fisher-score vector and calculate a so-called Fisher kernel [14]. In this case the feature representation can be viewed as how much the general model has to be adapted in order to accommodate each particular training example. It is possible to show a strong relationship between Fisher kernels on Markov chains and the n-gram/string kernels presented earlier [15].

1.3 Kernels for Trees and Graphs

Trees and graphs are other domains where structured kernels can be applied. A graph is a very general structural form which describes objects (or nodes) and relations between these objects called edges. Trees are a restricted class of graphs which do not contain cycles and have one node labeled as the root node. Furthermore, a tree may only contain one distinct path between any two nodes. These conditions make computing the tree kernel much simpler than graph kernels. A complete graph kernel would count all sub-graphs although this has been shown to be NP-hard [16] and so one must take another approach to build the feature vector, such as counting all walks in a graph. In this section we will introduce kernel methods for trees in 1.3.1 and graphs in section 1.3.2.

1.3.1 Tree Kernels

Tree structures are a very natural form for many problems. Trees can be used to express many types of data, for example parse trees of sentences [17]. The data only need to have some hierarchical structure. A tree structure is quite straightforward, it contains one *root* node with *edges* (or branches) that connects from a parent node to a child node. A *leaf* node is a node with no children. A *child* node is lower in the tree than it's parent: it is a descendant of its parent. Each tree induced by a node and it's children is called a *subtree*. It is clear that trees fit into the convolution kernel approach: the tree is a composition of subtrees or 'parts'. As a result, a kernel can be defined over a tree by calculating the intersection (or dot product) of the subtrees.

Collins et al. introduced kernel methods for tree structures in [17]. Let us define the feature representation of a tree, ϕ, as a composition of elements $\phi_i(T)$ indicating presence of subtree i in tree T. Then, $\phi(T_1)$ and $\phi(T_2)$ define the feature vector for all possible subtrees of tree T_1 and T_2 respectively. Therefore, a tree kernel is $\kappa(T_1, T_2) = \phi(T_1) \cdot \phi(T_2)$. In practice, one could define ϕ as a vector indicating all possible fragments from the set of trees given in the problem, although one does not need to restrict ϕ to this set as a general recursive formulation is computable via dynamic programming. Let the set of nodes of tree T_i be denoted by $N_i = \{n_1, n_2, \ldots, n_{|N|}\}$, then it is possible to define a tree kernel in the following way:

Definition 2 (Tree Kernel). *Let us define an indicator function $I_i(n_j)$ that returns 1 if sub-tree i is present at node n_j. The feature vector, ϕ, that can be stated as $\phi(T_i) = \sum_{n_1 \in N_1} \sum_i I_i(n_1)$, results in the following tree kernel:*

$$\kappa(T_1, T_2) = \sum_{n_1 \in N_1} \sum_{n_2 \in N_2} \sum_i I_i(n_1) I_i(n_2) = \sum_{n_1 \in N_1} \sum_{n_2 \in N_2} C(n_1, n_2) \quad (1.6)$$

The internal sum, given by $C(n_1, n_2)$, allows for a recursive definition. If n_1 and n_2 are different then $C(n_1, n_2)$ returns 0, while if they are leaf nodes and they are the same it returns 1. If they are not leaf nodes then:

$$C(n_1, n_2) = \prod_{j=1}^{nc(n_1)} 1 + C(ch(n_1, j), ch(n_2, j))$$

where $nc(n_1)$ is the children of n_1 and the i-th child is $ch(n_1, i)$. The kernel is calculated by building a table of size N_1 by N_2 and filling this table according to the recursive definitions and then summing. This has a quadratic time complexity of $O(|N_1||N_2|)$.

In order to calculate this kernel efficiently, Vishwanathan et al. [18] proposed using a suffix tree. Conceptually, their method works in a similar fashion to dot-products of sparse vectors. As most of the elements of a feature vector, ϕ, are zero, one can sort ϕ so that all non-zero entries are at the beginning. For strings, this can be done in practice by building a *suffix tree* which stores all prefixes of a string in a compact tree representation. To compute a tree kernel with this method, one must first convert a tree into its string representation. Here, brackets are used to denote its structure. For example, the string representation of a simple tree is [A [B C]]. The brackets denote the hierarchy: node A is the root and B and C are it's children. To obtain the same kernel as (1.6), one can use this string representation, and only count substrings that have matched brackets. The kernel $\kappa(T_1, T_2)$ is computable in linear time $O(|T_1| + |T_2|)$, where $|T_i|$ denotes the length of the string representation of tree i. This is the case as the matching statistics and suffix tree construction are computable in linear time; see [18] for more details.

1.3.2 Graph Kernels

Graphs are a very general data structure, such that strings and trees can be seen as special cases of graphs. They are a useful way to model many types of data. Research that involves learning from graphs can be categorized in two forms. Either each example is a node in a graph (such as a handwritten digit in a graph representing distances to other digits [19]) or each example is a graph. Here, we will describe the latter. One application that has a natural fit for this type of learning is molecular data. Here, a chemical compound is represented by its molecular graph (see figure 1.1). The vertices of the graph are atoms that are connected by edges labeled by their bond type (ex. single bond, double bond, etc).

A graph G is described with the following notation. Let $\mathcal{V}_i$, $\mathcal{E}_i$ be the set of vertices and edges respectively and let $|\mathcal{V}_i|$, $|\mathcal{E}_i|$ denote their size for graph G_i. The vertex set $\mathcal{V}$ is composed of elements $\{v_1, v_2, \ldots, v_{|\mathcal{V}|}\}$, while an edge from vertex v_i to v_j is described using the notation $(v_i, v_j) \in \mathcal{E}$.

We will be considering *labeled directed graphs* (see figure 1.1), where a set of labels, l, exist along with a function, label, that assigns labels to each vertex and edge. Graphs without labels are a special case where the same label is assigned to each vertex/edge. We will consider graphs which contain *directed* edges so an edge $(v_i, v_j) \not\Rightarrow (v_j, v_i)$. Molecular graphs are undirected so they must first be converted to directed graphs by adding two directed edges for every undirected edge.

The following definitions are crucial as they are used as features when comparing graphs. A *walk* is a sequence of vertices, $v_1, v_2, \ldots, v_{n+1}$, where

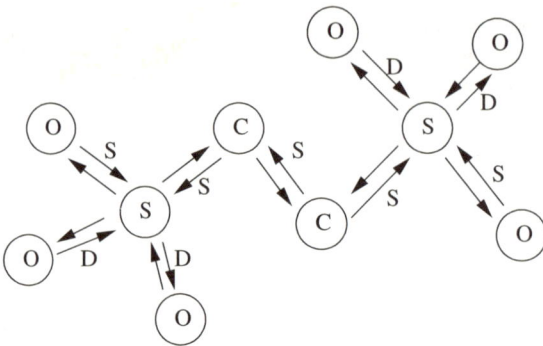

Fig. 1.1. This figure shows the molecular graph of a compound. It is represented as a labeled directed graph. Vertex labels list the atom type of C, O and S, being carbon, oxygen and sulphur respectively. The edge labels are listed as S and D for single and double bond.

$v_i \in \mathcal{V}$ and $(v_i, v_{i+1}) \in \mathcal{E}$. Here, the length of the walk is the number of edges, given by n in this case. A *path* is a walk with no repeated vertices, i.e. $v_i \neq v_j \Leftrightarrow i \neq j$. A *cycle* is a path with $(v_1, v_{n+1}) \in \mathcal{E}$.

Computational Issues

The cost of computing a kernel over a graph is much more prohibitive than trees. While in trees, when one randomly walks from a vertex, all paths will have a finite length. A graph may contain cycles and so one may traverse a path infinitely; therefore convergence conditions need to be found.

Gärtner analyzed the computational issues involved with computing a *complete graph* kernel in [16]. A complete graph kernel is one that is able to determine if two graphs are isomorphic. Consider a feature vector $\phi(G)$ that has a feature for every possible sub-graph (this would be analogous to example of the tree kernel presented earlier, where we considered all possible sub-trees). An inner product between these features can be used to identify whether a graph has a *hamiltonian path* [16]. A hamiltonian path is a path that visits each vertex in a graph exactly once. This problem is NP-hard and therefore a graph kernel using sub-graphs as features is not practical to compute. As a result, one must take a different approach that uses alternate features.

Approaches to Matching Graphs

In order to build a graph kernel by following the convolution kernel approach, one must enumerate all 'sub parts' of two graphs and return the cardinality of the intersection of these two sets. As described in the previous section, using the sub-graphs is not the best choice as it is too slow, therefore one must use other features to define the 'sub parts'. Several graph kernels have been developed using different features including walks, label distances, trees, and cycles. We will begin our discussion by describing a trivial symbolic graph kernel. Later, we will introduce several classes of graph kernels that are grouped based on their use of features. Finally, we will discuss an approach that does not use the classic convolution kernel approach, although produces a valid kernel.

A trivial symbolic graph kernel

A paper by Kashima [20] described several kernels including a trivial symbolic graph kernel that used counts of vertex labels as features. We define a feature vector of a graph $\phi(G)$ to contain features $\phi_i(G) = \sum_{v_j \in V} \texttt{label}(v_j) = l_i$, that indicate the number of times each unique label is found in graph G. The kernel $\kappa(G_1, G_2) = (\phi(G_1), \phi(G_2))$ counts the number of labels in common between two graphs although it does not incorporate any structural information.

1.3.3 Walk-based approaches to graph kernels

One of the most well-known approaches to computing a graph kernel uses random walks as features. This class of graph kernel has been termed (contiguous) label sequence graph kernel, although here we prefer to use the term (random) walk graph kernel. Two approaches exist for calculating this kernel: one that uses a marginalized kernel and one that uses a product graph approach. We will describe each of these and then show how they are equivalent. Finally, we will consider two popular extensions. One that includes non-matching walks as features, and another that removes tottering paths.

Counting random walks

The first approach we will consider uses walks as features and was introduced by Kashima [21]. This kernel counts all random walks up to infinite length by using a marginalized kernel approach. Essentially, several probabilities define a walk, determining the starting vertex, the transitions and the length. The starting probability p_s is defined with equal probability for all vertices, allowing the random walk to start from any vertex. A walk moves to another vertex with transition probability p_t which is defined with equal probability for all adjacent vertices. Finally, the length is determined by a stopping probability, p_q. The probability of a walk $\boldsymbol{w}$ of length l given a graph is then:

$$p(\boldsymbol{w}|G) = p_s(w_1) \prod_{i=2}^{l} p_t(w_i|w_{i-1}) p_q(w_l) \tag{1.7}$$

We must also define a kernel κ_z between two walks, which is the joint probability in the marginalized kernel. We use the notation $\boldsymbol{w}, \boldsymbol{w}'$ for walks from G_1, G_2 respectively.

$$\kappa_z(\boldsymbol{w}, \boldsymbol{w}') = \begin{cases} 0 & (l \neq l') \\ K_v(w_1, w_1') \prod_{i=2}^{l} K_e((w_{i-1}, w_i), (w'_{i-1}, w'_i)) K_v(w_i, w'_i) & (l = l') \end{cases} \quad (1.8)$$

where $K_e(w_i, w'_i)$ is any valid kernel between vertex labels (this could be as simple as an indicator function checking if vertex labels match), and similarly $K_e((w_i, w_{i+1}), (w'_i, w'_{i+1}))$ is any valid kernel defined over edge labels. We can now define the marginalized kernel and use (1.7) and (1.8) to solve.

$$\kappa(G_1, G_2) = \sum_{l=0}^{\infty} \sum_{\boldsymbol{w}_1} \sum_{\boldsymbol{w}_2} \kappa_z(\boldsymbol{w}_1, \boldsymbol{w}_2) p(\boldsymbol{w}_1|G_1) p(\boldsymbol{w}_2|G_2) \quad (1.9)$$

Since the path lengths considered span from 1 to ∞, one must find an equilibrium state. In order to this it is shown in [21] that terms can be rearranged into a recursive formula. An equilibrium is found by using a discrete-time linear system, that evolves with l, and is proven to converge.

Using product graphs

A second kernel [16] also counts walks up to infinite length. This kernel uses a product graph and finds convergence properties of the adjacency matrix for the product graph to calculate all walks. In order to derive this kernel, one must first define a product graph. The product graph created from $G_1 \times G_2$ is constructed with the following (see figure 1.2):

$\mathcal{V}_\times(G_1 \times G_2) = \{(v_1, v_2) \in \mathcal{V}_1 \times \mathcal{V}_2 : (\text{label}(v_1) = \text{label}(v_2))\}$
$\mathcal{E}_\times(G_1 \times G_2) = \{((u_1, u_2), (v_1, v_2)) \in \mathcal{V}^2(G_1 \times G_2) :$
$\quad (u_1, v_1) \in \mathcal{E}_1 \wedge (u_2, v_2) \in \mathcal{E}_2 \wedge (label(u_1, v_1) = label(u_2, v_2))\}$

We also define the adjacency matrix of the product graph as $E_\times$. The i, j-th element of $E_\times$ has an entry of 1 if vertex v_i is connected vertex v_j and is zero otherwise. One of the key properties used to compute this kernel is the following. If the matrix is taken to power n, E^n, each i, jth element now lists the number of walks of length n starting at vertex v_i and ending at v_j. The kernel can now be defined, using a sequence of weights λ ($\lambda_0, \lambda_1, \ldots$), resulting in the following kernel:

$$\kappa_\times(G_1, G_2) = \sum_{i,j=1}^{|\mathcal{V}_\times|} \left[\sum_{n=0}^{\infty} \lambda_n E_X^n \right]_{ij} \quad (1.10)$$

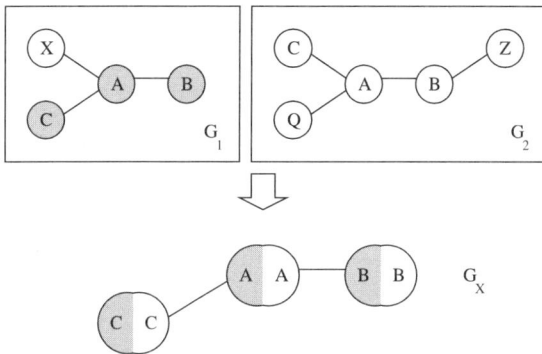

Fig. 1.2. Construction of a Product Graph - A product graph, $G_\times$, is constructed from two graphs, G_1 and G_2. Each vertex in G_1 is matched with vertices in G_2. Edges are added between product vertices if they exist in the original graphs. The shading indicates where each part of the product graph originated.

where $E_\times$ is constructed from $G_1 \times G_2$. The limit of this kernel can be computed by inverting the matrix $(\boldsymbol{I} - \gamma E_\times)^{-1}$, which takes $O(\mathcal{V}_\times^3)$. In order for the limit to exist, the geometric series implied by the inversion must converge and one must ensure to set λ appropriately such that $\lambda < 1/\min(\texttt{degree}(G_1), \texttt{degree}(G_2))$.

It is interesting to note that both methods presented in [21] and [16] are equivalent. Both methods calculate the number of walks up to infinite length by down-weighting longer paths. The only difference is the down-weighting scheme. A nice comparison between both methods is given in [22].

Now we will describe two popular extensions to the walk kernel: counting non-contiguous walks and non-tottering walks. The first extension we will consider is non-contiguous walks. This has some parallels with string kernels. This was introduced by Gärtner in [16] and is calculated using powers of product graph adjacency matrices as described previously. To define this kernel we now introduce an alternate product graph, G_o, that includes vertices with non-matching labels. G_o is constructed using:

$$\mathcal{V}_o(G_1 \times G_2) = \{(v_1, v_2) \in \mathcal{V}_1 \times \mathcal{V}_2\}$$
$$\mathcal{E}_o(G_1 \times G_2) = \{((u_1, u_2), (v_1, v_2)) \in \mathcal{V}^2(G_1 \times G_2) :$$
$$(u_1, v_1) \in \mathcal{E}_1 \wedge (u_2, v_2) \in \mathcal{E}_2 \wedge (label(u_1, v_1) = label(u_2, v_2))\}$$

To define this kernel, we must also define a parameter α ($0 \leq \alpha \leq 1$) that is used to penalize non-matching soft matches. α allows the kernel to choose the amount of relevance taken from the contiguous sequences $E_\times$ and non-contiguous E_o. The non-contiguous label sequence kernel is defined as:

$$\kappa_o(G_1, G_2) = \sum_{i,j=1}^{|\nu_\times|} \left[\sum_{n=0}^{\infty} \lambda_n ((1-\alpha)E_\times + \alpha E_o)^n \right]_{ij} \quad (1.11)$$

We note that the kernel only counts *anchored* walks, which begin and end with matching product vertices. It is calculated using the same method as the contiguous walk kernel, using an exponential or geometric convergence. With a geometric convergence the kernel is calculated as: $(I - \gamma((1-\alpha)E_\times + \alpha E_o))^{-1}$ $E_\times$ and E_o can be added as corresponding columns and rows from $E_\times$ are filled with zeros so that they are the same dimension.

The second extension was presented by Mahé in [23] and is used to remove tottering paths. A tottering path is one of the form $w = w_1, \ldots, w_n$ with $w_i = w_{i+2}$. Intuitively, it seems that a path that contains totters and passes one vertex after just visiting it will contain noisy or less useful information. It is calculated by augmenting the graphs with extra edges and nodes such that totters can no longer occur when the standard walk kernels are used.

1.3.4 Other approaches

Label Distances

A second class of graph kernels is created by using distances between labels as features, instead of matching walks. Two different methods for this have been suggested ([16],[24]) however the former is perhaps simpler and appears to be computationally faster, so this is the method we will present.

The kernel is based on *label pairs* or distances between vertices. In order to calculate this kernel, we must define the label matrix. The label matrix L is a a $V_\times \times l$ matrix, where l is the number of unique vertices.

The following property is crucial to calculating the distances between labels. By multiplying the adjacency matrix E by the label matrix L, the resulting matrix $[LE^n L^T]_{ij}$, has components that correspond to the number of walks of length n between vertices labeled ℓ_i and vertices labeled ℓ_j. The kernel is defined as:

$$\kappa(G_1, G_2) = \left\langle L \left(\sum_{i=0}^{\infty} \lambda_i E^i \right) L^T, L^2 \left(\sum_{j=0}^{\infty} \lambda_j (E^2)^j \right) (L^2)^T \right\rangle \quad (1.12)$$

The label matrix, L is order $l \times n$, where l is the number of unique vertex labels, and n is the number of vertices. The adjacency matrix, E is order $n \times n$, so the resulting feature, $[LE^n L^T]_{ij}$, is a matrix order $l \times l$. This kernel is very fast to compute (it is easy to compute explicit feature vectors beforehand) and also takes into account some "latent" information about the graph structure as walks of equal length are considered. However, the feature representation is not as rich as the walk-based kernels mentioned in the previous section, therefore it may not be as suitable for some tasks.

Trees

Another class of kernels uses tree patterns as features for graph kernels (this is distinct from the tree kernel for trees that we considered earlier). This was first presented in [25]. A tree pattern is one that is constructed by selecting a vertex in a graph as the root node and then recursively adding leaves to a node based on the structure of the graph. A tree kernel is defined using the following. First a function, ψ_t is defined that counts the number of times a tree-pattern of depth t occurs.

Definition 3 (Tree (graph) Kernel). *Let the kernel function that counts tree patterns between two graphs G_1 and G_2 be defined by:*

$$\kappa(G_1, G_2) = \sum_{t \in \tau} \lambda(t) \psi_t(G_1) \psi_t(G_2) \quad (1.13)$$

where τ is a set of trees of depths from 1 to ∞.

In practice, τ can be restricted to a finite depth size. As a graph is undirected, and furthermore almost all molecular graphs contain cycles, the depth of h will be infinite. The λ function sets trees of greater depth to have infinitely smaller value by setting $\lambda(t) = \alpha^t$ where α is between 0 and 1.

Cycles

A final class of kernels is one that combines both simple cycles and trees as features. It was first presented in [26]. This kernel used a combination of counting simple cycles and tree-based patterns as features for matching graphs. An algorithm is capable of calculating all the bi-connected components in linear time. A *simple cycle* is a cycle that contains no repeated vertices. It is computed by examining each bi-connected component, if it only contains one edge, then it is a bridge. A forest of trees is given by the bridges of the graph. If the bi-connected component contains more than one edge then it is a cycle. The kernel is the number of tree patterns and cycle patterns between two graphs.

Optimal assignment kernel

An alternative approach that uses a new type of kernel is the optimal assignment kernel. It does not use the convolution kernel approach, as is common with all the above, instead graphs are matched using a bipartite graph matching algorithm that is shown to result in a positive semi-definite matrix. A recent paper by Frohlich [27] introduced a new approach to define a kernel between two graphs. By defining two graphs as a bipartite graph, where the vertices of one graph are listed in one column of the bipartite graph and

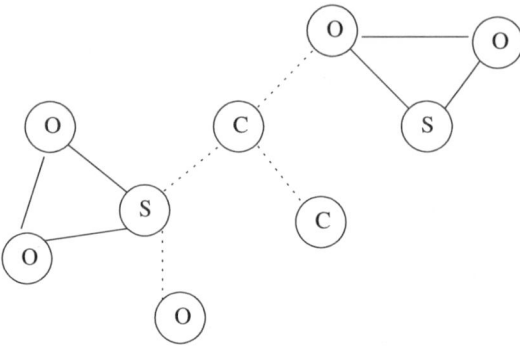

Fig. 1.3. A graph is shown along with its associated bridges, and cycles. The bridges are the edges given by dashed lines, while the cycles are given by solid lines.

connections are made where the highest similarity between vertices exist. The Hungarian method is used to solve this kernel (in $O(N^3)$ time), effectively matching areas between two graphs that are the most similar. Each vertex in the bipartite graph also weights neighbors in the graph with a λ weighting. This approach appears to be very promising as two similar molecular graphs usually have regions of the graph that are very similar.

1.4 Summary

In this chapter we have provided a brief review of kernel methods for strings, trees and graphs. The kernel paradigm provides a widely applicable way to embed symbolic information into a range of machine learning algorithms. Note that the application of these methods does not exclusively apply to the types of symbolic information considered in this chapter. Many of the techniques are general and can be applied (with modification) to arbitrary discrete structures. In this chapter however we have covered the structures that are currently most used and which are applicable to many of the currently popular applications in the problem domain.

The wealth of literature on structured kernels prevents us from covering all aspects in-depth, however we have presented some of the fundamental concepts and kernels which are currently being successfully applied. For the interested reader, further reference points to general frameworks (such as rational kernels or fisher kernels) have been provided where appropriate.

One common thread that links kernels for symbolic data is that the resulting feature representation is often explicitly known (so that those who design and apply these kernels can visualise the features that are used), however the features themselves never need to be represented explicitly. Therefore, kernel values can often be computed efficiently using tricks of linear algebra or dynamic programming. Note however that whilst these are powerful techniques, efficiency in this context means that these values can be computed in polynomial time with respect to the size of the structures. For large scale applications this can sometimes be too cumbersome, and an active area of research is in developing these types of kernels, which consider a variety of complex features, that can run in linear time.

References

1. Shawe-Taylor, J., Cristianini, N.: Kernel Methods for Pattern Analysis. Cambridge University Press (2004)
2. Herbrich, R.: Learning Kernel Classifiers: Theory and Algorithms (Adaptive Computation and Machine Learning). The MIT Press (2001)
3. Schoelkopf, B., Smola, A.J.: Learning with Kernels: Support Vector Machines, Regularization, Optimization, and Beyond. MIT Press, Cambridge, MA, USA (2001)
4. Haussler, D.: Convolution kernels on discrete structures. Technical Report UCSC-CRL-99-10, University of California, Santa Cruz (1999)
5. Watkins, C.: Dynamic alignment kernels. Technical Report CSD-TR-98-11, Royal Holloway, University of London (1999)
6. Lodhi, H., Shawe-Taylor, J., Cristianini, N., Watkins, C.: Text classification using string kernels. In Leen, T.K., Dietterich, T.G., Tresp, V., eds.: Advances in Neural Information Processing Systems 13, MIT Press (2001) 563–569
7. Leslie, C., Eskin, E., Noble, W.S.: The spectrum kernel: A string kernel for SVM protein classification. In: Proceedings of the Pacific Symposium on Biocomputing. (2002) 564–575
8. Teo, C.H., Vishwanathan, S.: Fast and space efficient string kernels using suffix arrays. In: In Proceedings of the Twenty Third International Conference on Machine Learning. (2006)
9. Lodhi, H., Saunders, C., Shawe-Taylor, J., Cristianini, N., C., W.: Text classification using string kernels. Journal of Machine Learning Research **2** (2002) 419–444
10. Leslie, C., Kuang, R.: Fast string kernels using inexact matching for protein sequences. Journal of Machine Learning Research **5** (2004) 1435–1455
11. Leslie, C.S., Eskin, E., Cohen, A., Weston, J., Noble, W.S.: Mismatch string kernels for discriminative protein classification. Bioinformatics **20**(4) (2004) 467–476
12. Kuang, R., Ie, E., Wang, K., Wang, K., Siddiqi, M., Freund, Y., Leslie, C.: Profile-based string kernels for remote homology detection and motif extraction. In: CSB '04: Proceedings of the 2004 IEEE Computational Systems Bioinformatics Conference (CSB'04), IEEE Computer Society (2004) 152–160
13. Cortes, C., Haffner, P., Mohri, M.: Rational kernels: Theory and algorithms. Journal of Machine Learning Research **5** (2004) 1035–1062
14. Jaakkola, T., Diekhaus, M., Haussler, D.: Using the fisher kernel method to detect remote protein homologies. 7th International Conference on Intelligent Systems for Molecular Biology (1999) 149–158

15. Saunders, C., Shawe-Taylor, J., Vinokourov, A.: String Kernels, Fisher Kernels and Finite State Automata. In Becker, S., Thrun, S., Obermayer, A., eds.: Advances in Neural Information Processing Systems 15. (2003)
16. Gärtner, T., Flach, P.A., Wrobel, S.: On graph kernels: Hardness results and efficient alternatives. In: Learning Theory and Kernel Machines, 16th Annual Conferences on Learning THeory and 7th Kernel Workshop, Proceedings. Volume 2843., Springer Verlag (2003) 129–143
17. Collins, M., Duffy, N.: Convolution kernels for natural language. Advances in Neural Information Processing Systems **volume 14** (2002)
18. Vishwanathan, S., Smola, A.: Fast kernels for string and tree matching. Advances in Neural Information Processing Systems **volume 15** (2003)
19. Zhu, X., Ghahramani, Z., Lafferty, J.: Semi-supervised learning using gaussian fields and harmonic functions. In: ICML. (2003)
20. Kashima, H., Inokuchi, A.: Kernels for graph classification. In: 1st ICDM Workshop on Active Mining (AM-2002), Maebashi, Japan (2002)
21. Kashima, H., Tsuda, K., Inokuchi, A.: Marginalized kernels between labeled graphs. In: Proceedings of the Twentieth International Conference on Machine Learning (ICML-2003), Washington DC, USA (2003)
22. Kashima, H., Tsuda, K., Inokuchi, A.: Kernels for graphs. In: Kernel Methods in Computational Biology. MIT Press (2004)
23. Mahé, P., Ueda, N., Akutsu, T., Vert, J.: Extensions of marginalized graph kernels. In: Proceedings of the 21st International Conference on Machine Learning (ICML-2004), Banff, Alberta, Canada (2004)
24. Borgwardt, K.M., Kriegel, H.: Shortest-path kernels on graphs. In: ICDM. (2005)
25. Ramon, J., Gartner, T.: Expressivity versus efficiency of graph kernels. In: First International Workshop on Mining Graphs, Trees and Sequences (held with ECML/PKDD03). (2003)
26. Horvath, T., Gärtner, T., Wrobel, S.: Cyclic pattern kernels for predictive graph mining. In: Proceedings of the International Conference on Knowledge Discovery and Data Mining. (2004)
27. H. Frohlich, J. K. Wegner, F.S., Zell, A.: Optimal assignment kernels for attributed molecular graphs. In: Int. Conf. on Machine Learning (ICML). (2005) 225–232

2

Comparing Sequence Classification Algorithms for Protein Subcellular Localization

Fabrizio Costa, Sauro Menchetti, and Paolo Frasconi

Machine Learning and Neural Networks Group
Dipartimento di Sistemi e Informatica
Università degli Studi di Firenze, Italy
Web: http://www.dsi.unifi.it/neural/

Summary. We discuss and experimentally compare several alternative classification algorithms for biological sequences. The methods presented in this chapter are all essentially based on different forms of statistical learning, ranging from support vector machines with string kernels, to nearest neighbour using biologically motivated distances. We report about an extensive comparison of empirical results for the problem of protein subcellular localization.

2.1 Introduction

Serial order is just one simple kind of relation, yet sequence learning can be seen as one very special and important case in relational learning. At least two interesting supervised learning problems can be defined on sequences: classification and translation. In the former problem, a category label is assigned to an input sequence. In the latter, an input sequence is mapped into an output sequence that may or may not have the same length. Sequence translation is perhaps the simplest instance of the structured output supervised learning problem [1, 2]. In this paper we focus on classification.

A very popular solution to sequence classification is based on kernel machines. In this approach, instances are implicitly mapped into a Hilbert space (called the feature space) where the kernel function plays the role of inner product. A linear decision function is usually computed in the feature space minimizing a regularized empirical loss functional.

Most kernel functions on sequential and relational data are simply specific instances of decomposition kernels [3]. An instance is first decomposed into the set of its parts, according to some given parthood relation. Instances are then compared by comparing their sets of parts. In most cases of practical interest, this reduces to counting the number of occurrences and multiplying the corresponding counts between two instances. In this way, features are directly associated with parts. The literature is replete with decomposition kernels specialized for different types of discrete data structures, including sequences [4, 5, 6], trees [7, 8], annotated graphs [9, 10] and more complex relational entities defined using higher order logic abstractions [11], ground terms [12], or Prolog execution traces [13]. Thanks to its generality, decomposition is an attractive and flexible approach for constructing similarity on structured objects based on the similarity of smaller parts. However, kernel design can be a difficult challenge when prior knowledge about the importance of each part is not available.

Indeed, one problem with the plain decomposition approach is the tradeoff between the expressivity and the relevance of the resulting features. In the case of sequence kernels, we can naturally use subsequences or substrings as parts. We can also implement the tradeoff between expressivity and relevance by limiting the maximum length of a subsequence or a substring to some integer k. However, since the number of features grows exponentially with k, we can easily end up with many irrelevant features even when fragments do not seem to be long enough to capture interesting behaviour in the data. The abundance of irrelevant parts may imply smaller margins [14] and can be often observed in terms of large-diagonal Gram matrices [15]. Weighted decomposition kernels (WDK), later described in Section 2.4, have been introduced to smoothen the expressivity-relevance tradeoff [16].

Symbolic sequences are commonly used in molecular biology to describe the primary structure of macromolecules such as proteins, DNA and RNA. Classification arises in different contexts, including family and superfamily classification [17] and protein subcellular localization. In this paper we will focus on the latter problem. Once proteins are synthesized, they are sorted to appropriate destinations in the cell. This process is extremely important for the functioning of both eukaryotic and prokaryotic organisms. The task of predicting subcellular localization from protein sequence has been addressed in several recent papers [18, 19, 20, 21, 22, 23, 24]. Here we report what we believe to be one of the most exhaustive experimental comparisons to date, testing alternative sequence classification algorithms on a large number of different data sets previously used in the literature.

2.2 Decomposition Kernels

We briefly review some concept about convolution kernels introduced in [3]. In the following, X will denote the instance space (e.g. a set of sequences). A Mercer kernel is a function $K : X \times X \mapsto R$ such that for any finite set of instances $\{x_1, \ldots, x_m\}$ the Gram matrix $G_{ij} = K(x_i, x_j)$ has all positive eigenvalues.

2.2.1 Decomposition structures

An R-decomposition structure [25] on a set X is a triple $\mathcal{R} = \langle \mathbf{X}, R, \mathbf{k} \rangle$ where $\mathbf{X} = (X_1, \ldots, X_D)$ is a D–tuple of non–empty sets; R is a finite relation on $X_1 \times \cdots \times X_D \times X$; $\mathbf{k} = (k_1, \ldots, k_D)$ is a D–tuple of positive definite kernel functions $k_d : X_d \times X_d \mapsto \mathbf{R}$. R is meant to capture parthood between a structure and its substructures. However, in this definition no special constraints are placed on R to guarantee that mereology axioms [26] are satisfied (one common mereological requirement is that parthood be a partial order). More simply, in this context $R(\mathbf{x}, x)$ is true iff $\mathbf{x}$ is a tuple of parts for x. In this case we say that $\mathbf{x}$ is a *decomposition* of x. For each $x \in X$, $R^{-1}(x) = \{\mathbf{x} \in \mathbf{X} : R(\mathbf{x}, x)\}$ denotes the multiset of all possible decompositions of x. The general form of a decomposition kernel is then obtained by comparing these multisets as follows:

$$K_\mathcal{R}(x, x') \doteq \sum_{\substack{\mathbf{x} \in R^{-1}(x) \\ \mathbf{x}' \in R^{-1}(x')}} \varkappa(\mathbf{x}, \mathbf{x}') \tag{2.1}$$

where we adopt the convention that summations over the members of a multiset take into account their multiplicity.

In order to compute $\varkappa(\mathbf{x}, \mathbf{x}')$, kernels on parts need to be combined. The resulting function is a valid kernel provided that the combination operator is closed with respect to positive definiteness. Haussler [3] proved that R-convolution kernels that use tensor product as a combination operator are positive definite. It is immediate to extend the result to decomposition kernels based on other closed operators. Interesting examples include direct sum:

$$(k_1 \oplus k_2)\left((x_1, x_2), (x_1', x_2')\right) = k_1(x_1, x_1') + k_2(x_2, x_2') \tag{2.2}$$

and tensor minimum:

$$(k_1 \oslash k_2)\left((x_1, x_2), (x_1', x_2')\right) = \min\{k_1(x_1, x_1'), k_2(x_2, x_2')\}. \tag{2.3}$$

Different components k_d can be even combined by different operators. Thus, denoting by op_i a valid operator, the most general form of kernel between decompositions is

$$\varkappa(\mathbf{x}, \mathbf{x}') = I_\mathcal{R}(x, x')k_1(x_1, x_1')op_1 k_2(x_2, x_2')op_2 \cdots op_{D-1} k_D(x_D, x_D').$$

where $I_\mathcal{R}(x, x') = 1$ if x and x' are decomposable in D parts and $I_\mathcal{R}(x, x') = 0$ otherwise. The traditional R-convolution kernel uses tensor product for all d:

$$K_{\mathcal{R}, \otimes}(x, x') \doteq \sum_{\substack{\mathbf{x} \in R^{-1}(x) \\ \mathbf{x}' \in R^{-1}(x')}} I_\mathcal{R}(x, x') \prod_{i=1}^{D} k_d(x_d, x_d'). \tag{2.4}$$

2.2.2 All-substructures kernels

Since decomposition kernels form a rather vast class, the relation R needs to be carefully tuned to different applications in order to characterize a suitable kernel. The "all-substructures kernels" are a popular family of decomposition kernels that just count the number of co-occurrences of substructures in two decomposable objects. In this case $D = 1$ and $\mathcal{R} = \langle X, R, \delta \rangle$, where $R(x_1, x)$ if x_1 is a substructure of x and δ is the *exact matching kernel*:

$$\delta(x_1, x'_1) = \begin{cases} 1 & \text{if } x_1 = x'_1 \\ 0 & \text{otherwise} \end{cases} \tag{2.5}$$

The resulting convolution kernel can also be written as

$$K_{\mathcal{R},\delta}(x, x') = \left| R^{-1}(x) \cap R^{-1}(x') \right|_{mset}$$

where $|\cdot|_{mset}$ denotes the multiset cardinality (in which each member is counted according to its multiplicity). Known kernels that can be reduced to the above form include the spectrum kernel on strings [5], the basic version (with no down-weighting) of co-rooted subtree kernel on trees [7], and cyclic pattern kernels on graphs [10]. All-substructure kernels became more and more sparse (i.e. diagonal dominant) as the size of parts in the decomposition grows [15]. This is because the chance of exact match between two parts decreases with their size.

In general, computing the equality predicate between x_1 and x'_1 may not be computationally efficient as it might require solving a subgraph isomorphism problem [27].

2.3 A Review of Kernels for Protein Sequences

2.3.1 Flattened Representations

Let us focus on strings on a finite alphabet Σ (e.g. in the case of protein sequences Σ consists of the 20 amino acid letters). In this case the instance space is $X = \Sigma^*$, the Kleene closure of Σ. One very simple decomposition structure is defined as $\mathcal{R} = \langle \Sigma, R, \delta \rangle$ where for $\sigma \in \Sigma$, $R(\sigma, x)$ is true iff σ occurs in x and $R^{-1}(x)$ is the multiset of characters that occur in x. The resulting all-substructures decomposition kernel merely counts the number of co-occurrences of letters in the two strings and does not take into account any serial order information. After normalization (dividing by string lengths) this kernel applied to protein sequences is simply the amino acid composition kernel:

$$k(x, x') = \frac{1}{|x| \cdot |x'|} \sum_{\sigma \in R^{-1}(x)} \sum_{\sigma' \in R^{-1}(x')} \delta(\sigma, \sigma') = \sum_{\sigma \in \Sigma} \frac{c(\sigma)}{|x|} \frac{c'(\sigma)}{|x'|} \tag{2.6}$$

where $c(\sigma)$ and $c'(\sigma)$ are the number of occurrences of σ in x and x', respectively, and $|\cdot|$ denotes string length. In spite of its simplicity, the above kernel has been shown to be effective for predicting protein subcellular localization [21].

2.3.2 Probability Kernels

An interesting alternative interpretation of the Hua and Sun [21] approach is that comparing the amino acid composition of two protein sequences is the same as comparing two multinomial probability distributions. This interpretation leads to the more general idea of defining probability kernels for sequences. The first we consider is derived from the so-called *probability product kernel* [28] whose associated feature space consists of probability distributions. To compute the feature mapping $\phi(x)$, a simple generative model $p(\lambda)$ is fitted to example x. The kernel between two examples is then evaluated by integrating the product of the two corresponding distributions:

$$k(x, x') = \langle p, p' \rangle = \int p(\lambda)^\rho p'(\lambda)^\rho d\lambda$$

where ρ is a positive constant (one could use e.g., $\rho = 1$ or alternatively $\rho = 1/2$, yielding the so-called Bhattacharyya kernel). By taking a non-parametric approach, $p(\lambda)$ may be replaced by a set of empirical histograms $[c_l(1), \ldots, c_l(n)]$, one for each attribute $l = 1, \ldots L$. In this case $c_l(j)$ is the observed count of the j-th bin associated with attribute j, yielding

$$k(x, x') = \sum_{l=1}^{L} \sum_{j=1}^{n} \frac{c_l(j)^\rho}{z_l} \frac{c'_l(j)^\rho}{z'_l} \quad (2.7)$$

being $z_l = \sum_j c_l(j)$ a normalization factor. In the case of protein amino acid composition, $L = 1$ and it is natural to choose $n = 20$ bins, one for each amino acid letter. Note that when $\rho = 1$ Equation 2.7 reduces to the amino acid composition kernel of Equation 2.6 using string lengths as normalization factors.

The second probability kernel we consider is the so-called *histogram intersection kernel* introduced in [29] for image processing applications. When applied to two strings of the same length, a histogram intersection kernel counts the number of characters in one string that are also present in the second string and can be written as

$$k(x, x') = \sum_{\sigma \in \Sigma} \min\{c(\sigma), c'(\sigma)\}. \quad (2.8)$$

The minimum operator can also be applied after normalizing the empirical counts, yielding

$$k(x, x') = \sum_{\sigma \in \Sigma} \min\left\{\frac{c(\sigma)}{z}, \frac{c'(\sigma)}{z'}\right\}. \quad (2.9)$$

2.3.3 The Fisher Kernel

The Fisher kernel [17] (which can be seen as closely related to probability kernels) induces a feature space directly associated with the parameter space of a generative model. Interestingly, the Fisher kernel was originally introduced for solving a sequence classification problem in the context of protein remote homology. In [17], hidden Markov models (HMM) were used as the underlying generative model. An HMM, however, can only reliably model the generative distribution of protein that share some sequence similarity and cannot be trained on arbitrary data sets containing unrelated sequences, such as subcellular localization data sets. Forming proper sequence sets (e.g. by running sequence search algorithms on genomic data) for training a set of HMMs is out of the scope of the present paper. Therefore, Fisher kernels are not further considered here.

2.3.4 Spectrum and Mismatch Spectrum Kernels

The r-spectrum of a string x is the set of all substrings of x whose length is r. Clearly the r-spectrum is a decomposition of x according to the relation $R_r(z, x)$ iff z is a substring of x and $|z| = r$. The r-spectrum kernel between two strings x and x' is then defined as

$$k_r(x, x') = \left| R_r^{-1}(x) \cap R_r^{-1}(x') \right|_{mset}. \quad (2.10)$$

It can be optionally normalized e.g. by dividing by the two string lengths. Like the Fisher kernel, the spectrum kernel was originally introduced for sequence classification in the context of protein remote homology [5]. Unlike the Fisher kernel, it has a purely discriminant nature, which makes it immediately suitable for solving the problem of subcellular localization problem studied in this paper. Note that similar algorithmic ideas had been experimented long before the explosion of interest of kernel machines in bioinformatics. Wu et al. [30], for example, used string spectra in combination with singular value decomposition (to reduce dimensionality) and neural networks in a problem of protein family classification.

The (r, m)-mismatch spectrum of a string x is the set of all substrings of x whose length is r and that differ from an r-substring of x by at most m characters [31]. In this case, we have a decomposition relation $R_{(r,m)}(z, x)$ that is true iff z is a substring of x, $|z| = r$, and there exists a substring y of x such that the mismatch between y and z is at most m characters. The (r, m)-mismatch spectrum kernel between two strings x and x' is then defined as

$$k_{(r,m)}(x, x') = \left| R_{(r,m)}^{-1}(x) \cap R_{(r,m)}^{-1}(x') \right|_{mset}. \quad (2.11)$$

2.4 Weighted Decomposition Kernels

2.4.1 Definitions

A weighted decomposition kernel (WDK) aims to address the sparsity problem that arises with all-substructures kernels when the size of parts grows. A WDK is characterized by the following decomposition structure:

$$\mathcal{R} = \langle (S, Z_1, \ldots, Z_D), R, (\delta, \kappa_1, \ldots, \kappa_D) \rangle$$

where $S, Z_1, \ldots, Z_D \subset X$, $R(s, z_1, \ldots, z_D, x)$ is true iff $s \in S$ is a special part type of x called the *selector* and $\mathbf{z} = (z_1, \ldots, z_D) \in Z_1 \times \cdots \times Z_D$ is a tuple of parts of x called the *contexts* of occurrence of s in x. This setting results in the following general form of the kernel:

$$K(x, x') = \sum_{\substack{(s, \mathbf{z}) \in R^{-1}(x) \\ (s', \mathbf{z}') \in R^{-1}(x')}} \delta(s, s') \sum_{d=1}^{D} \kappa_d(z_d, z'_d) \tag{2.12}$$

where the direct sum between kernels over parts can be replaced by the tensor product (Equation 2.2) or by the tensor minimum operator (Equation 2.3).

Compared to all-substructures kernels, which simply count the number of substructures, a WDK weights different matches between substructures (selectors) according to contextual information in which they occur. Note that in our definition of WDK, selectors are compared by the exact matching kernel. Therefore, choosing large selectors can give raise to the same sparsity problems encountered with all-substructures kernels. However, using a WDK we can achieve a better control of sparsity by making selectors small and, at the same time, recovering information about larger parts through the contexts in which a selector occurs. For this reason, it makes sense to compare contexts by kernels based on *soft* matches that return a non-zero value even when contexts are not exactly identical. A natural solution for achieving this goal is to use a probability kernel for each of the $\kappa_d, d = 1, \ldots, D$.

2.4.2 Weighted Decomposition Kernels on Strings

When applied to strings, both the selector and the contexts can be chosen to be substrings. Suppose for simplicity that $D = 1$ (single context). Given a string $x \in \Sigma^*$, two integers $1 \leq t \leq |x|$ and $r \geq 0$, let $x(t, r)$ denote the substring of x spanning string positions from $t - \lfloor \frac{r}{2} \rfloor$ to $t + \lceil \frac{r}{2} \rceil$. Then, let us introduce the relation $R_{(r,l)}$ depending on two integers $r \geq 0$ (the selector length) and $l \geq 0$ (the context length) defined as follows:

$$R_{(r,l)} = \{(s, z, x) : x \in \Sigma^*, s = x(t, r), z = x(t, l), t \leq |x|\}.$$

The resulting WDK is:

$$K(x, x') = \sum_{t=1}^{|x|} \sum_{\tau=1}^{|x'|} \delta(x(t, r), x'(\tau, r)) \kappa(x(t, l), x'(\tau, l)).$$

Intuitively, when applied to protein sequences, this kernel computes the number of common r-mers[1] each weighed by the similarity κ between the amino

---

[1] In the biochemistry parlance, a k-mer is just a k-gram or substring of length k in a biological sequence.

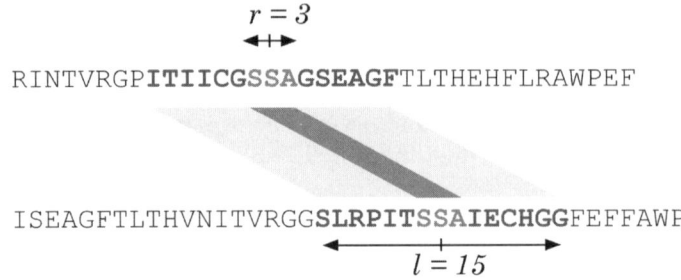

Fig. 2.1. The application of a WDK to protein sequences. The match between the two substrings "SSA" in the example is weighted by the context kernel. Using the HIK of Equation 2.9 the weight highlighted in the picture is 0.048889.

acid composition of their environments — measured, for example, by one of the probability kernels as defined in Equation 2.7 or Equation 2.8. Figure 2.1 shows an example. In this example, if $l = 3$ or if $\kappa(\cdot, \cdot) \equiv 1$, then the WDK reduces to the r-spectrum kernel. Note that although the above equation seems to imply a complexity of $O(|x||x'|)$, more efficient implementations are possible, as discussed below.

2.4.3 Weighted Decomposition Mismatch Kernels on Strings

We obtain a natural extension of the WDK on strings extending the match between selectors so to include the case of m mismatching symbols. The decomposition relation, parallelling what happens in the (r, m)−mismatch spectrum kernel, now becomes $R_{(r,l,m)}(z, x)$ that is true iff z is a substring of x, $|z| = r$, and there exists a substring y of x such that the mismatch between y and z is at most m characters and which identifies substrings $x(t, l)$ of x spanning string positions from $t - \lfloor \frac{l}{2} \rfloor$ to $t + \lceil \frac{l}{2} \rceil$ with $t \leq |x|$. When working with protein sequences, the mismatch assumption implies that we compare each r−mer and its context with all possible mutations up to m different residues.

2.4.4 Algorithms and Complexity

The computational efficiency of a decomposition kernel depends largely on the cost of constructing and matching substructures. In general, exact matching of substructures might lead to intractability. This is the case, for example, when instances are general graphs [27]. On the other hand, the spectrum kernel on strings can be computed in linear time by taking advantage of a suffix tree data structure [5].

Intractability or high complexity can be intrinsically avoided in a WDK provided that some reasonable design guidelines are kept:

1. selectors should be small, i.e. they should have constant size with respect to the size of the entire instance;
2. each context should be computable efficiently given the selector;
3. simple probability kernels (e.g. HIK) should be used for matching contexts;

Examples of suitable selectors include: short substrings in the case of sequences, production rules in the case of parse trees, individual vertices or adjacent vertices in the case of graphs, atoms in the case of logical ground terms. Note that contexts may be large, still efficiency can be achieved because probability distributions and not whole substructures are matched. In practice, in order to define a proper probability kernel for contexts, we need to further characterize the nature of the data types under consideration. One reasonably general case is when each context consists of atomic parts that are characterized by discrete-valued attributes. Under these assumptions, forming histograms of various attributes trivially reduces to the enumeration of atomic parts. Strings, parse trees, and attributed graphs all naturally fit these assumptions.

One efficient algorithm for computing WDK in the case of discrete-valued attributes and histogram kernels is based on the idea of bucketing. For the sake of simplicity, suppose there is only a single context ($D = 1$) and note that Equation 2.12 can be rewritten in this case as

$$K(x, x') = \sum_{s \in S(x,x')} \sum_{z \in Z_x(s)} \sum_{z' \in Z'_x(s)} \kappa(z, z') \qquad (2.13)$$

where $S(x, x')$ is the set of selectors that occur in both x and x' and $Z_x(s)$ [$Z'_x(s)$] is the bucket of histograms associated with the contexts in which the selector occurs in x [x'].

Before kernel calculation, each instance x is preprocessed to build a dictionary data structure that maps each selector s to its bucket of histograms $Z_x(s)$. Assuming the set of selectors S possesses a total order (for example, a lexicographic order can be used in the case of strings or graphs where a canonical order is defined on adjacency lists), the dictionary can be implemented with balanced search trees. In this case, the dictionary for a given example can be constructed in time $O(t \log t + T_c)$, being $t = |R^{-1}(x)|$ and T_c an upper bound on the time for calculating all the context histograms. When dealing with sequences, context histograms can be updated in $O(1)$ moving along the sequence e.g. from left to right. Therefore, the time for filling in all buckets $Z_x(s)$, for all s, is $T_c = O(|x| + l)$, being l the size of each context and $|x|$ the length of the sequence[2]. As shown in procedure WEIGHTEDDE-COMPOSITIONKERNEL, the outer summation over selectors in Equation 2.12

---

[2] Note that in this analysis we are interested in the scaling with respect to the instance size (e.g. string length). Therefore we are regarding as constants the number of attributes and the number of attribute values. For example in the case of protein sequence these constants are 1 (the only attribute is the amino acid type) and 20 (the number of amino acids), respectively.

is computed by traversing the dictionary associated with x and x' from the smallest to the largest selector:

WEIGHTEDDECOMPOSITIONKERNEL($M_x, M_{x'}$)
  $k \leftarrow 0$
  $i \leftarrow begin[M_x]$
  $i' \leftarrow begin[M_{x'}]$
  while $i \neq$ NIL and $i' \neq$ NIL
    do $s \leftarrow key[i]$
       $s' \leftarrow key[i']$
       if $s > s'$
         then $i' \leftarrow$ SUCC$[i']$
       elseif $s < s'$
         then $i \leftarrow$ SUCC$[i]$
       else
           $Z_x(s) \leftarrow value[i]$
           $Z'_x(s) \leftarrow value[i']$
           $k = k +$ BUCKETKERNEL$(Z_x(s), Z'_x(s))$
           $i \leftarrow$ SUCC$[i]$
           $i' \leftarrow$ SUCC$[i']$
  return k

In the above code, M_x and $M_{x'}$ are the dictionaries associated with x and x', respectively. Field $begin[M_x]$ is an iterator pointing to the first item of M_x. Procedure SUCC returns the successor of an iterator (or NIL if no successor exist) and can be implemented to run in amortized constant time. Fields $key[i]$ and $value[i]$ are the selector and the associated set of histograms, respectively, for the item pointed by i. Procedure BUCKETKERNEL is in charge of computing the two inner summations in Equation 2.13. If the best case, if each selector occurs at most a constant number of times in a given instance (with respect to the size of the instance), then BUCKETKERNEL clearly runs in constant time. In the worse case, all histograms end up in the same bucket (i.e. there is only one distinct selector) and the running time of BUCKETKERNEL is $O(t^2)$. Thus, the above strategy leads to a complexity reduction of the kernel computation between two instances of the same size ranging from $O(t^2)$ up to $O(t)$, depending on indexing sparseness. Note that depending on the choice of κ in Equation 2.13, complexity can be further reduced. Since a kernel is a bilinear operator, we have

$$\sum_{\substack{z \in Z_x(s) \\ z' \in Z'_x(s)}} \kappa(z, z') = \sum_{\substack{z \in Z_x(s) \\ z' \in Z'_x(s)}} \langle \phi(z), \phi(z') \rangle_\kappa = \left\langle \sum_{z \in Z_x(s)} \phi(z), \sum_{z' \in Z'_x(s)} \phi(z') \right\rangle_\kappa.$$
(2.14)

Therefore, complexity can be further reduced if the feature space representations $\phi(z)$ can be efficiently computed, as in the case of the amino acid composition kernel of Equation 2.6.

In the case of weighted decomposition mismatch kernels we cannot resort to Mismatch Tree Data Structures [5] for an efficient kernel computation since each r−mer bears additional information in the context. We therefore have to explicitly instantiate every possible selector resulting from m mismatches and associate the same context to each one of them. In order to speed up the computation we can approximate the expansion of the mismatched selector by instantiating only those r−mer that are actually present in the sequence. Without resorting to the approximation the number of buckets will increase by a factor of $|\Sigma|^m$, that is an exponential increase with respect to the number of allowed mismatch. The approximation gives an upper bound on the increase factor limiting it to the size of the sequence (there can't be more r−mers in a sequence of length $|x|$ than $|x| - r$), though in this case the computational complexity becomes soon dominated by the generation and search phase of r−mers. In this case it might become more convenient to confront each r−mer of one sequence with all the r−mers of the other sequence and determine each time whether one is a m−mismatch of the other. In practice only small values of allowed mismatches, $m = 1, 2$, guarantee acceptable running times.

2.5 Protein sequences and protein functions

In bioinformatics, the study of the relationship between the sequential nature of proteins and their three-dimensional structure and biological function is of key importance. On the one hand it is important to gain information on how new functions can arise from changes in sequence and structure. A second reason is that large scale genome sequencing projects have increased the gap between the number of proteins of known sequence but of unknown function.

It is generally accepted that we can infer structural and functional similarity on the basis of common protein ancestry and that high sequence similarity is a clear sign of common ancestry. The relationship between the conservation of the pair sequence-structure has been widely explored in literature (and it has become widely accepted and used in everyday sequence analysis and database annotation practice). It is known that when two proteins have high (above experimentally determined thresholds) sequences similarity we can infer that the two proteins have a similar 3D structure (though the converse does not hold as we can have proteins with similar structure but with very little sequence similarity). Unfortunately, trying to directly use these thresholds to infer similarity for protein functions is not a viable approach [32].

The conservation of the pair sequence-function is somewhat more problematic, as 1) the definition of protein function is controversial (as we can identify different levels at which the action of a protein takes place i.e. chemical, biochemical or cellular, and these levels are then related in complex ways), 2) machine readable data on functional aspects of proteins is available limited only to a few types of functions and 3) the conservation level is highly dependent on the specific protein function. In the literature, several approaches have been used to classify protein functions, ranging from enzymatic activity as described by the Enzyme Commission [33], to the usage of SWISSPROT keywords [32], to the more comprehensive Gene Ontology (GO) [34]. Subcellular localization is the destination compartment where the mature protein is sorted after synthesis. Since localization is strongly correlated with protein function, we can approximate function prediction by assuming that proteins that share the same working environment also share the same macro-function and classify them accordingly. This approach is less fine-grained and not hierarchical but offers the advantage of having a clear definition and it can be easily measured.

It is known in the literature [35] that if we take into consideration the relationship between the sequence similarity and the structure/function we do observe a transition-phase behaviour, i.e. there exist a sharp transition between the so called "safe" or "trivial" zone and the "impossible" zone; the first one characterized by proteins with high alignment that share the same function and the second characterized by proteins that have levels of sequence similarity so low that it is not possible to determine if they have the same function considering only the sequence information. A small interval in the sequence similarity axis separates the two areas and is called the "problematic" or "twilight" zone. The sharpness of the transition strongly depends on the actual procedure adopted to compute the similarity score between two sequences but different approaches have consistently shown the same transition phenomenon [22]. An important feature of the mentioned zone diagram concerns the coverage: the vast majority of protein pairs belong to the impossible zone and those that are highly similar (safe zone) represent only a small fraction. Bearing this in mind it is still possible to devise a simple approach to function inference that relies on finding homologue proteins (i.e. proteins with a high sequence similarity) and assign their known function or localization to given unknown test proteins. Nair & Rost [22] showed that using measures like BLAST [36] and HSSP distance [37, 38], it is possible to infer subcellular localization with high accuracy. More specifically if we require a 90% accuracy then we need a 70% similarity threshold. Unfortunately only 20-10% (depending on the cell compartment) test proteins could be classified using the known localization of 1K unique proteins from the SWISSPROT database [39]. The result is nonetheless interesting and it shows that subcellular localization is highly conserved. This information is important as the actual biological procedure for protein localization depends on the specific destination compartment but mainly relies on a signal mechanism, i.e. the presence of short terminal sequences of 20-30 aminoacids (most often N-terminal[3]) that allow proteins to cross cell compartments until they reach the final destination. It has been experimentally demonstrated how the absence of such signals alters the final position of the protein (or conversely that replacing the signal provokes

---

[3] The N-terminal end refers to the extremity of a protein or polypeptide terminated by an amino acid with a free amine group (NH2).

a misplacement of the protein). Despite this mechanism being sensitive only to a small part of the protein sequence, it is shown that similar proteins are still located in the same compartments and therefore that the similarity based approach is viable. Instead of finding a close enough neighbour to a given test protein sequence (as in the KNN approach), we propose to learn the notion of similarity in a supervised framework using weighted decomposition kernels for sequences as defined in Sec.2.4 and a support vector technique.

2.6 Experimental setting

2.6.1 Redundancy Issues

Whilst the size of protein structure and sequence data banks is continuously growing there still remains a non-homogeneity issue as some protein families are heavily represented whereas other have just a single entry. Using directly all this redundant data can therefore lead to overestimation problems, possibly masking otherwise observable regularities. A desiderata is then that all protein types (defined in evolutionary terms for example) need be represented. A common solution is to guarantee that no pair of proteins should have more than a given level of sequence similarity and that the number of proteins should be maximal. The similarity measure is application specific though measures like BLAST [36] or HSSP distance [37, 38] are the most common choices. BLAST returns a score based on subsequences that have a good match and whose alignment is statistically significant while HSSP distance is based on the percentage of residues identical between two proteins and the length of the alignment. In [40, 41] a simple greedy algorithm to reduce redundancy is presented. The algorithm builds the binary adjacency matrix of the sequences similarity graph, that is, vertices are sequences and edges exist between sequences if the similarity measure is above a given threshold. The algorithm then proceeds iteratively removing vertices belonging to the most populated adjacency list until these become empty: the remaining sequences form the final non redundant set.

2.6.2 Data Sets

In the following sections we report information on the size and the number of classes of several datasets used in literature for sub-localization experimentation. On seven datasets no redundancy reduction was employed (SubLoc Eukaryotic, SubLoc Prokaryotic, Archea, Gram Positive and Negative Bacteria, Fungi, Plant). On four dataset (Lockey NR1K, Lockey NR3K, LocTree Plant and Non-Plant) the [40, 41] redundancy reduction strategy was employed to avoid biases.

Table 2.1. Number of sequences within each subcellular localization category for the Reinhardt and Hubbard data set. The data set is available at http://www.bioinfo.tsinghua.edu.cn/SubLoc/.

Species	Subcellular localization	Size
Prokaryotic	Cytoplasmic	688
	Periplasmic	202
	Extracellular	107
Eukaryotic	Nuclear	1,097
	Cytoplasmic	684
	Mitochondrial	321
	Extracellular	325

Reinhardt & Hubbard's Data Set.

This data set was generated by [18] extracting two sets of 997 sequences for prokaryotic organisms and 2,427 sequences for eukaryotic organisms from the SWISSPROT database release 33.0 [39]. Admissible sequences needed to be complete proteins (as the database lists also fragments of protein sequences), not contain ambiguities (such as residues denoted by X within the sequence) and had to have reliable localization annotation determined directly by experimentation (excluding database entry annotated with keywords such as POSSIBLE, PROBABLE, BY SIMILARITY). No transmembrane protein was included as this particular kind of localization can be determined very reliably by other methods [42, 43]. Redundancy was tolerated up to a 90% sequence identity for any pair of sequences in their respective set. The three prokaryotic localization categories are cytoplasmic, periplasmic and extracellular, while the four eukaryotic categories are nuclear, cytoplasmic, mitochondrial and extracellular; the number of sequences within each subcellular localization category is reported in Table 2.1.

Archaea, Gram Bacteria, Fungi, Plant.

This data set was generated by [23] and comprises protein localization for different specific taxonomic groups ranging from bacteria to fungus to plants. No specific redundancy reduction procedure was adopted. The number of sequences within each subcellular localization category is reported in Table 2.2.

LOCKey Data Set.

This data set was generated by Nair and Rost (02) extracting two sets of 1,161 and 3,146 sequences from the SWISSPROT database release 40 [39]. Sequences where the localization was annotated with words POSSIBLE, PROBABLE, SPECIFIC PERIODS, BY SIMILARITY and proteins with multiple annotations and localizations were discarded. To eliminate any bias the set was reduced by [40, 41] algorithm using a similarity notion based on the HSSP distance [37, 38]. The procedure guarantees that any pair of 100 residues long sequences have less than 35% pairwise identical residues. The number of sequences within each of the 10 subcellular localization category is reported in Table 2.3.

Nair & Rost Eukaryotic Sequences.

This data set was generated by [24] extracting two sets of sequences one for non-plant eukaryotic proteins (1,505 sequences) and one for plant organisms (304 sequences) from the SWISSPROT database release 40 [39]. Sequences annotated with words such as MEMBRANE, POSSIBLE, PROBABLE, SPECIFIC PERIODS, BY SIMILARITY and proteins with multiple annotations and localizations were discarded. In order to find a balance between biased data (that would yield over estimated accuracy results) and too small data set (that would yield incorrect estimates) the set was reduced by [40, 41] algorithm using a similarity notion based on the HSSP distance [37, 38]. The procedure guarantees that any pair of 250 residues long sequences have less than 25% pairwise identical residues. The number of sequences within each of the 10 subcellular localization category is reported in Table 2.4.

Table 2.2. Number of sequences within each subcellular localization category for the Lu et al. [23] data set, available at http://www.cs.ualberta.ca/~bioinfo/PA/Subcellular/experiments/intro.html.

Subcellular localization	Archaea	Gram Positive	Gram Negative	Fungi	Plant
Chloroplast					1,899
Cytoplasmic	404	930	1861	395	477
Endoplasmic reticulum				64	64
Extracellular	5	252		171	127
Golgi				52	35
Lysosomal					
Mitochondria				406	307
Nucleus				621	168
Peroxisomal			385	64	29
Vacuole				19	82
Membrane	62	340		302	135
Wall	6	19	46		
Inner membrane			432		
Outer membrane			197		
Total	477	1,541	3,147	2,094	3,293
# of classes	4	4	6	9	10

Table 2.3. Number of sequences within each subcellular localization category for the Nair and Rost data set. The data set is available at the site http://www.cs.ualberta.ca/~bioinfo/PA/Subcellular/experiments/intro.html

Subcellular localization	NR3K	NR1K
Chloroplast	197	94
Cytoplasmic	544	136
Endoplasmic reticulum	108	14
Extracellular	724	334
Golgi	52	22
Lysosomal	49	7
Mitochondria	467	190
Nucleus	922	352
Peroxisomal	55	8
Vacuole	28	4
Total	3,146	1,161

2.6.3 Experimental Results

We have compared several approaches ranging from k-NN methods that use traditional similarity measures based on alignment scores such as BLAST [36] and PSI-BLAST [44] to SVM methods based on a variety of kernels for sequences. In the following we report experimental tests on fair basis between all these methods over the two sets of redundant and non-redundant sequence databases. The learning task is a multi-classification with a number of classes varying from 5 to 10. When using support vector machines we employed a one-vs-all multiclass learning strategy. The experimental setting uses a 4-fold validation technique to determine, with a grid search, the best parameters combination. The only parameter that was not optimized was the context size l for the WDK that was set to 15 in all experiments (this value was

Table 2.4. Number of sequences within each subcellular localization category for the Plant/Non-plant Nair and Rost data set. The data set is available at http://cubic.bioc.columbia.edu/results/2005/LOCtree-data.

Subcellular localization	NonPlant	Plant
Number of classes	5	6
Chloroplast		103
Cytoplasmic	336	77
Extracellular	361	22
Mitochondria	201	50
Nucleus	560	32
Organelles	47	20
Total	1,505	304

suggested as adequate by previous experiments). The final results are reported micro-averaging the multiclass accuracy over a 5-fold experiment. In the case of composition, spectrum and WDK we have also tested the hypothesis that the most relevant part of the localization signal would be encoded in the N-terminal leading sequence of the protein: instead of using all the aminoacidic sequence we have considered only the first 20-50 residues where most probably the localization signal is placed to infer the final cell compartment for the whole protein.

Alignment k-NN.

Proteins that have a significant biological relationship to one another often share only isolated regions of sequence similarity. For identifying relationships of this nature, the ability to find local regions of optimal similarity is advantageous over global alignments that optimize the overall alignment of two entire sequences. BLAST (Basic Local Alignment Search Tool) [36] is an alignment tool that uses a measure of local similarity to score sequence alignments in such a way as to identify regions of good local alignment. We perform a K-NN classification using BLAST score as distance measure. We tested both a plain majority voting and a weighted majority voting strategy to determine the final class. In the first case we assign the class of the majority of the k neighbours, while in the second case we weight each vote with the actual BLAST score. The parameters search space includes the number of neighbours $k \in 1, 3, 5, 7$ and the E-Value (Expectation value) $\in 1, 10$, where the expectation value is the number of different alignments with scores equivalent to or better than the given threshold that are expected to occur in a database search by chance (the lower the E-value, the more significant the score). High E-values have been used to guarantee that at least one similar sequence could be determined even for the small training data sets used. In this working conditions the best k value was 1, as adding information from more unreliable neighbours was not beneficial to the accuracy of the prediction. This result is expected as in data sets which are not controlled for redundancy the single best homologue protein alone is sufficient to determine the correct class.

PSI-BLAST (Position-Specific Iterated BLAST) [44] is a tool that produces a position-specific scoring matrix constructed from a multiple alignment of the top-scoring BLAST responses to a given sequence. When a profile is used to search a database, it can often detect subtle relationships between proteins that are distant structural or functional homologues. This information is used to determine evolutionary relatedness often not detected by a BLAST search with a sample sequence query. A second experiment was therefore carried out using PSI-BLAST to obtain sequence profiles for each test protein in our experiments. Note that profiles alone have been computed performing multiple-alignment against the vast and comprehensive NR Database (composed by all Non-Redundant GenBank CDS translations, Protein Data Bank, SWISSPROT, Protein Information Resources, Protein Research Foundation). The

Table 2.5. Micro-averaged 5-fold multi-class accuracy and standard deviation results for Trivial, BLAST and PSI-BLAST distance based k-NN. C # is the number of classes. Last four data sets have low redundancy.

Data set	Size	C #	Trivial	BLAST	PSI-BLAST
SubLocEu	2,427	4	45.2	85.1±1.5	85.7±1.5
SubLocPro	997	3	69.0	89.2±1.2	91.1±3.0
Archaea	477	4	84.7	98.3±0.9	97.7±1.4
Fungi	2,094	9	29.7	75.9±1.1	74.0±1.1
GN	3,174	6	58.6	92.7±0.8	89.7±0.8
GP	1,541	4	60.4	92.1±0.7	88.1±1.2
Plant	3,293	10	57.7	93.8±1.2	91.6±1.3
NR1K	1,161	10	30.3	62.1±2.4	63.2±4.3
NR3K	3,146	10	29.3	61.5±1.8	61.9±1.4
Plant	304	6	33.9	26.8±2.8	29.3±1.7
NonPlant	1,505	5	37.2	53.9±2.2	53.7±1.6

classification procedure and parameters exploration follows the same strategy as in the previous case using two iterations as commonly done in literature. Once again, mainly due to the small size of the training data sets, the best parameters are $k = 1$ and E-value= 1.

Experimental results are reported in Table 2.5 where as a baseline we also include the results obtained using a trivial predictor that always outputs the most frequent class.

Amino Acid Composition Kernel.

An SVM multi-class experiment has been performed over the 7 redundant and the 4 non redundant sequence data sets. Three experimental settings have been tested: 1) straightforward application of the amino acid composition kernel of Equation 2.6, 2) the composition with a Gaussian kernel and 3) the composition kernel over the first N-terminal leading residues. The grid search for parameters explored variation on the regularization parameter C in the range $\{0.01, 0.1, 1, 10, 50, 100\}$, on the Gaussian variance parameter g in a range $\{0.01, 0.1, 1, 10, 50, 100, 150, 200, 250, 300\}$ and on the length of the leading N-terminal sequence nt in the range $\{20, 50\}$. Experimental results are reported in Table 2.6.

Spectrum Kernel.

An SVM multi-class experiment has been performed over the 7 redundant and the 4 non redundant sequence data sets. Three experimental settings have been tested: 1) straightforward application of the spectrum kernel of Equation 2.10, 2) the spectrum with a Gaussian kernel and 3) the spectrum kernel over the first N-terminal leading residues. The grid search for parameters explored variation on the regularization parameter C in the range $\{0.01, 0.1, 1, 10, 50, 100\}$, on the size of the r−mer r in a range $\{2, 3, 4\}$, on the Gaussian variance parameter g in a range $\{0.01, 0.1, 1, 10, 50, 100, 150, 200, 250, 300\}$ and on the length of the leading N-terminal sequence nt in the range $\{20, 50\}$. Experimental results are reported in Table 2.7.

Weighted Decomposition Kernel.

An SVM multi-class experiment has been performed over the 7 redundant and the 4 non redundant sequence data sets. Three experimental settings have been tested: 1) straightforward application of the WDK kernel of Equation 2.13, 2) WDK with a Gaussian kernel and 3) WDK over the first N-terminal leading residues. The grid search for parameters explored variation on the regularization parameter C in the range $\{0.01, 0.1, 1, 10, 50, 100\}$, on the size of the r−mer r in a range $\{2, 3, 4\}$, on the Gaussian parameter g in a range $\{0.01, 0.1, 1, 10, 50, 100, 150, 200, 250, 300\}$ and on the length of the leading N-terminal sequence nt in the range $\{20, 50\}$. Experimental results are reported in Table 2.8.

Weighted Decomposition Mismatch Kernel.

We tested the mismatch WDK 2.4.3 on the SubLoc Eukaryotic and Prokaryotic datasets. The grid search for parameters explored variation on the regularization parameter C in the range $\{0.01, 0.1, 1, 10, 50, 100\}$, on the size of the

Table 2.6. Micro-averaged 5-fold multi-class accuracy and standard deviation results and parameters for Composition, Composition + Gaussian, and Composition over Partial N-terminal sequence kernels. Parameters are C: regularization coefficient, g: Gaussian variance, and nt: length of N-terminal sequence. Last four data sets have low redundancy.

Data set	Composition		G-Composition			N-Composition			
	C	Accuracy	g	C	Accuracy	nt	g	C	Accuracy
SubLocEu	50	69.2±1.2	10	10	81.3±1.1	50	10	10	73.7±1.1
SubLocPro	10	88.5±2.4	10	1	92.2±2.0	50	10	1	83.2±1.6
Archaea	100	98.3±0.5	1	50	98.5±0.9	50	1	50	96.4±1.2
Fungi	10	57.5±1.4	10	10	70.3±1.5	50	10	1	59.7±0.8
GN	100	83.8±1.0	10	10	90.8±0.5	50	10	10	83.9±0.8
GP	50	86.0±1.0	10	1	89.1±1.1	50	10	10	83.9±2.4
Plant	100	64.7±1.0	10	10	88.3±0.6	50	10	10	83.7±0.9
NR1K	10	52.6±2.6	1	10	60.2±2.8	50	10	1	53.4±3.2
NR3K	10	44.8±2.0	10	1	56.3±1.1	50	10	0.1	48.2±0.6
Plant	50	52.9±2.0	10	1	58.2±4.9	50	0.1	50	53.6±4.7
NonPlant	10	54.6±1.9	1	50	61.7±2.4	50	1	10	57.9±3.6

Table 2.7. Micro-averaged 5-fold multi-class accuracy and standard deviation results and parameters for Spectrum, Spectrum with Gaussian, and Spectrum over Partial N-terminal sequence kernels. Parameters are C: regularization coefficient, r: size of the r-mers, g: Gaussian parameter, and nt: length of N-terminal sequence. Last four data sets have low redundancy.

Data set	Spectrum			G-Spectrum				N-Spectrum			
	r	C	Accuracy	r	g	C	Accuracy	nt	r	C	Accuracy
SubLocEu	4	10	83.3±2.3	2	1	50	83.2±0.3	50	3	50	77.0±1.4
SubLocPro	3	10	89.3±3.4	2	1	10	90.5±2.3	50	3	10	83.1±1.9
Archaea	2	10	97.9±0.7	2	0.1	50	98.1±0.8	50	2	10	96.8±1.2
Fungi	3	1	74.5±0.2	3	1	10	74.5±1.3	50	3	1	62.4±0.9
GN	3	10	91.7±0.4	3	0.01	100	91.4±0.3	50	3	10	87.1±0.9
GP	3	10	91.8±0.8	3	0.01	100	92.1±1.2	50	3	10	87.3±1.0
Plant	3	50	92.0±0.4	3	0.1	100	92.1±0.5	50	3	10	89.4±1.2
NR1K	3	10	55.8±1.8	2	1	50	60.7±3.3	50	2	0.01	53.3±1.8
NR3K	3	1	51.2±1.5	2	1	1	53.7±1.6	50	3	0.1	45.6±0.8
Plant	2	0.01	46.8±6.2	2	1	0.01	47.7±3.0	50	2	1	52.9±6.9
NonPlant	2	1	57.7±1.3	2	1	1	60.0±2.8	50	2	1	54.9±1.9

r−mer r in a range $\{2, 3, 4\}$. The allowed mismatch was $m = 1$. Using the approximation mentioned in Section 2.4.4 the micro-averaged 5-fold multi-class accuracy was 84.6 for the SubLoc Eukaryotic and 90.0 for the Prokaryotic set, thus yielding a result not statistically different from that obtained with the plain WDK. The method is however much more computationally expensive

Table 2.8. Micro-averaged 5-fold multi-class accuracy and standard deviation results and parameters for WDK, WDK with Gaussian, and WDK over Partial N-terminal sequence kernels. Parameters are C: regularization coefficient, r: size of the selector, g: Gaussian parameter, and nt: length of N-terminal sequence. Last four data sets have low redundancy.

Data set	WDK			G-WDK				N-WDK			
	r	C	Accuracy	r	g	C	Accuracy	nt	r	C	Accuracy
SubLocEu	2	50	84.6±0.9	2	1	10	85.5±0.6	50	3	10	81.3±1.5
SubLocPro	2	10	89.4±1.5	2	1	10	90.2±1.5	50	2	10	84.2±2.3
Archaea	2	10	97.2±0.9	2	0.1	50	97.2±0.9	50	2	10	96.8±1.2
Fungi	3	10	77.3±0.6	2	1	10	77.2±1.5	50	3	10	66.4±1.7
GN	2	10	91.5±1.0	2	0.1	50	91.6±0.8	50	3	10	88.5±1.0
GP	2	10	92.3±0.5	2	0.01	100	91.5±0.7	50	3	10	88.4±1.5
Plant	3	10	92.7±0.3	3	0.01	100	92.7±0.4	50	3	10	90.3±1.2
NR1K	3	10	59.5±1.9	2	1	10	61.1±4.5	50	2	1	58.0±1.8
NR3K	2	1	53.5±1.7	2	1	10	55.6±1.2	50	2	1	51.8±0.8
Plant	2	0.1	49.0±2.1	2	0.01	100	50.3±2.3	50	2	1	51.3±4.1
NonPlant	2	1	59.8±2.3	2	0.01	100	61.4±2.4	50	2	1	59.8±2.6

as the mismatched selectors are being made explicit and therefore no further experiments on other data sets have been carried out.

2.6.4 Hierarchical Localization

It has been suggested [24] that better classification results could be achieved in inferring the final localization of a protein by mimicking the cellular sorting mechanism. The idea is to identify the reasons why proteins are transported. We should identify signal and transit peptides and nuclear localization signals, rather than characterize the proteins in their final location. In addition localization errors should not be equally weighted since some cellular compartments are more similar than others, e.g. the endoplasmic reticulum is more similar to the extra-cellular environment than to the nuclear due to the biological sorting machinery. The suggested approach consists in moving from a multi-classification problem to a hierarchical-classification problem where a hierarchical ontology of localization classes is modelled onto biological pathways. In this way mistakes are penalized by construction in a different way if they are about the same pathway or not. Two hierarchies are suggested, one for proteins belonging to non-plant organisms and one for plant organisms. In the non-plant case we have a first distinction between proteins belonging to the secretory pathway and the intra-cellular environment, that is extracellular plus organelles proteins and nuclear plus cytoplasmic and mitochondrial proteins. At a second level we then discriminate between extracellular and organelles proteins, and between nuclear and cytoplasmic plus mitochondrial proteins. Finally we discriminate between cytoplasmic and mitochondrial proteins. In the plant case the chloroplast are further discriminated from the mitochondrial proteins (see Figure 2.2). Classification is performed by training different models to discriminate the various categories. The testing phase takes place by performing several binary decisions on the predicted output of the appropriate model at various nodes of the hierarchy.

Hierarchical Weighted Decomposition Kernel.

In [24] the final SVM predictor LOCTree comprises a variety of information on the protein ranging from the amino acid composition to the 50 N-terminal residues, from the output of SignalP neural network system to the profile information resulting from the alignment to the NR data set. We compared a WDK approach against the LocTree system for the hierarchical-classification task. Note that at each level we have a different number of classification tasks ranging from 2 at level 0 to 5 at level 3; that is at the final level we report the classification accuracy in one of the 5 possible classes. The grid search for parameters explored variation on the regularization parameter C in the range $\{0.01, 0.1, 1, 10, 50, 100\}$, on the size of the k−mer k in a range $\{2, 3, 4\}$, on the Gaussian parameter g in a range $\{0.01, 0.1, 1, 10, 50, 100, 150, 200, 250, 300\}$. Best results were obtained for Plant data set with $k = 2$, $C = 100$ and $g = .01$

Fig. 2.2. Hierarchical ontology of localization classes modelled onto biological pathways.

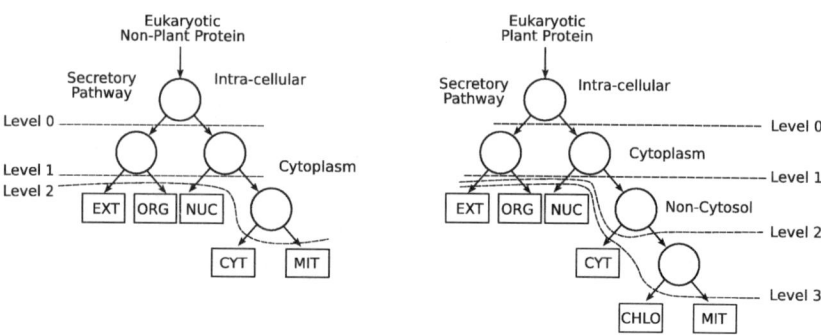

while for Non-Plant data set with $r = 2$, $C = 10$ and $g = 1$. Comparative experimental results are reported in Table 2.9. Confronting the final classification result with the WDK compounded with the Gaussian kernel for the Plant (50.3±2.3) and NonPlant (61.4±2.4) data sets, we observe no significant difference let alone improvement. We believe the better performance obtained by the LocTree system is to be attributed to the different features used rather than to the original problem being cast into a hierarchical-classification task.

2.7 Discussion

We report the best result obtained by each method (be it plain, compounded with a Gaussian or specialized only on the leading N-terminal sequence) in Figure 2.3 for redundant and non-redundant data sets. Even though there is no clear all-times winner we can draw some conclusions on the properties of the several string classification methods applied to the sub-cellular localization task. We note that alignment k-NN methods are quite sensitive to

Table 2.9. Multiclass accuracy for WDK and LocTree. Hierarchy levels and classes are explained in Fig.2.2.

	Plant			NonPlant		
Level	N Class	WDK	LocTree	N Class	WDK	LocTree
0	2	88.2	94±4	2	84.0	89±2
1	4	77.6	88±5	4	67.6	78±4
2	5	55.3	77±7	5	60.3	74±6
3	6	45.7	70±3	-	-	-

Fig. 2.3. Comparison on redundant (a) and non-redundant (b) data sets of best results obtained by each method among the plain, compounded with a Gaussian or specialized only on the leading N-terminal sequence setting.

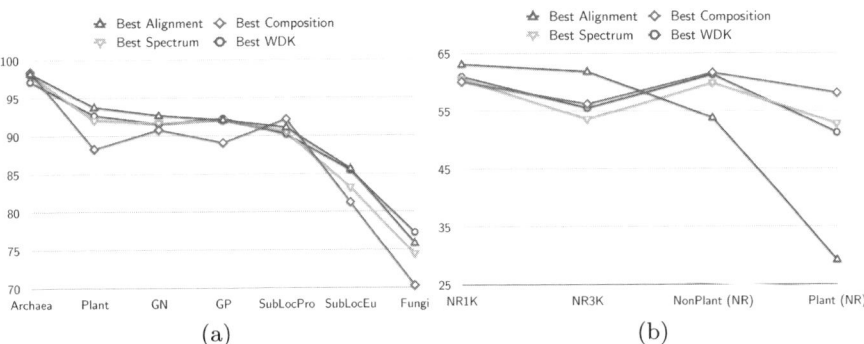

redundancy in the data set. This is to be expected since good performance in k-NN methods rely on the presence of train examples "close" to test sequences: if those examples are forcefully removed from the data set the induction becomes inevitably less accurate and reliable. A desired property of a protein classification method is that it should be robust to the various definitions (including thresholds values) of redundancy. This can make alignment k-NN methods less interesting despite their occasional superior performance.

Composition kernel SVM methods are more stable and show interesting performance levels, despite the simplicity of the approach, especially when dealing with non-redundant data. Probably this happens when (lacking better information from similar examples) the cellular compartment is predicted resorting to global properties. For example when we want to discriminate between the secretory pathway and the intra-cellular environment it can be enough to estimate the average hydrophobicity of the whole sequence. When, however, the redundancy level of the data set increases, the composition alone is unable to capture finer information on the similarity or common ancestry between proteins, and the classification performance falls short of other methods such as spectrum or weighted decompositional kernels.

We finally observe how the WDK approach generally outperforms the spectrum kernel and performs reasonably well over non-redundant data, matching the top results of alignment methods in redundant conditions.

Acknowledgments

This research is supported by EU STREP APrIL II (contract no. FP6-508861) and EU NoE BIOPATTERN (contract no. FP6-508803).

References

1. Taskar, B., Chatalbashev, V., Koller, D., Guestrin, C.: Learning structured prediction models: A large margin approach. In: Twenty Second International Conference on Machine Learning (ICML05), Bonn, Germany (2005)
2. Tsochantaridis, I., Hofmann, T., Joachims, T., Altun, Y.: Support vector machine learning for interdependent and structured output spaces. In: International Conference on Machine Learning (ICML04). (2004)
3. Haussler, D.: Convolution kernels on discrete structures. Technical Report UCSC-CRL-99-10, University of California, Santa Cruz (1999)
4. Lodhi, H., Saunders, C., Shawe-Taylor, J., Cristianini, N., Watkins, C.: Text classification using string kernels. J. Mach. Learn. Res. **2** (2002) 419–444
5. Leslie, C.S., Eskin, E., Noble, W.S.: The spectrum kernel: A string kernel for svm protein classification. In: Pacific Symposium on Biocomputing. (2002) 566–575
6. Cortes, C., Haffner, P., Mohri, M.: Rational kernels: Theory and algorithms. J. of Machine Learning Research **5** (2004) 1035–1062
7. Collins, M., Duffy, N.: Convolution kernels for natural language. In: NIPS 14. (2001) 625–632
8. Vishwanathan, S., Smola, A.: Fast kernels on strings and trees. In: Advances in Neural Information Processing Systems 2002. (2002)
9. Kashima, H., Tsuda, K., Inokuchi, A.: Marginalized kernels between labeled graphs. In: Proceedings of ICML'03. (2003)
10. Gärtner, T.: A survey of kernels for structured data. SIGKDD Explor. Newsl. **5**(1) (2003) 49–58
11. Gärtner, T., Lloyd, J., Flach, P.: Kernels and distances for structured data. Mach. Learning **57**(3) (2004) 205–232
12. Passerini, A., Frasconi, P.: Kernels on prolog ground terms. In: Int. Joint Conf. on Artificial Intelligence (IJCAI'05), Edinburgh (2005)
13. Passerini, A., Frasconi, P., De Raedt, L.: Kernels on prolog proof trees: Statistical learning in the ILP setting. Journal of Machine Learning Research **7** (2006) 307–342
14. Ben-David, S., Eiron, N., Simon, H.U.: Limitations of learning via embeddings in euclidean half spaces. J. of Mach. Learning Research **3** (2002) 441–461
15. Schölkopf, B., Weston, J., Eskin, E., Leslie, C.S., Noble, W.S.: A kernel approach for learning from almost orthogonal patterns. In: Proc. of ECML'02. (2002) 511–528
16. Menchetti, S., Costa, F., Frasconi, P.: Weighted decomposition kernels. In: Proc. Int. Conf. on Machine Learning (ICML'05). (2005)
17. Jaakkola, T., Diekhans, M., Haussler, D.: A Discriminative Framework for Detecting Remote Protein Homologies. J. of Comp. Biology **7**(1–2) (2000) 95–114
18. Reinhardt, A., Hubbard, T.: Using neural networks for prediction of the subcellular location of proteins. Nucleic Acids Research **26**(9) (1998) 2230–2236
19. Chou, K.C., Elrod, D.: Prediction of membrane protein types and subcellular locations. Proteins **34** (1999) 137–153
20. Emanuelsson, O., Nielsen, H., Brunak, S., von Heijne, G.: Predicting subcellular localization of proteins based on their n-terminal amino acid sequence. J Mol. Biol. **300** (2000) 1005–1016
21. Hua, S., Sun, Z.: Support Vector Machine for Protein Subcellular Localization Prediction. Bioinformatics **17**(8) (2001) 721–728

22. Nair, R., Rost, B.: Sequence conserved for subcellular localization. Protein Science **11** (2002) 2836–2847
23. Lu, Z., Szafron, D., Greiner, R., Lu, P., Wishart, D.S., Poulin, B., Anvik, J., Macdonell, C., Eisner, R.: Predicting subcellular localization of proteins using machine-learned classifiers. Bioinformatics **20**(4) (2004) 547–556
24. Nair, R., Rost, B.: Mimicking cellular sorting improves prediction of subcellular localization. J Mol Biol **348**(1) (2005) 85–100
25. Shawe-Taylor, J., Cristianini, N.: Kernel Methods for Pattern Analysis. Cambridge Univ. Press (2004)
26. Varzi, A.: Parts, wholes, and part-whole relations: the prospects of mereotopology. Knowledge and Data Engineering **20** (1996) 259–286
27. Gärtner, T., Flach, P., Wrobel, S.: On graph kernels: Hardness results and efficient alternatives. In Schölkopf, B., Warmuth, M.K., eds.: Proc. of COLT/Kernel '03. (2003) 129–143
28. Jebara, T., Kondor, R., Howard, A.: Probability product kernels. J. Mach. Learn. Res. **5** (2004) 819–844
29. Odone, F., Barla, A., Verri, A.: Building kernels from binary strings for image matching. IEEE Transactions on Image Processing **14**(2) (2005) 169–180
30. Wu, C., Berry, M., Shivakumar, S., McLarty, J.: Neural networks for full-scale protein sequence classification: Sequence en coding with singular value decomposition. Machine Learning **21**(1) (1995) 177–193
31. Leslie, C., Eskin, E., Cohen, A., Weston, J., Stafford Noble, W.: Mismatch string kernels for discriminative protein classification. Bioinformatics **20**(4) (2004) 467–476
32. Devos, D., Valencia, A.: Practical limits of function prediction. Proteins: Structure, Function, and Genetics **41** (2000) 98–107
33. Webb, E.C.: Enzyme nomenclature 1992 : recommendations of the nomenclature committee of the international union of biochemistry and molecular biology on the nomenclature and classification of enzymes. San Diego : Published for the International Union of Biochemistry and Molecular Biology by Academic Press (1992)
34. Lewis, S., Ashburner, M., Reese, M.: Annotating eukaryote genomes. Current Opinion in Structural Biology **10**(3) (2000) 349–354
35. Doolittle, R.: Of URFs and ORFs: a primer on how to analyze derived amino acid sequences. University Science Books, Mill Valley California (1986)
36. Altschul, S., Gish, W., Miller, W., Myers, E., Lipman, D.: A basic local alignment search tool. J Mol. Biol. **215** (1990) 403–410
37. Rost, B.: Twilight zone of protein sequence alignment. Protein Engineering **12**(2) (1999) 85–94
38. Sander, C., Schneider, R.: Database of homology-derived protein structures and the structural meaning of sequence alignment. Proteins **9**(1) (1991) 56–68
39. Boeckmann, B., Bairoch, A., Apweiler, R., Blatter, M.C., Estreicher, A., Gasteiger, E., Martin, M.J., Michoud, K., O'Donovan, C., Phan, I., Pilbout, S., Schneider, M.: The swiss-prot protein knowledgebase and its supplement trembl in 2003. Nucleic Acids Res **31**(1) (2003) 365–370
40. Hobohm, U., Scharf, M., Schneider, R., Sander, C.: Selection of representative protein data sets. Protein Science **1** (1992) 409–417
41. Mika, S., Rost, B.: Uniqueprot: creating sequence-unique protein data sets. Nucleic Acids Res. **31**(13) (2003) 3789–3791

42. Liò, P., Vannucci, M.: Wavelet change-point prediction of transmembrane proteins. Bioinformatics **16**(4) (2000) 376–382
43. Chen, C., Rost, B.: State-of-the-art in membrane protein prediction. Applied Bioinformatics **1**(1) (2002) 21–35
44. Altschul, S., Madden, T., Schaffer, A., Zhang, J., Zhang, Z., Miller, W., Lipman, D.: Gapped blast and psi-blast: a new generation of protein database search programs. Nucleic Acids Res **25**(17) (1997) 3389–3402

3

Mining Structure-Activity Relations in Biological Neural Networks using *NeuronRank*

Tayfun Gürel[1,2], Luc De Raedt[1,2,4] and Stefan Rotter[1,3]

[1] Bernstein Center for Computational Neuroscience, Freiburg, Germany
[2] Institute for Computer Science, Albert Ludwigs University of Freiburg, Germany
[3] Institute for Frontier Areas of Psychology and Mental Health, Freiburg, Germany
[4] Department of Computer Science, Katholieke Universiteit Leuven, Belgium
guerel@informatik.uni-freiburg.de, LucDeRaedt@cs.kuleuven.be, stefan.rotter@biologie.uni-freiburg.de

Summary. Because it is too difficult to relate the structure of a cortical neural network to its dynamic activity analytically, we employ machine learning and data mining to learn structure-activity relations from sample random recurrent cortical networks and corresponding simulations. Inspired by the *PageRank* and the *Hubs & Authorities* algorithms for networked data, we introduce the *NeuronRank* algorithm, which assigns a source value and a sink value to each neuron in the network. Source and sink values are used as structural features for predicting the activity dynamics of biological neural networks. Our results show that *NeuronRank* based structural features can successfully predict average firing rates in the network, as well as the firing rate of output neurons reflecting the network population activity. They also indicate that link mining is a promising technique for discovering structure-activity relations in neural information processing.

3.1 Introduction

Important functions of our brain are mediated by the operation of complex neuronal networks. The relation between structure and function of the various types of networks has been subject of many theories and intense computational modeling. Fundamental questions, however, remain unanswered: How important is the structure of a network for its function? Are certain structural features essential for a particular function? Can one and the same structure support different functions? Can different structures support the same function? How does the repeated usage of a network change its structure and its function, respectively? How does the interaction between networks determine the function of the whole system? These and other related questions are of central interest in the neurosciences.

3.1.1 Link mining and networked data

Neural networks in the brain have, at the structural level, the same format as social networks, food webs, citation networks, the Internet, or networks of biochemical reactions: they can be represented as large graphs, linking many interacting elements to each other. The best studied example of such networks is probably the world wide web. Web pages and hyperlinks between them can be considered as vertices and directed edges, respectively, of a huge graph. Similarly, collections of scientific publications exhibit network characteristics, as publications cite each other. Social networks, or the transmission of epidemic diseases are further examples for networks with nodes representing human beings.

Link mining [10] [11] is a recent direction within the field of data mining [24] that aims at discovering knowledge and models from 'networked' data, which is essentially empirical data about a particular network structure and its dynamics. Recent contributions include Kleinberg's [14] Hubs & Authorities algorithm, which is able to detect authoritative sources of information on the web by exploiting its link structure and Page and Brin's [18] *PageRank* algorithm underlying the Google search engine, which successfully predicts the relevance of a web page to the user and ranks the page. The interest for rating and ranking has also extended to classification of the linked entities. From Chakrabarti's et al. [4] simple iterative relaxation labeling algorithm to Taskar's et al. [22] complex Relational Markov Networks, many approaches used link information successfully for classification of web pages, research articles, and similar entities. Domingos' [9] viral marketing method assigned network values to human beings in a social network in order to optimize marketing costs by focusing the marketing on the influential customers. These are just a few examples of the tasks studied within the rapidly developing area of link mining.

3.1.2 Neuronal networks of the mammalian cortex

The cortex of mammals ('gray matter') is the locus of many brain functions related to sensory perception, voluntary movement control and 'higher' cognitive processes. Despite this broad spectrum of different functions, the structure of cortex is roughly the same everywhere, both across modalities and across species. It is clear that the limited information in the genome cannot represent all details of the sophisticated network layout directly, but it could encode general rules for the formation of neuronal connectivity during development and learning. In order to find these rules, neuroanatomists have characterized both the common and the distinguishing features of the various networks statistically [2]. In a nutshell, a local volume of $1\,\mathrm{mm}^3$ cortical tissue contains roughly 10^5 neurons, of which 80 % are excitatory, the rest is inhibitory. The degree of convergence/divergence in the network is enormous:

each neuron contacts about 10 % of all the other cells, yielding a recurrent network of unprecedented complexity.

What kind of activity dynamics is implied by such networks? Sparsely coupled random recurrent networks of integrate-and-fire neurons have established themselves as computational models here. They have been studied both analytically and with the help of numerical simulations [3]. These networks may display different types of activity, depending on certain structural features, and on the characteristics of external input. Among other things, irregularity of individual spike trains and synchrony across neurons were identified as meaningful and easy-to-handle descriptors of collective dynamic states. Asynchronous-irregular (AI) states of network activity, in fact, are regarded the most interesting ones, since they have also been demonstrated in electrophysiological experiments *in vivo*.

More widespread cortical networks spanning distances of several millimeters, say, are not any more well characterized as a random graph. Long-range connections are exceedingly sparse as compared to local couplings. Consequently, viewed on a large scale, cortex is dominated by local connections, enhanced by a relatively small number of long-range connections. Networks with this specific characteristic are known as 'small world' networks. They can be highly clustered, like regular lattices, yet have small characteristic path lengths, like random graphs [23]. Such networks were also shown to exhibit somewhat different dynamical states than neural networks of totally random connectivity [1].

The structure of large networks may be characterized in terms of small subgraph patterns, called network motifs. Figure 3.1 depicts all possible motifs involving three nodes. Motif counts can be thought of as statistical summary information for network connectivity. Song et al. [21] have shown that certain connectivity motifs of pyramidal neurons in layer 5 rat visual cortex are in fact overrepresented as compared to random networks. Relations between motif statistics and activity dynamics of networks have also been investigated [19]. The results indicate that the abundance or absence of particular network motifs may contribute to the stability of certain networks arising in biology, e.g. the regulatory transcription network of the bacterium *E. coli*, or the neural connectivity map of the worm *C. elegans*.

3.1.3 Motivation and scope of this work

We started to investigate structure-function relations in the cortex by systematically exploring the relation between network structure and activity dynamics in cortical network models. The analysis of Brunel [3] and others showed how the complex dynamics of a random-topology cortical network is determined by various structural parameters. In particular, the influence of the relative strength of the inhibitory synaptic couplings in the network and the role of external inputs was elucidated. The question how structural variations contribute to variations in activity dynamics, however, was not tackled in

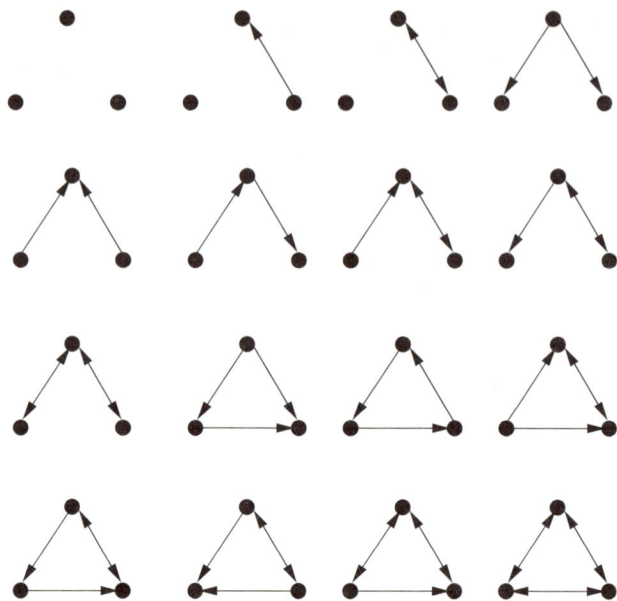

Fig. 3.1. Network motifs of three nodes. Any directed unlabeled graph of 3 nodes is isomorphic to exactly one of these motifs. Figure is adapted from Song et al. [21]

this work. Motivated by the finding that structural variations indeed influence the network dynamics [1] [19], we aimed at an automated discovery of structure-activity relations.

In this article[5] we investigate the applicability of link mining techniques to reveal structure-activity relations in biological neural networks. In particular, we are interested in learning a function that maps structural features of neural networks to activity-related features. We introduce the *NeuronRank* [12] algorithm, which yields structural features describing the level to which neurons are *functionally* excitatory and/or inhibitory within a recurrent network. *NeuronRank* is inspired by the *Hubs & Authorities* algorithm, and is shown to yield good predictions of network activity.

We proceed by giving an overview of our approach in Section 3.2. In Section 3.3, we present our network model. We explain how we analyze the network activity in Section 3.4. We introduce our key contribution, the *NeuronRank* algorithm, in Section 3.5. We describe our structural feature extraction methodology in Section 3.6. We finally present our experimental results in Section 3.7 and our conclusions in Section 3.8.

---

[5] This work is the significantly extended version of "Ranking neurons for mining structure-activity relations in biological neural networks: *NeuronRank*", Neurocomputing (2006), doi:10.1016/j.neucom.2006.10.1064

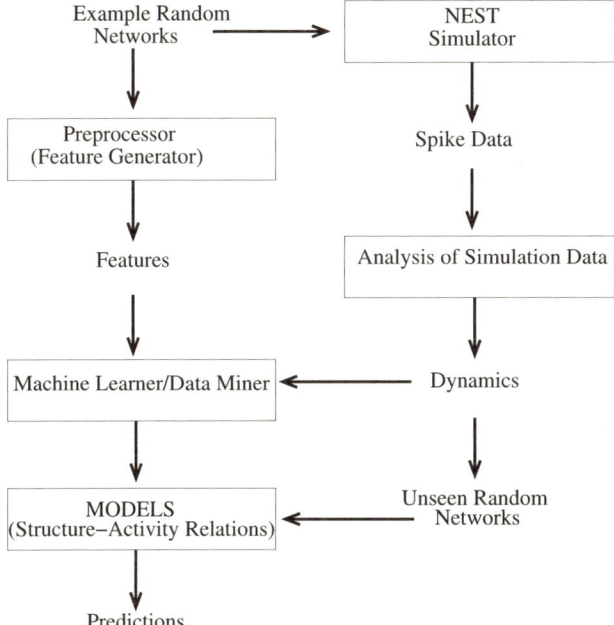

Fig. 3.2. Mining structure-activity relations in biological neural networks: overview of methods.

3.2 Overview of the method

Aiming at discovering structure-activity relations in cortical networks, we focus here on the following specific question: Which structural features of recurrent random networks are effective to predict the characteristics of their activity? Our approach is to employ link mining algorithms to find such features, and to use machine learning methods to achieve an effective prediction based on these features by learning the desired mappings from a set of examples. Figure 3.2 depicts a schematic overview of our approach.

Various structural features of specific networks were extracted, based on graph-theoretical descriptors and on the new *NeuronRank* algorithm. To assess the activity dynamics exhibited by these networks, we performed numerical simulations and measured the mean firing rates and other characteristic parameters based on spike statistics describing the activity dynamics. Several 'examples' consisting of a set of values for structural features and the corresponding activity features were then processed by machine learning algorithms, which generated statistical models for predicting the dynamics of unseen networks based on their specific structural features. We assessed the quality of these models by determining their predictive power.

3.3 The network models

3.3.1 Neuron model

We performed our simulations using the leaky integrate-and-fire neuron, which is regarded as a simple and moderately realistic model for spiking neurons, popular in the field of computational neuroscience [3]. The model describes the dynamics of the neuronal membrane potential based on the following differential equation

$$\tau \dot{V}(t) = -V(t) + RI(t),$$

where τ is the membrane time constant and $I(t)$ is the current reflecting synaptic input to the neuron. The equivalent circuit describing the neuronal membrane is shown in Figure 3.4. Synaptic currents are induced by the spikes of the presynaptic neurons as follows

$$RI(t) = \tau \sum_i J_i \sum_k \delta(t - t_i^k - D),$$

where τ is the membrane time constant, J_i is the efficacy of the synapse with neuron i, t_i^k are the spike times of neuron i, D is a fixed transmission delay, and δ is the Dirac delta-function.

Each spike in a presynaptic neuron i increases or decreases the membrane potential by J_i. The synaptic efficacy J_i is either positive or negative depending on whether the synapse is excitatory or inhibitory, respectively. The

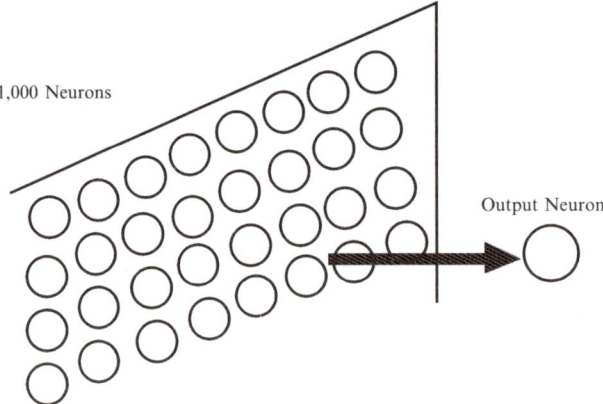

Fig. 3.3. Setup of the numerical simulations. We simulated recurrent cortical networks of 1,000 neurons. Each neuron in the network received external input in the form of an excitatory Poisson spike train with a total mean rate slightly above the threshold for sustained activity. All neurons in the network projected to a single 'readout' neuron, which did not receive any external inputs.

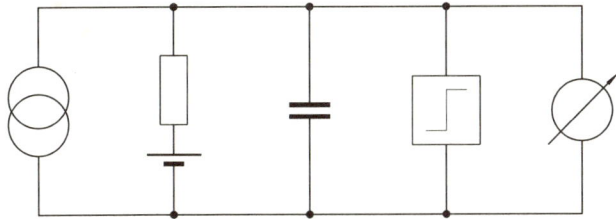

Fig. 3.4. Equivalent circuit for the integrate-and-fire neuron. In this model, spikes are generated by a simple threshold mechanism. The voltage between the lower and the upper nodes corresponds to the membrane potential.

capacitor is passively discharging through the leak resistor, with time constant τ. Whenever the voltage across the capacitor hits the threshold θ, it is instantly discharged to a value of U_{reset}, and a spike is sent to all postsynaptic neurons.

The following parameters were used in all simulations: membrane time constant $\tau = 20\,\text{ms}$, membrane leak resistance $R = 80\,\text{M}\Omega$, spike threshold $U_\theta = 20\,\text{mV}$, reset potential $U_{\text{reset}} = 10\,\text{mV}$, refractory period $2\,\text{ms}$. Synaptic currents were modeled as delta-pulses, delayed by $D = 1.5\,\text{ms}$ with respect to the inducing action potential.

3.3.2 Types of networks

We considered three types of networks to explore structure-activity relations:

Sparse random networks with identical synapses. We created recurrent networks of $n = 1{,}000$ integrate-and-fire neurons, in accordance with basic statistical features of the neocortex regarding neuron types and synaptic connectivity [2]. Each of the $n(n-1)$ potential synapses was established with probability 0.1, independently of all the others. Neurons were inhibitory with probability 0.2, and excitatory otherwise. Excitatory neurons made only synapses with a positive efficacy on their axons, inhibitory neurons had only synapses with negative efficacies (Dale's principle). The amplitude of all excitatory postsynaptic potentials was $0.1\,\text{mV}$, all inhibitory postsynaptic potentials had an amplitude of $-0.6\,\text{mV}$.

Small-world networks. We also generated small-world networks, starting from a locally connected ring graph with connectivity 0.1, and then randomly and independently rewiring all existing links with probability 0.1 [23]. The relative abundance of excitatory and inhibitory neurons was the same as in the random network case, and Dale's principle was obeyed. Inhibitory neurons were 6 times as efficient as excitatory neurons, as before.

Sparse random networks with normally distributed synaptic efficacies. We also experimented with random networks with normally distributed inhibitory and excitatory weights. We generated random networks as in the first setting, but then randomized synaptic efficacies by adding zero mean Gaussian perturbations.

All network simulations were performed using the NEST simulator ([7] [17], http://www.nest-initiative.org).

3.4 Activity-related features

We studied dynamical activity 'states' in the networks, based on the spike firing characteristics of the neurons, in particular the mean firing rate and the firing rate of a readout neuron.

Mean firing rate. The networks were simulated for 1.2 s biological time. Spike counts and spike rates were determined and averaged over all neurons

$$\nu_{\mathrm{avg}} = \frac{1}{n}\sum_{i=1}^{n}\frac{1}{T}\int_{0}^{T}\sum_{k}\delta(t-t_i^k)\,dt,$$

where n is the number of neurons in the network, T is the duration of simulation and t_i^k is the time of the k-th spike in the i-th neuron.

Firing rate of a readout neuron. The cortex is composed of many interacting local networks. It is, therefore, a relevant question how the activity of a local network affects other neurons or networks it is connected to. Here we considered the case of a single readout neuron that receives input from all neurons of a network (Figure 3.3). We were particularly interested in how the firing rate of the readout neuron depends on the structural variations of the network it receives input from. The firing rate of the readout neuron was defined as

$$\nu_{\mathrm{out}} = \frac{1}{T}\int_{0}^{T}\sum_{k}\delta(t-t^k)\,dt,$$

where T is the duration of simulation and t^k is the time of the k-th spike of the readout neuron.

3.5 The *NeuronRank* Algorithm

The *NeuronRank* algorithm, which is introduced below, assigns a *source value* α_i and a *sink value* ω_i to each neuron i, based only on structural information. The source value α_i of a neuron encodes the net effect on the network induced by a spike in that neuron. As a rule, excitatory neurons will have positive source values, whereas inhibitory neurons will have negative source values. Exceptions to this rule, however, may exist. Namely, if an excitatory neuron excites many inhibitory neurons, it may attain a negative source value. On the other hand, if an inhibitory neuron inhibits many other inhibitory neurons, it may attain a positive source value. The absolute source value of a neuron is an indicator for its total impact on network activity. The sink value w_i, on the other hand, encodes the sensitivity of a neuron for activity somewhere else in the network. Neurons with higher sink values tend to be excited more by other neurons and therefore tend to have higher firing rates.

In a recurrent network, the source value of a neuron depends on the source values of all other neurons. In other words, the vector of all source values in a network recursively depends on itself, and the same holds for the vector of sink values. The *NeuronRank* algorithm, which we introduce below, finds a consistent set of source and sink values in a network. It iteratively updates the source value of a neuron according to the source values of its postsynaptic nodes. If A denotes the weighted adjacency matrix of the network

$$A_{ij} = \begin{cases} 1 & \text{for an excitatory synapse } j \to i \\ -g & \text{for an inhibitory synapse } j \to i \\ 0 & \text{otherwise,} \end{cases}$$

where $g > 0$ is a number that encodes the relative impact of inhibitory couplings relative to excitatory ones. The update rule for the row vector of source values $\alpha = (\alpha_1, \ldots, \alpha_n)$ is given by

$$\alpha \leftarrow \alpha A$$

starting with initial values $\alpha_i \pm 1$ depending on whether neuron i is excitatory or inhibitory. In contrast to source values, the sink value of a neuron is updated according to the sink values of its presynaptic nodes. The update rule for the column vector of sink values $w = (w_1, \ldots, w_n)^T$ is therefore given by

$$w \leftarrow A w,$$

starting with initial values $w_i = 1$ for all neurons. In each step of the iteration both α and w are normalized to unit length, and the iteration stops upon convergence. The detailed algorithm is depicted in Table 3.1.

3.6 Structural features

3.6.1 *NeuronRank* features

Upon convergence of the *NeuronRank* algorithm, statistical summary information about the source and sink values in a network is passed on to data mining algorithms. We considered in particular mean and variance of the source values, as well as mean and variance of the sink values of all neurons. In addition, sum, mean and variance were computed separately for excitatory and inhibitory neurons, respectively. This yielded a set of total 16 *structural features*.

Table 3.1. The NeuronRank algorithm.

Input: A directed labeled (inhibitory/excitatory) recurrent network N, represented by a weighted adjacency matrix A.
Output: Source (α) and sink (ω) values of all nodes in N

for each node i in N
  $\omega_i \leftarrow 1$
  if i is excitatory
    $\alpha_i \leftarrow 1$
  else if i is inhibitory
      $\alpha_i \leftarrow -1$
    endif
  endif
endfor
repeat
  $\alpha \leftarrow \alpha A$
  $\omega \leftarrow A \omega$
  normalize α and ω such that $\sum_i \alpha_i^2 = 1$ and $\sum_i \omega_i^2 = 1$
until convergence
return α and ω

3.6.2 Network motifs

Network motifs [16] are subnetworks involving a small number of neurons. Formally, a network motif of n-th order is an isomorphic set of unlabeled directed graphs $G = (V, E)$, where $V = \{v_1, v_2, ...v_n\}$ is the set of nodes and E is the set of edges. Consider a network with N nodes. Each of the $\binom{N}{n}$ different subnetworks with n nodes is then isomorphic to exactly one order n motif. The motif count k_G is the number of subnetworks that are isomorphic to one particular G. Examples of network motifs are shown in figure 3.1. Network motifs can be defined and counted in labeled networks as well, e.g. in neural networks where neurons have different types. In this case, label matching is part of equivalence checking.

We implemented an algorithm, which counts second order and third order motifs in a network. Motif counting was performed both with and without considering neuron types. Note that there are 16 third-order network motifs without considering the types, and 93 third-order network motifs considering the neuron types.

3.6.3 Cluster coefficient

We also computed the average cluster coefficient [23] of each network and used it as a feature for machine learning algorithms. The cluster coefficient was originally defined on undirected graphs, here we use the following straightforward extension to directed graphs

$$C = \frac{1}{N} \sum_n \frac{E_n}{k_n(k_n - 1)},$$

where k_n is the number of neighbors of neuron n (either presynaptic or postsynaptic), and E_n is the total count of synapses among the neighbors. The latter number is bounded by $k_n(k_n - 1)$. The coefficient $\frac{E_n}{k_n(k_n-1)}$, therefore, yields a measure of clustering from the perspective of neuron n. The average C of these coefficients for all neurons yields a measure for the cliquishness of the network [23].

3.7 Experimental results

In order to tackle the question whether structural features, including those obtained from the *NeuronRank* algorithm, are good predictors for the network activity, we applied well-known machine learning algorithms implemented in the WEKA workbench for data mining [24] to the four structure-activity relation prediction tasks sketched below. The machine learning algorithms employed were J48, a decision tree learner, the K2 algorithm [5] for learning Bayesian Networks, and Support Vector Machines [6] for predicting the class variable of the networks (e.g. 'high firing rate' or 'low firing rate'). J48 is the java version of C4.5 decision tree learning algorithm [20], which induces a tree in a top-down manner using information gain as its heuristic. C4.5 supports attributes with numerical values and performs tree post-pruning. K2 is a hill climbing based learning algorithm for Bayesian networks, which can deal with missing data. K2 uses a special ordering of the attributes to guide its search when adding arcs to the Bayesian network. In all experiments listed below, we report on the predictive accuracy of these algorithms using 10-fold cross-validation.

Predicting the average firing rate in sparse random networks with identical synapses. We generated 330 random networks and performed numerical simulations of their activity dynamics. For each network, we measured the firing rate averaged across all neurons. Firing rates above the median were labeled as 'high', below the median as 'low'. The task then was to predict the firing rate correctly ('high' vs. 'low'), based on the features extracted by motif counting (cc: clustering coefficient, inh: inhibitory neuron count, 3m: third order motifs, 2ie: second order motifs with signs, 3ie: third order motifs with signs) and by the *NeuronRank* algorithm (mean, variance and sum of source and sink values). Note that running *NeuronRank* took 25.2 seconds per network on average, whereas counting third order type-specific motifs took 255.2 seconds per network on average (Pentium-4 3.2 GHz processor, SuSE Linux 10.0 operating system). The results are shown in Table 3.2.

Predicting the firing rate of a readout neuron. In the numerical simulations, a readout neuron was added to each of the networks described above. This neuron received input from all neurons in the network, but no external input (see also Figure 3.3). We considered the same structural features as in the previous setting (ignoring the readout neuron) and trained the machine learning algorithms on examples to predict the firing rate of the readout neuron as 'low' or 'high' on unseen networks. The results are shown in Table 3.2.

Predicting the firing rate in small-world networks. We generated small world networks as explained in Section 3.3 and simulated their activity with low-frequency Poisson input. We used the same structural features and discretized the firing rate into 'high' and 'low', as described above. Prediction accuracies are shown in Table 3.3.

Predicting the average firing rate in sparse random networks with continuous synaptic efficacies. The same features, discretization, and input were applied as in the first experiment. However, normally distributed synaptic efficacies were imposed. The task was again to predict the firing rate as either 'high' or 'low'. Table 3.4 shows the outcome of this experiment.

The results of our experiments clearly demonstrate the predictive power of the *NeuronRank* features. For all experiments, *NeuronRank* features contributed significantly to the accuracy of the prediction, so we can regard them as excellent indicators of network activity. *NeuronRank* features, when used together with inhibitory neuron count (inh), was the best of all feature combinations for activity prediction. 3-rd order type-specific motif counts (3ie), also used together with inhibitory neuron count (inh), had the closest prediction accuracy to the *NeuronRank* features in general. This is not surprising since 3ie feature set contains the most detailed and the most complete summary of the network structure among motif counting based features. Some machine learning algorithms, such as decision trees (J48) and Bayesian networks (K2), are more sensitive to large numbers of features than support vector machines. This explains why the performance gain for SVMs when working with 3-rd order type specific motifs was typically larger.

Note that *NeuronRank* based best features (inh +source values+sink values) outperformed motif counting based best features (inh +3ie) in prediction. The difference was statistically significant according to sign test applied to those two feature sets with a p-value of 0.0032 (see Table 3.5).

3.8 Conclusions

We showed that it is possible to model certain aspects of the activity dynamics in random cortical networks by employing data mining and machine learning techniques. Furthermore, we demonstrated that *NeuronRank*, which is related to the *Hubs & Authorities* and *PageRank*, can successfully extract structural features that are relevant for predicting certain aspects of network activity. This indicates that learning algorithms can relate the neural structure (which is in the format of networked data) to certain features of

Table 3.2. Accuracy of prediction in sparse random networks using three well known machine learning algorithms: K2 Bayesian Network algorithm, J48 decision tree learner, and Support Vector Machines (SVM) using first and second order polynomial kernels. Note that the presented SVM results are the better ones from both types of kernels. cc: clustering coefficient, inh: inhibitory neuron count in the network, 2ie/3ie: second/third order motifs respecting node types, 3m: third order motifs ignoring node types

Features	Average Firing Rate			Readout Firing Rate		
	BN-K2	J48	SVM*	BN-K2	J48	SVM*
cc	48.5%	48.5%	49.4%	50.6%	50.6%	50.0%
inh	87.0%	87.0%	89.1%	89.7%	89.7%	90.9%
inh + cc	87.0%	86.7%	88.8%	89.7%	89.1%	90.1%
inh + 2ie	89.7%	91.5%	88.8%	91.2%	91.2%	91.5%
inh + 3m	87.0%	85.8%	88.8%	89.7%	88.5%	91.5%
inh + 3ie	86.7%	90.1%	93.3%	92.1%	91.5%	93.4%
inh + source values	92.7%	94.8%	94.2%	94.5%	93.0%	95.8%
inh + sink values	93.0%	93.0%	94.5%	91.8%	95.2%	95.8%
inh + source + sink values	92.1%	93.0%	94.8%	92.7%	93.6%	95.5%
source values	92.4%	93.0%	93.3%	92.4%	92.1%	93.0%
sink values	90.9%	92.4%	92.1%	92.1%	93.0%	93.0%
source + sink values	92.1%	93.3%	93.6%	92.4%	93.0%	94.2%

neural activity. Structural features can be successfully extracted by link mining methods. Our results indicate that employed link mining methods in the discovery of structure-activity and structure-function relations is useful. Information extraction tools, which are designed for networked data, can help neuroscientists to gain insight to computations performed by sophisticated neural structures. Since recent simulation results show how specific neural structures result in precise firing patterns in cortical neural networks [13] [8], employing link mining algorithms would be interesting and promising to discover structural features that predict further functional aspects such as precise spike timing. Building on our experiences with simulated activity data, we are currently adapting our algorithms to discover structure-activity relations in

Table 3.3. Accuracy of prediction in small-world networks. Abbreviations denote the same features and algorithms as in Table 3.2.

	Average Firing Rate		
Features	BN-K2	J48	SVM*
cc	50.0%	50.0%	52.5%
inh	84.8%	84.0%	85.8%
inh + cc	84.8%	85.8%	85.5%
inh + 2ie	87.0%	87.8%	89.5%
inh + 3m	84.8%	84.8%	85.8%
inh + 3ie	86.0%	87.3%	90.8%
inh + source values	84.5%	87.5%	91.5%
inh + sink values	80.5%	87.8%	90.3%
inh + source + sink values	83.3%	88.5%	91.5%
source values	78.5%	82.8%	87.5%
sink values	65.8%	82.8%	81.0%
source + sink values	83.3%	88.5%	91.5%

biological neuronal networks, like cell cultures grown on multi-electrode arrays [15].

Acknowledgments

This work was funded by the German Federal Ministry of Education and Research (BMBF grant 01GQ0420 to BCCN Freiburg).

References

1. Mauricio Barahona and Louis M. Pecora. Synchronization in small-world systems. *Physical Review Letters*, 89 : 054101, 2002.
2. Valentino Braitenberg and Almut Schüz. *Cortex: Statistics and Geometry of Neuronal Connectivity.* Springer-Verlag, Berlin, 2nd edition, 1998.

Table 3.4. Accuracy of prediction in networks with normally distributed synaptic efficacies. Abbreviations denote the same features and algorithms as in Table 3.2.

	Average Firing Rate		
Features	BN-K2	J48	SVM*
cc	48.5%	48.5%	49.1%
inh	86.4%	86.4%	87.5%
inh + cc	86.4%	86.1%	88.2%
inh + 2ie	87.6%	89.7%	90.6%
inh + 3m	86.4%	85.2%	88.5%
inh + 3ie	87.0%	89.4%	93.6%
inh + source values	88.5%	89.4%	94.5%
inh + sink values	89.7%	89.4%	91.8%
inh + source + sink values	89.4%	90.1%	94.5%
source values	86.7%	89.4%	90.1%
sink values	87.3%	89.4%	88.8%
source + sink values	87.6%	89.1%	90.6%

Table 3.5. Comparison of the best *NeuronRank* based feature set (inh+source values +sink values) to the best motif counting based feature set (inh+3ie) according to sign test. *NeuronRank* features were significantly better at binary prediction task

# NeuronRank better	# 3ie better	p value
11	1	0.0032

3. Nicolas Brunel. Dynamics of sparsely connected networks of excitatory and inhibitory spiking neurons. *Journal of Computational Neuroscience*, 8(3): 183–208, 2000.
4. Soumen Chakrabarti, Byron E. Dom, and Piotr Indyk. Enhanced hypertext categorization using hyperlinks. In Laura M. Haas and Ashutosh Tiwary, editors, *Proceedings of SIGMOD-98, ACM International Conference on Management of Data*, pages 307–318, Seattle, US, 1998. ACM Press, New York, US.

5. Gregory F. Cooper and Edward Herskovits. A bayesian method for the induction of probabilistic networks from data. *Machine Learning*, 9(4):309–347, 1992.
6. Nello Cristianini and John Shawe-Taylor. *An Introduction to Support Vector Machines and Other Kernel-based Learning Methods.* Cambridge University Press, March 2000.
7. Markus Diesmann and Marc-Oliver Gewaltig. NEST: An environment for neural systems simulations. In Theo Plesser and Volker Macho, editors, *Forschung und wisschenschaftliches Rechnen, Beiträge zum Heinz-Billing-Preis 2001*, volume 58 of *GWDG-Bericht*, pages 43–70. Ges. für Wiss. Datenverarbeitung, Göttingen, 2002.
8. Markus Diesmann, Marc-Oliver Gewaltig, and Ad Aertsen. Stable propagation of synchronous spiking in cortical neural networks. *Nature*, 402(6761):529–533, 1999.
9. Pedro Domingos. Mining social networks for viral marketing. *IEEE Intelligent Systems*, 20(1):80–82, 2005.
10. Lise Getoor. Link mining: a new data mining challenge. *SIGKDD Explorations*, 5(1):84–89, 2003.
11. Lise Getoor and Christopher P. Diehl. Link mining: a survey. *SIGKDD Explorations*, 7(2):3–12, 2005.
12. Tayfun Gürel, Luc De Raedt, and Stefan Rotter. Ranking neurons for mining structure-activity relations in biological neural networks: Neuronrank. *Neurocomputing*, doi:10.1016/j.neucom.2006.10.1064, 2006.
13. Eugene M. Izhikevich. Polychronization: Computation with spikes. *Neural Computation*, 18(2):245–282, February 2006.
14. Jon M. Kleinberg. Authoritative sources in a hyperlinked environment. *Journal of the ACM*, 46(5):604–632, 1999.
15. Shimon Marom and Goded Shahaf. Development, learning and memory in large random networks of cortical neurons: lessons beyond anatomy. *Quarterly Reviews of Biophysics*, 35(1):63–87, February 2002.
16. R. Milo, S. Shen-Orr, S. Itzkovitz, N. Kashtan, D. Chklovskii, and U. Alon. Network motifs: simple building blocks of complex networks. *Science*, 298(5594):824–827, October 2002.
17. Abigail Morrison, Carsten Mehring, Theo Geisel, Ad Aertsen, and Markus Diesmann. Advancing the boundaries of high connectivity network simulation with distributed computing. *Neural Computation*, 17(8):1776–1801, 2005.
18. Lawrence Page, Sergey Brin, Rajeev Motwani, and Terry Winograd. The pagerank citation ranking: Bringing order to the web. Technical report, Stanford Digital Library Technologies Project, 1998.
19. Robert J J. Prill, Pablo A A. Iglesias, and Andre Levchenko. Dynamic properties of network motifs contribute to biological network organization. *Public Library of Science, Biology*, 3(11), October 2005.
20. Ross J. Quinlan. *C4.5: Programs for Machine Learning (Morgan Kaufmann Series in Machine Learning)*. Morgan Kaufmann, January 1993.
21. Sen Song, Sjöström Per, Markus Reigl, Sacha Nelson, and Dmitri Chklovskii. Highly nonrandom features of synaptic connectivity in local cortical circuits. *Public Library of Science, Biology*, 3(3):0507–05019, 2005.
22. Benjamin Taskar, Pieter Abbeel, and Daphne Koller. Discriminative probabilistic models for relational data. In Adnan Darwiche and Nir Friedman, editors, *Uncertainty in AI*, pages 485–492. Morgan Kaufmann, 2002.

23. D. J. Watts and S. H. Strogatz. Collective dynamics of 'small-world' networks. *Nature*, 393(6684):440–442, June 1998.
24. Ian H. Witten and Eibe Frank. *Data Mining: Practical Machine Learning Tools and Techniques, Second Edition (Morgan Kaufmann Series in Data Management Systems)*. Morgan Kaufmann, June 2005.

4

Adaptive Contextual Processing of Structured Data by Recursive Neural Networks: A Survey of Computational Properties

Barbara Hammer[1], Alessio Micheli[2], and Alessandro Sperduti[3]

[1] Institute of Computer Science, Clausthal University of Technology, Julius Albert Straße 4, Germany. hammer@in.tu-clausthal.de
[2] Dipartimento di Informatica, Università di Pisa, Largo Bruno Pontecorvo 3, 56127 Pisa, Italy. micheli@di.unipi.it
[3] Dipartimento di Matematica Pura ed Applicata, Università di Padova, Via Trieste 63, 35121 Padova, Italy. sperduti@math.unipd.it

Summary. In this section, the capacity of statistical machine learning techniques for recursive structure processing is investigated. While the universal approximation capability of recurrent and recursive networks for sequence and tree processing is well established, recent extensions to so-called contextual models have not yet been investigated in depth. Contextual models have been proposed to process acyclic graph structures. They rely on a restriction of the recurrence of standard models with respect to children of vertices as occurs e.g. in cascade correlation. This restriction allows to introduce recurrence with respect to parents of vertices without getting cyclic definitions. These models have very successfully been applied to various problems in computational chemistry. In this section, the principled information which can be processed in such a way and the approximation capabilities of realizations of this principle by means of neural networks are investigated.

4.1 Introduction

In many real world application domains, entities are compound objects and the main computational task consists in discovering relations among the different parts of the objects or with respect to some other entity of interest. Very often, the problem formulation includes both discrete (e.g. symbols) and numerical entities and/or functions. For example, in Chemistry and Biochemistry, chemical compounds are naturally represented by their molecular graph, where information attached to each vertex is a symbol denoting the name of the involved atom/atoms, and where a major task is to correlate the chemical structures with their properties (e.g. boiling point, chemical reactivity, biological and pharmaceutical properties, etc.). The acronyms QSPR and QSAR are used in these fields to denote the analysis of the quantitative relationships between structure and properties or biological activities.

Learning methods able to deal with both discrete information, as the structure of a compound object, and numerical relationships and functions, such as Recursive Neural Networks (RNNs) [1], offer an opportunity to directly approach QSPR/QSAR analysis when the knowledge of the mathematical relationships among data is poor or absent and when data are affected by noise and errors, as it is typical when they are experimentally acquired.

In particular, the RNN approach for QSPR/QSAR is characterized by the direct treatment of variable-size structured data (in the form of hierarchical data), which is a vehicle of information much richer than the traditional flat vectors of descriptors used in the traditional QSPR/QSAR approaches. Moreover, RNN provide an adaptive encoding of such input structures, i.e. the encoding of the structures is a function learned according to the task and the training data. The use of an adaptive model for structures avoids both the need for an a priori definition of a flat molecular representation (by vectors of descriptors) or the need of the prior definition of a similarity measure for the structured data.

The success of the approach has been proven for some datasets both by comparison with respect to state-of-art QSPR/QSAR approaches and, recently, by empirical comparison versus a Support Vector Machine (SVM) with tree kernel [2]. The flexibility of the structure-based approach allows for the treatment of different problems and data sets: Recent applications showing the feasibility of the approach exploiting RNN in a variegate set of QSPR and QSAR analysis problems can be found in [3, 4, 5, 6, 7, 8, 9]. The RNN approach can be further exploited to deal with tasks for which the lack of prior knowledge on the problem makes the direct use of structural information for the prediction task appealing.

These features of RNN can be exploited also in other structured domains, such as XML document categorization, or Natural Language Processing tasks, just to name a few.

In this chapter, we present some of the major RNN models with simple recursive dynamics and with enlargements by contextual connections. The aim of the chapter is to review the computational properties of these models and to give a comprehensive and unitary view of both the classes of data that can be used for learning, e.g. sequences, trees, acyclic graphs, and various modifications (such as positional graphs), and the classes of transductions that can be computed and approximated.

In Section 4.2 we present the different types of structural transductions which are of interest. In particular, causal and contextual transductions are considered. Mathematical models for implementing these types of transductions and amenable to neural realizations are discussed in Section 4.3. In Section 4.4, we discuss the computational properties of the models presented in the previous section, relating each other with respect to both the class of data they are able to process and the class of transductions they can realize with respect to their functional dependencies, disregarding computational issues.

In Section 4.5, the universal approximation ability of the models is discussed based on these results on the functional dependency of recursive mechanisms.

4.2 Structural Transductions

A *transduction* $T : G \to O$ is a mapping between a structured domain (SD) G and a discrete or continuous output-space O. Various classes of data can be considered in a SD. Vectors constitute the simplest data structures corresponding to a flat domain. In fact, each vector can be seen as a single vertex with *label* $l(v)$ composed by a tuple of attributes, which can represent both numerical or categorical information[4].

In a general discrete structure, a graph $g \in G$, we have a set of labeled vertices $Vert(g)$ connected by a set of edges $Edg(g)$. Discrete structures of interest are hierarchical structures, such are sequences (where a total order on the vertices is defined), rooted positional trees (where a distinctive vertex is the root, a partial topological order holds among the vertices, and a position is assigned to each edge leaving from each vertex) and Directed Positional Acyclic Graphs (DPAG), the generalization of this tree structure to acyclic graphs. Positional trees with bounded out-degree K are also called *K-ary trees*. In the following we use the term *tree* referring to the class of labeled K-*ary* trees.

A supersource for DPAGs is defined as a vertex s, with zero in-degree, such that every vertex in the graph can be reached by a directed path starting from s. In the case of trees, the supersource is always defined by its root node. Rooted positional trees are a subclass of DPAG, with supersource, formed by prohibiting cycles in the undirected version of the graph. In DPAG the position is specified by a positional index assigned to each entering and leaving edge from a node v. Formally, in a DPAG we assume that for each vertex $v \in Vert(g)$ with a set of edges denoted by $edg(v)$, two injective functions $P_v : edg(v) \to [1, 2, ..., in]$ and $S_v : edg(v) \to [1, 2, ..., out]$ are defined on the edges entering and leaving from v. For a DPAG, if an edge (u, v) is present in $edg(g)$, then u is a *parent* (*predecessor*) of v (denoted by $pa_j[v]$ for $P_u((v, u)) = j$) and v is a *child* (or *successor*) of u (denoted by $ch_j[v]$ for $S_u((u, v)) = j$). If there exists a path from u to v then u is an *ancestor* of v and v is a *descendant* of u. As before, we assume a restriction k of the number of children and parents per node. Often, we assume $k = 2$ for simplicity.

Other subclasses of DPAG can be considered for special purposes, e.g the class of Directed Bipositional Acyclic Graphs (DBAGs), where a constraint is added on the positions such that $P_u((v, u)) = S_v((v, u))$ for each edges in $Edg(g)$, i.e. the enumeration of children and parents is the same.

---

[4] In the following the attribute vectors are assumed to be numeric: symbolic labels from a finite alphabet can be represented by a coding function as numerical vectors; for the sake of presentation, symbols are retained in the graphical representation of structures to label the vertices in a concise way.

Over a SD various classes of transductions can be defined. Of specific interest to learning applications are the following classes:

Adaptive transductions: the similarity measures on structures is learned for the task at hand, e.g. using a trainable RNN.

Contextual transductions: for each vertex the response depends on the whole information represented in the structured data, according to its topology.

Causal transductions (over hierarchical data structures): for each vertex v the response for the vertex v only depends on v and its descendant vertices (i.e the vertices in $Vert(g)$ which can be reached from v by following a directed path).

I/O-isomorphic and Supersource transductions: in an Input/Output isomorphic transduction (I/O-isomorphic transduction), for each vertex of the input graph the system emits an output, i.e. the system produces a structured output with a topology isomorphic to the original input graph. In a supersource transduction the response of the system is given in the form of a scalar value only in correspondence of the supersource of the input graph.

We always assume stationarity of the transduction, i.e. the transfer function is fixed for the whole structure. In Figure 4.1 we show an example of a supersource transduction and an I/O-isomorphic transduction. These examples provide instances of contextual transductions that cannot be computed by pure causal and stationary transductions. We will show through the paper the effects that the causality assumption and its relaxation induce on the computational capabilities of the learning models and on the classes of data that can be treated.

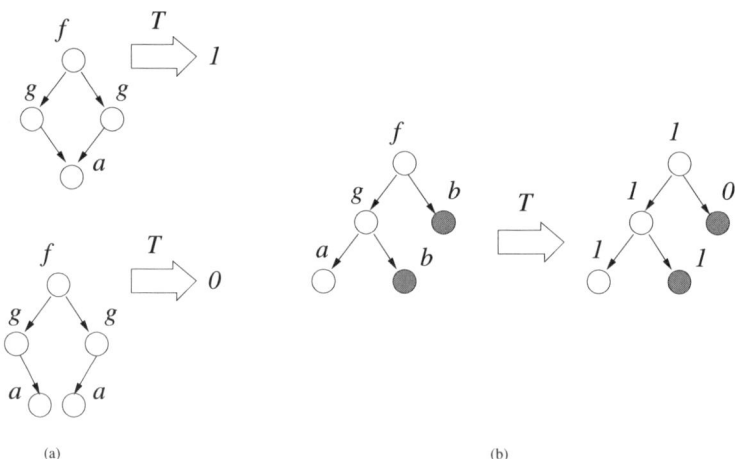

Fig. 4.1. Example of supersource (a) and IO-isomorphic (b) contextual transductions.

Notice that a causal IO-isomorphic transduction T_1 can be implemented via a suitable supersource transduction T_2. In fact, for each vertex v in the input structure g, the output label generated by T_1 in correspondence of that vertex, can be reproduced by a transduction T_2 that takes as input the subgraph in g rooted in v (and including all and only all its descendants) and returns the output label generated by T_1 for v. This implementation scheme, however, does not work for contextual IO-isomorphic transductions since the subgraph rooted in v does not constitute a complete context for v.

Causality is a natural property of e.g. transductions for time series where future events depend on past events but not vice versa. For the processing of chemical data, the assumption of causality is usually not valid since chemical structures do not possess a distinguished direction of causality of the constituents.

4.3 Models

Here we present formal models that can be used to implement causal and contextual models. The main issue consists in the way in which the constituents of a recursive structure are processed within the context posed by the involved structural elements. Different kinds of dependencies, i.e. causalities of the constituents can be accounted for.

4.3.1 Causal Models

A (stationary) causal IO-isomorph transduction, and thus a supersource transduction, can be implemented by resorting to a recursive state representation, i.e by defining a state (vector) $\boldsymbol{x}(v) \in X$ associated with each vertex v of a DPAG g and two functions f^s and f^o such that for $v \in \mathit{Vert}(g)$

$$\boldsymbol{x}(v) = f^s(\boldsymbol{l}(v), \boldsymbol{x}(ch_1[v]), \ldots, \boldsymbol{x}(ch_K[v])) \qquad (4.1)$$
$$\boldsymbol{y}(v) = f^o(\boldsymbol{x}(v)) \qquad (4.2)$$

where $\boldsymbol{x}(ch_1[v]), \ldots, \boldsymbol{x}(ch_K[v])$ is the $K(= out)$-dimensional vector of states for the vertices children of v. In the above definition, f^s is referred to as the *state transition function* and f^o is referred to as the *output function*. The base case of the recursion is defined by $\boldsymbol{x}(nil) = \boldsymbol{x_0}$. The output variable $\boldsymbol{y}(v)$ provides the value of $T(g)$ for both the class of supersource or IO-isomorphic transductions. Stationarity assumption is made by the invariance of f^s and f^o over the vertices of the structure.

The above equations define a very general recursive processing system. For example, Mealy and Moore machines are finite-state machines that can be described as an instance of such framework, where the input domain is limited to sequences, and X to a finite set. Extension of the input domain to trees leads to the definition of *Frontier-to-Root tree Automata* (FRA) (or bottom-up tree automata).

Compact and graphical representations of the symbolic transformation of the state variables can be obtained by the shift operator. Specifically, given a state vector $\boldsymbol{x}(v) \equiv [x_1(v), \ldots, x_m(v)]^t$, we define structural shift operators $q_j^{-1} x_i(v) \equiv x_i(\text{ch}_j[v])$ and $q_j^{+1} x_i(v) \equiv x_i(\text{pa}_j[v])$. If $\text{ch}_j(v) = nil$ then $q_j^{-1} x_i(v) \equiv x_0$, the null state. Similarly, if $\text{pa}_j(v) = nil$ then $q_j^{+1} x_i(v) \equiv x_0$. Moreover, we consider a vector extension $\boldsymbol{q}^e x_i(v)$ whose components are $q_j^{-e} x_i(v)$ as defined above and $e \in \{-1, +1\}$. We also denote by $q_j^e \boldsymbol{x}$ the componentwise application of q_j^e to each $x_i(v)$ for $i = 1, \ldots, m$, and by $\boldsymbol{q}^e \boldsymbol{x}$ the componentwise application of $\boldsymbol{q}^e$ to each $x_i(v)$ for $i = 1, \ldots, m$ [10].

The recursive processing system framework provides an elegant approach for the realization of adaptive processing systems for learning in SD, whose main ingredients are:

- an input domain of structured data, composed in our case by a set of labeled DPAGs with supersource;
- a parametric transduction function T, i.e. in the recursive processing system used to compute T, the functions f^s and f^o are made dependent on tunable (free) parameters;
- a learning algorithm that is used to estimate the parameters of T according to the given data and task.

A specific realization of such approaches can be found in the family of recursive neural network (RNN) models [1].

The original recursive approach is based on the causality and stationarity assumptions. Since RNNs implement an adaptive transduction, they provide a flexible tool to deal with hierarchical structured data. In fact, the recursive models have proved to be *universal approximators* for trees [11]. However, the causality assumption put a constraint that affects the computational power for some classes of transductions and when considering the extension of the input domain to DPAGs. In particular, the context of the state variables computed by RNN for a vertex is limited to the descendant of such vertices. For instance, for sequential data the model computes a value that only depends on the left part of the sequence. The concept of *context* introduced by Elman [12] for recurrent neural networks (applied to time series) is to be understood in the restrictive sense of "past" information. Of course this assumption might not be justified in spatial data. For structures the implications are even more complex as we will see in the following of the paper.

RNN Architectures

Recursive neural networks (RNNs) are based on a neural network realization of f^s and f^o, where the state space X is realized by $\mathbb{R}^m$. For instance, in a fully connected Recursive Neural Network (RNN-fc) with one hidden layer of

m hidden neurons, the output $\boldsymbol{x}(v) \in \mathbb{R}^m$ of the hidden units for the current vertex v, is computed as follows:

$$\boldsymbol{x}(v) = f^s(\boldsymbol{l}(v), \boldsymbol{q}^{-1}\boldsymbol{x}(v)) = \boldsymbol{\sigma}(\boldsymbol{W}\boldsymbol{l}(v) + \sum_{j=1}^{K} \hat{\boldsymbol{W}}^j q_j^{-1}\boldsymbol{x}(v)), \quad (4.3)$$

where $\boldsymbol{\sigma}_i(\boldsymbol{u}) = \sigma(u_i)$ (sigmoidal function), $\boldsymbol{l}(v) \in \mathbb{R}^n$ is a vertex label, $\boldsymbol{W} \in \mathbb{R}^{m \times n}$ is the free-parameters (weight) matrix associated with the label space and $\hat{\boldsymbol{W}}^j \in \mathbb{R}^{m \times m}$ is the free-parameters (weight) matrix associated with the state information of jth children of v. Note that the bias vector is included in the weight matrix $\boldsymbol{W}$. The stationarity assumption allows for processing of variable-size structure by a model with fixed number of parameters.

Concerning the output function g, it can be defined as a map $g : \mathbb{R}^m \to \mathbb{R}^z$ implemented, for example, by a standard feed-forward network.

Learning algorithms can be based on the typical gradient descendent approach used for neural networks and adapted to recursive processing (Back-Propagation Through Structure and Real Time Recurrent Learning algorithms [1]).

The problem with this approach is the apriori definition of a proper network topology. Constructive methods permit to start with a minimal network configuration and to add units and connections progressively, allowing automatic adaptation of the architecture to the computational task. Constructive learning algorithms are particularly suited to deal with structured inputs, where the complexity of learning is so high that it is better to use an incremental approach.

Recursive Cascade Correlation (RCC) [1] is an extension of Cascade Correlation algorithms [13, 14] to deal with structured data.

Specifically, in RCC, recursive hidden units are added incrementally to the network, so that their functional dependency can be described as follows:

$$x_1(v) = f_1^s\left(\boldsymbol{l}(v), \boldsymbol{q}^{-1} x_1(v)\right)$$
$$x_2(v) = f_2^s\left(\boldsymbol{l}(v), \boldsymbol{q}^{-1}[x_1(v), x_2(v)]\right) \quad (4.4)$$
$$\vdots$$
$$x_m(v) = f_m^s\left(\boldsymbol{l}(v), \boldsymbol{q}^{-1}[x_1(v), x_2(v), ..., x_m(v)]\right)$$

The training of a new hidden unit is based on already frozen units. The number of hidden units realizing the $f^s()$ function depends on the training process.

The computational power of RNN has been studied in [15] by using hard threshold units and frontier-to-root tree automata. In [16], several strategies for encoding finite-state tree automata in high-order and first-order sigmoidal RNNs have been proposed. Complexity results on the amount of resources needed to implement frontier-to-root tree automata in RNNs are presented in [17]. Finally, results on function approximation and theoretical analysis of learnability and generalization of RNNs (referred as folding networks) can be found in [18, 11, 19].

4.3.2 Contextual Models

Adaptive recursive processing systems can be extended to contextual transductions by proper "construction" of contextual representations. The main representative of this class of models is the *Contextual Recursive Cascade Correlation* (CRCC), which appeared in [20] applied to the sequence domain, and subsequently extended to structures in [21, 10]. A theoretical study of the universal approximation capability of the model appears in [22].

The aim of CRCC is to try to overcame the limitations of causal models. In fact, due to the causality assumption, a standard RNN cannot represent in its hypothesis space contextual transductions. In particular for a RNN the context for a given vertex in a DPAG is restricted to its descending vertices. In contrast, in a contextual transduction the context is extended to all the vertices represented in the structured data.

As for RCC, CRCC is a constructive algorithm. Thus, when training hidden unit i, the state variables $x_1, \ldots, x_{i-1}$ for all the vertices of all the DPAGs in the training set are already available, and can be used in the definition of x_i. Consequently, equations of causal RNN, and RCC in particular (Equation 4.4), can be expanded in a contextual fashion by using, where possible, the variables $q^{+1} x_i(v)$. The equations for the state transition function for a CRCC model are defined as:

$$
\begin{aligned}
x_1(v) &= f_1^s\Big(l(v), q^{-1} x_1(v)\Big) \\
x_2(v) &= f_2^s\Big(l(v), q^{-1}[x_1(v), x_2(v)], q^{+1}[x_1(v)]\Big) \\
&\vdots \\
x_m(v) &= f_m^s\Big(l(v), q^{-1}[x_1(v), x_2(v), ..., x_m(v)], q^{+1}[x_1(v), x_2(v), ..., x_{m-1}(v)]\Big)
\end{aligned}
\tag{4.5}
$$

Hence, in the CRCC model, the frozen state values of both the children and parents of each vertex can be used as input for the new units without introducing cycles in the dynamics of the state computing system.

Following the line developed in Section 4.3.1 it is easy to define neural network models that realize the CRCC approach. In fact, it is possible to realize each f_j^s of Equation 4.5 by a neural unit associating each input argument of the f_j^s with free parameters $\mathcal{W}$ of the neural unit. In particular, given a vertex v, in the CRCC model each neural unit owns:

- connections associated to the input label of v, and connections associated to the state variables (both the output of already frozen hidden units and

the current j-th hidden unit) of the children of the current vertex v (as for the RNN/RCC model presented in Section 4.3.1);
- connections associated to the state variables of the *parents* of v which are already *frozen* (i.e., the output of already frozen hidden units computed for the parents of v). The connections associated to *parents* add new parameters to the model that are fundamental to the contextual processing.

The major aspect of this computation is that increasing context is taken into account for large m. In fact, the hidden unit for each layer takes directly a local context (parents and children) of each vertex as input and progressively, by composition of context developed in the previous steps, it extends the context involving other vertices, up to the vertices of the whole structure. The context window follows a nested development; as an example, let consider a vertex $v' = pa[v]$ and the presence of 3 hidden units. Following the definition given in eq. 4.5, $x_2(v')$ depends on $x_1(pa[v'])$ and $x_3(v)$ depends on $x_2(pa[v]) = x_2(v')$. Thus, we obtain that also $x_3(v)$ depends on $x_1(pa[v'])$, which means that $x_3(v)$ includes information by $pa[v'] = pa[pa[v]]$ in its "context window".

In particular, adding new hidden units to the CRCC network leads to an increase of the "context window" associated to each vertex v. Hence, the size of the context window can grow during model training and we do not need to fix it prior to learning. Given a sufficient number of hidden units the information included in x_i grows to all the context of the structured data according to the structure topology. A formal analysis of the context development of the CRCC model is presented in [10], showing the extension of the power capability allowed by CRCC for tasks that cannot be computed by causal models. In particular, such results include contextual IO-isomorph transductions and supersource transductions on DPAG.

Beside these results, there are other interesting cases that deserve to be discussed for supersource transductions that both a causal and contextual models can compute. In fact, relaxing the causal assumption can be useful also for supersource transductions whenever the meaning of a sub-structure depends on the "context" in which it is found. In such way it is possible to consider in which position within a larger structure the given substructure does occur. In order to show the relevance of a contextual processing for this issue, let us consider an example of sub-structure encoding, as shown in Figure 4.2. In this figure, the internal encoding of some fragments (subgraphs) is reported as represented in a 2-dimensional feature space F. Each point in the F space represents the numerical code developed by the CRCC model for each fragment in the input structured domain. This can be practically obtained by projecting the $\boldsymbol{x}(v)$, usually belonging to a m-dimensional space into a 2-dimensional space, e.g. by Principal Component Analysis (PCA), that can be easily visualized. In such case, a causal mapping yields an unique code for each occurrence of the fragment. By a contextual mapping, each fragment can be represented in different ways depending on the context (position). Hence, clearly, CRCC achieves a more expressive substructure encoding. A

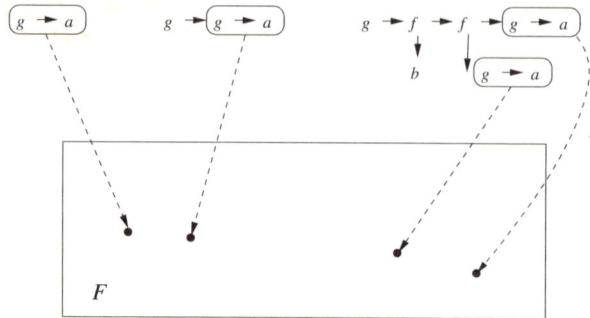

Fig. 4.2. Different representations of the "g(a)" structural fragment in the projection of the X space (denoted by F) by a contextual model.

PCA plot and analysis of the internal representation of CRCC developed for alkanes data can be found in [10].

4.4 Classes of Functions

Each (neural) model $\mathcal{M}$ is characterized by the class of transductions it can compute, each specific transduction being defined by one (or more) specific instantiation(s) of the weight values. In this section we will try to characterize these classes of functions, with specific focus on the domain of structures they are able to discriminate. We proceed by considering the general form of the equations associated to a model, i.e. the dependency constraints defined by both the function f^s generating the state space and the output function f^o. This means that we are not interested in the specific realization of f^s and f^o, but only on which information they take in input.

A complete formal characterization of this idea can be obtained by reasoning at syntactic level over the algebraic term generated by a specific input vertex.

Let start by considering models where the output generated at each input vertex v only depends on the state representation for that vertex, i.e., $y(v) = f^o(x(v))$. Members of this class of models can differentiate in the way they use the information stored in the state space, however all of them produce an output for each vertex v by using only the state information associated to v, i.e. $x(v)$. This means that we can focus on the dependency constraints defined by f^s. As a concrete example of how this can be done, let consider eq. (4.1) and the binary tree $t_1 \equiv l_4(l_3, l_2(l_1, \xi))$, where l_i represent the labels associated to the tree's vertices and the symbol ξ represents the absence of a child. Then, abstracting from the specific values assumed by the labels, and defined $x_0 = f^s(\xi)$, the algebraic term generated by applying eq. (4.1) to the root of t is $f^s(l_4, f^s(l_3, x_0, x_0), f^s(l_2, f^s(l_1, x_0, x_0), x_0))$, while the algebraic term generated for the right child of the root is $f^s(l_2, f^s(l_1, x_0, x_0), x_0)$.

From a conceptual point of view, this corresponds to generate, for each input structure, an isomorphic structure, where the information attached to each vertex is the algebraic term generated for the corresponding vertex in the input structure. In the following we will refer with $term(\boldsymbol{x}(v))$ the algebraic term corresponding to the expansion of $\boldsymbol{x}(v)$.

Definition 1 (Collision Relation). *Let consider a model $\mathcal{M}$ which for each vertex v associates the state representation $\boldsymbol{x}(v)$. Then we define the collision relation $\bowtie_{\mathcal{M}}$ over two vertices v_1, v_2, as follows:*

$$v_1 \bowtie_{\mathcal{M}} v_2 \Leftrightarrow term(\boldsymbol{x}(v_1)) = term(\boldsymbol{x}(v_2))$$

This relation is very useful to characterize the class of input structures that can be discriminated by the given model. In fact, if two different vertices are in relation, this means that they cannot be discriminated since they are mapped into the same representation, regardless of the specific implementation of f^s.

We can thus exploit the collision relation to define equivalence classes over vertices.

Definition 2 (Collision Equivalence Classes). *Given a model $\mathcal{M}$, with $Vert_{\mathcal{M}}$ we define the partition of the set of vertices into equivalence classes according to the corresponding collision relation $\bowtie_{\mathcal{M}}$.*

Definition 3 (Transduction Support). *Given a model $\mathcal{M}$, and a transduction T defined over a set of structures S, with set of vertices $Vert(S)$, we say that T is supported by $\mathcal{M}$ if each equivalence class belonging to $Vert(S)_{\mathcal{M}}$ only contains vertices for which T returns the same value (or no value).*

Definition 4 (Model Completeness). *A model $\mathcal{M}$ is said to be complete with respect to a family of transductions $\mathcal{T}$ defined over a set of structures S, if for all $T \in \mathcal{T}$, T is supported by $\mathcal{M}$.*

Theorem 1 (RNN/RCC Completeness for Supersource Transductions on Trees). *RNN/RCC supports all the supersource transductions defined on the set of trees (and sequences).*

Proof. First of all note that RNN/RCC is stationary, so it always returns the same value if the same subtree is presented. Moreover it is causal in such a way that the term generated for a vertex v, together with the label attached to it, contains a subterm for each descendant of v. Descendants which correspond to different subtrees will get a different term. A model is incomplete if there exist two trees t_1 and t_2, $t_1 \neq t_2$, and a supersource transduction T for which $T(t_1) \neq T(t_2)$, while $term(supersource(t_1)) = term(supersource(t_2))$. Let assume this happens. Since $t_1 \neq t_2$ there exists at least one path from the root of the two trees which ends into a vertex v with associated labels that are different. But this means that also $term(supersource(t_1))$ and $term(supersource(t_2))$ will differ in the position corresponding to v, contradicting the hypothesis that $term(supersource(t_1)) = term(supersource(t_2))$.

Theorem 2 (RNN/RCC Incompleteness for I/O-isomorphic Transductions on Trees). *RNN/RCC cannot support all the I/O-isomorphic transductions defined on the set of trees.*

Proof. This can be proved quite easily by observing that any I/O-isomorphic transduction which assigns a different value to different occurrences of the same subtree within a given tree (see, for example, figure 4.1.(b)) cannot be supported. This is due to the fact that the term representation for the roots of the two identical subtrees will belong to the same equivalence class, while the transduction assigns different values to these vertices.

Corollary 1 (Sequences). *RNN and RCC can support only I/O-isomorphic transductions defined on prefixes of sequences.*

Proof. As for subtrees, the roots of all the identical prefixes of a set of sequences belong to the same equivalence class. I/O-isomorphic transductions that depend on subsequences not included in the prefix can assign different values to such prefixes.

Theorem 3 (Characterization of Causal Model Collision Classes on DPAGs). *Let consider any causal and stationary model $\mathcal{M}$ and DPAGs defined on the set of vertices S. Let t denote any tree defined on S. Let $visit(g)$ be the tree obtained by a complete recursive visit of the DPAG g starting from its supersource. Let consider the DPAG $\hat{g}$ with the smallest number of vertices for which $visit(\hat{g}) = t$. Then, each vertex $v \in \hat{g}$ will belong to a different equivalence class of $S_{\mathcal{M}}$. Moreover, the set of equivalence classes to which vertices of $\hat{g}$ belong to, is exactly the same set of equivalence classes to which vertices of any graph g such that $visit(g) = t$ belong to.*

Proof. Basic idea: for causal models it is possible to turn/rewrite a DPAG into an associated/equivalent tree, where shared vertices are replicated according to all the possible positions that they can have in the DPAG (*unsharing* process, see Figure 4.3). Since this construction is possible for *any* DPAG, the class of DPAG is absorbed into the class of trees for *causal* models.

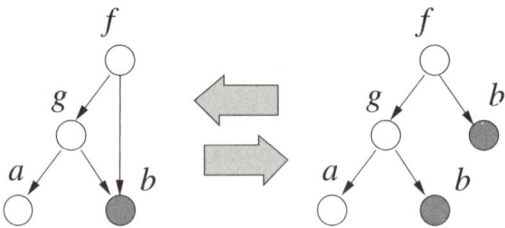

Fig. 4.3. Example of unsharing process for a DPAG: causal models allow to rewrite a DPAG as an equivalent tree.

4 Processing structured data by RNNs   79

Corollary 2 (RNN/RCC Incompleteness on DPAGs). *RNN/RCC cannot support all the supersource transductions defined on DPAGs. The same is true for I/O-isomorphic transductions.*

Interestingly we can derive also the following result.

Corollary 3 (Classes of structures). *The set of transductions on DPAGs supported by causal and stationary models (e.g. RNN/RCC) can equivalently be defined on the class of trees.*

Proof. Since the unsharing construction is possible for *any* DPAG, for each vertex of a DPAG g it is possible to find in its equivalence class a vertex belonging to a tree t obtained by applying the unsharing construction to g, i.e. $visit(g) = t$. Thus, any transduction defined on a DPAG g can be computed with equal results on the equivalent tree t.

The above result stresses the fact that causal and stationary models look at input structures as if they were trees, i.e. for causal and stationary models the class of DPAG is absorbed into the class of trees.

From a computational point of view the unsharing construction helps in understanding that it is better to share all the vertices that can be shared. In fact, in this way there are fewer vertices to process while the computational power of the model stays the same. This observation is at the basis of the optimization of the training set suggested in [1].

The problem with causal (and stationary) models is that they only exploit information concerning a vertex and its descendants. In this way it is not possible for that models to discriminate between occurrences of the same substructure ν in different contexts, i.e. different sets of parents for the supersource of ν.

The CRCC model tries to embed this idea in its transition state function (eq. 4.5) where information about the context of a vertex v is gathered through factorization of the state space into a sequence of state variables that progressively increase the visibility above the ancestors of v.

In [10] the concept of context window has been introduced to study the computational properties of CRCC. More precisely, the context window of a state variable for a vertex is defined as the set of all the state variables (defined over all possible vertices) that (directly or indirectly) contribute to its determination. Here we use a more precise definition of the context since the term associated to a vertex state vector not only involves the terms generated by all the relevant state variables mentioned above, but it also relates these terms each other through the state transition function(s).

Theorem 4 (CRCC More Powerful than Causal Models). *The set of supersource transductions defined on DPAGs supported by CRCC is strictly larger than the set of transductions supported by causal (and stationary) models (e.g., RCC/RNN). The same is true for I/O-isomorphic transductions.*

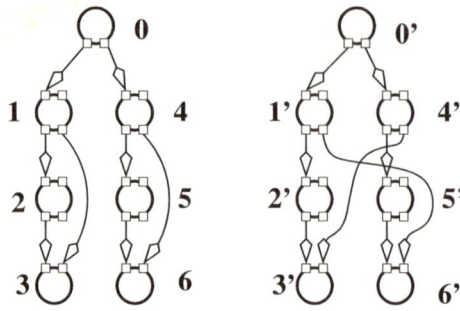

Fig. 4.4. Example of two graphs which are mapped to identical values with CRCC networks for a non-compatible enumeration of parents. Thereby, the enumeration of parents and children is indicated by boxes at the vertices, the left box being number one, the right box being number two.

Proof. Let consider a tree t which does not contain two or more vertices with the same set of children and the same set of siblings and the same parent. Then, for causal models the supersources of t and any DPAG g such that $visit(g) = t$ will belong to the same equivalence class. This will not be the case for CRCC. In fact, for any such DPAG g_1, given any other such DPAG $g_2 \neq g_1$, there will exists at least one vertex v_1 belonging to g_1 for which either the set of children or the set of siblings or the set of parents will be different from the ones of any vertex v_2 belonging to g_2. This means that when considering CRCC's eq. 4.5 for vertices v_1 and v_2, the corresponding terms will be different as well as the terms corresponding to the supersources. Thus all the supersources of involved DPAGs will belong to different equivalence classes. Moreover, vertices that are in different equivalence classes for causal models will remain in different equivalence classes for CRCC. Thus there will be additional supersource and I/O-isomorphic transductions that CRCC can support with respect to causal models.

Unfortunately it has been demonstrated in [22] that CRCC is not able to support all the supersource transductions on DPAGs. This is due to the fact that there exist high symmetric DPAGs for which the term generated by CRCC is the same (see Figure 4.4).

Thus, we have

Theorem 5 (Incompleteness of CRCC for transductions defined on DPAGs). *CRCC cannot support all the supersource transductions defined on DPAGs. The same is true for I/O-isomorphic transductions.*

There is, however, a restricted class of DPAGs, called DBAGs, where completeness can be achieved.

Definition 5 (DBAGs). *Assume D is a DPAG. We say that the ordering of the parents and children is compatible if for each edge (v, u) of D the equality*

4 Processing structured data by RNNs 81

$$P_u((v,u)) = S_v((v,u))$$

holds. *This property states that if vertex v is enumerated as parent number i of u then u is enumerated as child number i of v and vice versa. We recall that this subset of DPAGs with compatible positional ordering is called* rooted directed bipositional acyclic graphs *(DBAGs)*.

First, the causal models retain the incompleteness limitations on DBAGs.

Theorem 6 (RNN/RCC Incompleteness for Transductions Defined on DBAGs). *RNN/RCC cannot support all the supersource transductions defined on DBAGs. The same is true for I/O-isomorphic transductions.*

Proof. Follows directly from Theorem 3 applied to the subclasses of DBAGs.

On the contrary, CRCC is complete for supersource and I/O-isomorphic transductions defined on DBAGs.

Theorem 7 (CRCC Completeness for Supersource Transductions on DBAGs). *The set of supersource and I/O-isomorphic transductions defined on DBAGs is supported by CRCC.*

Proof. The property that the ordering of parents and children is compatible has the consequence that a unique representation of DBAGs is given by an enumeration of all paths in the DBAG starting from the supersource. While this fact is simple for tree structures, it is not trivial for graph structures which include nodes with more than one parent. For simplicity, we restrict the argumentation to the case of at most two children and parents, the generalization to more children and parents being straightforward. Now the proof proceeds in several steps.

Step 1: Assume v is a vertex of the DBAG. Consider a path from the supersource to the vertex v. Assume this path visits the vertices $v_0, v_1, \ldots, v_n = v$, v_0 being the supersource. Then a *linear path representation* of the path consists of the string $s = s_1 \ldots s_n \in \{0,1\}^*$ with $s_i = 0 \iff v_i$ is the left child of v_{i-1}. Note that the absolute position of a vertex in the DBAG can be recovered from any given linear path representation. The dynamics of a CRCC as given by equation 4.5 allows to recover all paths leading to v starting from the supersource for large enough number of the context unit. More precisely, assume the maximum length of a path leading to v is h. Then all linear path representations to v can be recovered from the functional representation given by $x_{h+2}(v)$ as defined in equation 4.5.

This can easily be seen by induction: for the supersource, the only linear path representation is empty. This can be recovered from the fact that $\boldsymbol{q}^{+1}[x_1(v)]$ corresponds to an initial context for the root. For some vertex v with maximum length h of a path leading to v, we can recover the linear path representations of the parents from $\boldsymbol{q}^{+1}[x_1(v), \ldots, x_{h+1}(v)]$ by induction. Further, it is clear from this term whether v is the left or right child from the considered parent because of the compatibility of the ordering. Thus we obtain the linear path representations of all paths from the supersource to v by extending the linear path representation to the parents of v by 0 or 1, respectively.

Step 2: Assume h is the maximum length of a path in a given DBAG. Assume v is a vertex. Then one can recover the following set from $x_{2h+2}(v)$: $\{(l(v'), B(v')) \mid v'$ is a vertex in the DBAG, $B(v')$ is the lexicographically ordered set of all linear path representations of the supersource to $v'\}$.

This fact is obvious for an empty DBAG. Otherwise, the term $x_{h'}(v_0)$, v_0 being the supersource and $h' \geq h + 2$ can easily be recovered from $x_{2h+2}(v)$. Because of Step 1, we can extract $(v_0, B(v_0))$ from this term, and we obtain the representations $x_{h'}(v')$ of all children v' of v_0. Thus, we can continue this way, adding the terms $(v', B(v'))$ for the children. Note that a vertex might be visited more than once, which can easily be detected due to the lexicographic ordering of $B(v')$.

Step 3: For any given DBAG, the set $\{(l(v'), B(v')) \mid v'$ is a vertex in the DBAG, $B(v')$ is the lexicographically ordered set of all linear path representations of the supersource to $v'\}$ is unique. This can be seen as follows: we can introduce a unique identifier for every vertex v in the DBAG, and assign its label $l(v)$. The supersource is uniquely characterized because the only path to the supersource is empty. Starting from the supersource we can add all connections (together with a positioning) to the DBAG in the following way: we consider the shortest linear path representation for which not yet all connections are contained in the constructed graph. Assume the final vertex of this path is v. Obviously, all connections of this path but the last one are already contained in the graph since we considered the shortest path. Thus, we can uniquely locate the parent of v in the graph and insert the edge together with the positioning.

Thus, the functional dependence as defined by CRCC is unique for every DBAG and vertex v contained in the DBAG if the context number is large enough. Therefore, the set of supersource and I/O-isomorphic transductions defined on DBAGs is supported by CRCC.

Note that these arguments hold also if the functional form of CRCC is restricted to the function

$$x_m(v) = f_m^s\left(l(v), \boldsymbol{q}^{-1}[x_m(v)], \boldsymbol{q}^{+1}[x_{m-1}(v)]\right)$$

since all relevant information used in the proof is contained in this representation.

It should be noticed that the condition imposed by DBAGs is only a sufficient condition to assure the CRCC completeness for transductions defined on the classes of DPAGs, which includes DBAGs. In fact, as introduced in [22], the limitation occurs only for high symmetric cases.

Thus, the overall picture as shown in Fig. 4.5 results if the functional form of the definitions is considered.

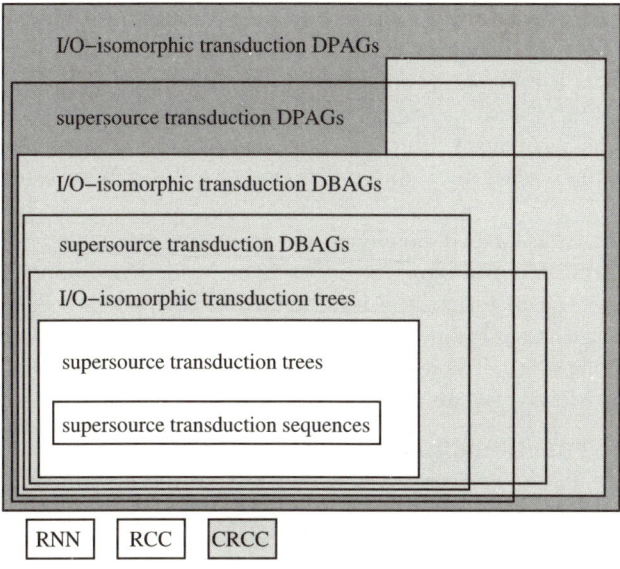

Fig. 4.5. Classes of functions supported by RNN, RCC, and CRCC because of the functional form

4.5 Approximation capability

An important property of function classes used for learning functions based on examples is its universal approximation ability. This refers to the fact that, given an unknown regularity, there exists a function in the class which can represent this regularity up to a small distance ϵ. Obviously, the universal approximation capability is a condition for a function class to be successful as a learning tool in general situations. There exist various possibilities to make this term precise. We will be interested in the following two settings:

Definition 6 (Approximation completeness). *Assume X is an input set. Assume Y is an output set. Assume $l : Y \times Y \to \mathbb{R}$ is a loss function. Assume F and G are function classes of functions from X to Y. F is* approximation complete *for G in the maximum norm if for every $g \in G$ and $\epsilon > 0$ there exists a function $f \in F$ such that $l(f(x), g(x)) \leq \epsilon$ for all $x \in X$.*

Assume X is measurable with probability measure P. Assume all compositions of functions in F and G with l are (Borel-)measurable. F is approximation complete *for G in probability if for every $g \in G$ and $\epsilon > 0$ there exists a function $f \in F$ such that $P(x \mid l(f(x), g(x)) > \epsilon) \leq \epsilon$.*

We are interested in the cases where X consists of sequences, trees, DPAGs, or DBAGs, and Y is either a real-vector space for supersource transductions or the same structural space for I/O-isomorphic transductions. We equip X with the σ-algebra induced by the sum of structures of a fixed form, whereby structures of a fixed form are identified with a real-vector space by concatenating all labels. We choose l as the Euclidean distance if Y is a real-vector space, and $l(s, s') = \max\{\|\boldsymbol{l}(v) - \boldsymbol{l}(v')\| \mid v$ and v' are vertices of s and s', respectively, in the same position$\}$ for structures s and s' of the same form. Since we only consider I/O-isomorphic transductions, it is not necessary to define a loss function for structures of a different form.

So far, the functional dependence of various models which process structured data has been considered, neglecting the form of the transition function f^s. Obviously, the fact that the functional form supports a certain class of functions is a necessary condition for the universal approximation ability of the function class, but it is not sufficient. Due to the limitation of the respective functional form, we can immediately infer the following results:

Theorem 8 (Functional incompleteness). *RNN/RCC is not approximation complete in the maximum norm or in probability for I/O-isomorphic transductions on trees.*

RNN/RCC is not approximation complete in the maximum norm or in probability for I/O-isomorphic or supersource transductions on DPAGs or DBAGs.

CRCC is not approximation complete in the maximum norm or in probability for I/O-isomorphic or supersource transductions on DPAGs.

For all other cases, the functional form as defined in the models above provides sufficient information to distinguish all different input structures.

However, it is not clear whether this information can be stored in the finite dimensional context space provided by neural networks, and whether these stored values can be mapped to any given desired outputs up to ϵ.

First, we consider approximation in the maximum norm. This requires that input structures up to arbitrary depth can be stored in the context neurons and mapped to desired values. This situation includes the question of the computation capability of recurrent neural networks, i.e. the mapping of binary input sequences to desired binary output values. The capacity of RNNs with respect to classical mechanisms has been exactly characterized in the work provided by Siegelmann and coworkers [23, 24, 25, 26]: RNNs with semi-linear activation function are equivalent to so-called non-uniform Boolean circuits. Although this capacity subsumes Turing capability, it does not include all functions, hence RNNs and alternatives cannot approximate every (arbitrary) function in the maximum norm.

The publication [11] provides a general direct diagonalization argument which can immediately be transferred to our settings which shows the principled limitation of the systems:

Theorem 9 (Approximation incompleteness for general functions in max-norm). *Assume the transition function f^s of a RNN/RCC or CRCC* network is given as the composition of a finite number of fixed continuous nonlinear functions and affine transformations. Consider input structures with label 1, i.e. only structural differences are important. Then, there exist functions in the class of supersource transductions from sequences, trees, and DBAGs which cannot be approximated by a RNN/RCC or CRCC network in the maximum norm.

Proof. For any number i there exists at most a finite number of neural networks with at most i neurons and the specified structure. For any fixed architecture f with i neurons and a fixed input structure s_i with i vertices the set $\{y \mid y$ is output of a network with architecture f and input structure s_i with weights absolutely restricted by $i\}$ is compact. Therefore we can find a value y_i which is more than ϵ away from the union of these sets for all architectures with i neurons. Obviously, the mapping $s_i \mapsto y_i$ cannot be approximated in the maximum norm by any network with a finite number of neurons and finite weights.

For restricted function classes, approximation results can be found: it has been shown e.g. in [27] that RNNs with sequential input can approximate finite state automata up to arbitrary depth. Similarly, it has been shown in [28] that RNNs with tree structured inputs can approximate tree automata up to arbitrary depth. Thus, the following holds:

Theorem 10 (Completeness of RNN for finite automata in max-norm). *The class of RNN networks (with sigmoidal activation function or similar) is approximation complete in the maximum norm for the class of supersource transductions computed by finite state automata or tree automata, respectively.*

Note that RCC restricts the recurrence of RNNs since recurrent connection cannot go back to already frozen units due to the specific training mode of CC. The effects of the limitation of recurrence on the capacity of the models has been discussed in various approaches including [29, 30]. In particular, [30] shows the following

Theorem 11 (Incompleteness of RCC for finite automata in max-norm). *The class of RCC networks cannot approximate every finite state automaton in the maximum norm.*

Thus, considering approximation in the maximum norm, RCCs are strictly weaker than RNNs on sequences and the same transfers to tree structured inputs, DPAGS, and also CRCC as depicted in Figure 4.6.

In practice, however, approximation of functions up to some reasonable depth is usually sufficient. Thus, in practice, approximation completeness in probability is interesting. For such investigations, a number of properties of the transition functions are relevant:

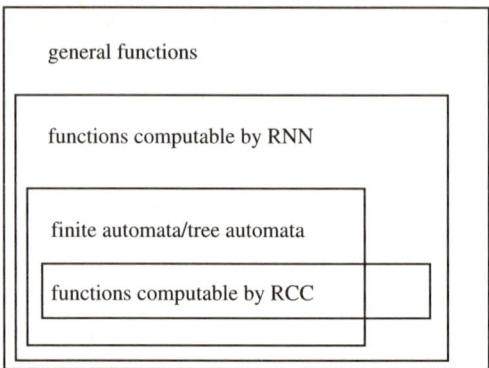

Fig. 4.6. Approximation capability by RCC and RNN networks when considering approximation in the maximum norm

- We assume that the function f^s is implemented by a neuron which constitutes a composition of an affine transformation and the sigmoidal function σ as shown in Equation 4.3.
- σ is the logistic function which has the following properties: it is a squashing function, i.e. $\lim_{x\to\infty} \sigma(x) = 1$ and $\lim_{x\to-\infty} \sigma(x) = 0$. σ is continuously differentiable at 0 with nonvanishing derivative $\sigma'(0)$, therefore, one can approximate uniformly the identity by $x \approx (\sigma(\epsilon \cdot x) - \sigma(0))/\epsilon \cdot \sigma'(0)$ for $\epsilon \to 0$.
- We only consider continuous supersource transductions starting from structures to $[0,1]$, the codomain of σ, and I/O-isomorphic transductions where labels are restricted to $[0,1]$, such that the restriction of the function to one label is continuous.

The following result has been shown in [11] which extends well known approximation results of recurrent networks for sequences such as [31] to tree structures:

Theorem 12 (Completeness of RNNs in probability). *Consider RNNs with the logistic activation function σ. Then every continuous supersource transduction from tree structures can be approximated arbitrarily well in probability by a RNN network.*

Obviously, at most causal I/O-isomorphic transductions can be approximated by RNNs. Note that approximation of causal transduction reduces to the approximation of all substructures of a given structure, thus, it follows immediately that Theorem 12 also holds for causal I/O-isomorphic transductions.

We now consider RCC and CRCC models. A general technique to prove approximation completeness in probability has been provided in [22]:

1. Design a unique linear code in a real-vector space for the considered input structures assumed a finite label alphabet. This needs to be done for every finite but fixed maximum size.

2. Show that this code can be computed by the recursive transition function provided by the model.
3. Add a universal neural network in the sense of [32] as readout; this can be integrated in the specific structure of CC.
4. Transfer the result to arbitrary labels by integrating a sufficiently fine discretization of the labels.
5. Approximate specific functions of the construction such as the identity or multiplication by the considered activation function σ.

Steps 3-5 are generic. They require some general mathematical argumentation which can be transferred immediately to our situation together with the property that σ is squashing and continuously differentiable (such that the identity can be approximated). Therefore, we do not go into details concerning these parts but refer to [22] instead.

Step 1 has already been considered in the previous section: we have investigated whether the functional form of the transition function, i.e. the sequence $term(\boldsymbol{x}(v))$, v being the supersource resp. root vertex, allows to differentiate all inputs. If so, it is sufficient to construct a representation of the functional form within a real-vector space and to show that this representation can be computed by the respective recursive mechanism. We shortly explain the main ideas for RCC and CRCC. Technical details can be found in [22].

Theorem 13 (Completeness of RCC for sequences in probability).
Every continuous supersource transduction from sequences to a real-vector space can be approximated arbitrarily well by a RCC network in probability.

Proof. We only consider steps 1 and 2: Assume a finite label alphabet $L = \{l_1, \ldots, l_n\}$ is given. Assume the symbols are encoded as natural numbers with at most d digits. Adding leading entries 0, if necessary, we can assume that every number has exactly d digits. Given a sequence $(l_{i_1}, \ldots, l_{i_t})$, the real number $0.l_{i_t} \ldots l_{i_1}$ constitutes a unique encoding. This encoding can be computed by a single RCC unit because it holds

$$0.1^d \cdot l_{i_t} + 0.1^d \cdot 0.l_{i_{t-1}} \ldots l_{i_1} = 0.l_{i_t} \ldots l_{i_1}$$

This is a simple linear expression depending on the input l_{i_t} at time step t and the context $0.l_{i_{t-1}} \ldots l_{i_1}$ which can be computed by a single RCC unit with linear activation function resp. approximated arbitrarily well by a single neuron with activation function σ.

As beforehand, this result transfers immediately to continuous *causal I/O-isomorphic* transductions since this relates to approximation of prefixes.

For tree structures, we consider more general multiplicative neurons which also include a multiplication of input values.

Definition 7 (Multiplicative neuron). *A multiplicative neuron of degree d with n inputs computes a function of the form*

$$x \mapsto \sigma \left(\sum_{i=1}^{d} \prod_{j_1,\ldots,j_i \in \{1,\ldots,n\}} w_{j_1\ldots j_i} \cdot x_{j_1} \cdot \ldots \cdot x_{j_i} + \theta \right)$$

(including an explicit bias θ).

It is possible to encode trees and more general graphs using recursive multiplicative neurons as shown in the following.

Theorem 14 (Completeness of RCC with multiplicative neurons for trees in probability). *Every continuous supersource transduction from tree structures to a real-vector space can be approximated arbitrarily well by a RCC network with multiplicative neurons in probability.*

Proof. As beforehand, we only consider steps 1 and 2, the representation of tree structures and its computation in a RCC network. As beforehand, we restrict to two children per vertex, the generalization to general fan-out being straightforward. Assume a finite label alphabet $L = \{l_1, \ldots, l_n\}$ is given. Assume these labels are encoded as numbers with exactly d digits each. Assume l_f and l_ϵ are numbers with d digits not contained in L. Then a tree t can uniquely be represented by $r(t) = l_\epsilon$ if t is empty and $r(t) = l_f l(v) r(t_1) r(t_2)$ (concatenation of these digits), otherwise, where $l(v)$ is the root label and t_1 and t_2 are the two subtrees. Note that this number represents the functional sequence $term(\boldsymbol{x}(v))$, v being the root vertex, which arises from a recursive function and uniquely characterizes a tree structure.

It is possible to compute the real numbers $0.r(t)$ and $(0.1)^{|r(t)|}$, $|r(t)|$ being the length of the digit string $r(t)$, by two multiplicative neurons of a RCC network as follows:

$$(0.1)^{|r(t)|} = 0.1^{2d} \cdot (0.1)^{|r(t_1)|} \cdot (0.1)^{|r(t_2)|}$$

for the neuron computing the length (included in the RCC network first) and

$$0.r(t) = 0.1^d \cdot l_f + 0.1^{2d} \cdot l(v) + 0.1^{2d} \cdot 0.r(t_1) + 0.1^{2d} \cdot (0.1)^{|r(t_1)|} \cdot 0.r(t_2)$$

for the neuron computing the representation. Note that the second neuron depends on the output of the first, but not vice versa, such that the restricted recurrence of RCC is accounted for. The neurons have a linear activation function which can be approximated arbitrarily well by σ.

As beforehand, this result can immediately be transferred to continuous *causal* I/O-isomorphic transductions since these are equivalent to supersource transductions on subtrees.

Contextual processing adds the possibility to include information about the parents of vertices. This extends the class of functions which can be computed by such networks. We obtain the following completeness result:

4 Processing structured data by RNNs

Theorem 15 (Completeness of CRCC with multiplicative neurons for DBAGs in probability). *Every continuous I/O-isomorphic or super-source transduction on DBAGs can be approximated by a CRCC network with multiplicative neurons in probability.*

Proof. We restrict to fan-in and fan-out 2, the generalization to larger fan-in and fan-out being straightforward. Note that, according to Theorem 7, the functional form given by CRCC allows a unique representation of DBAGs, provided the number of the hidden neurons is large enough (depending on the maximum length of the paths of the DBAG). Thus, it is sufficient to show that a unique numeric representation of the functional form of CRCC can be computed by a CRCC network. Thereby, as remarked in the proof of Theorem 7, we can restrict the functional form to

$$x_m(v) = f_m^s\Big(l(v), q^{-1}[x_m(v)], q^{+1}[x_{m-1}(v)]\Big)$$

Assume the label alphabet is given by $L = \{l_1, \ldots, l_n\}$ represented by numbers with exactly d digits. Assume $l_($, $l_)$, $l_,$, l_f, and l_ϵ are numbers with exactly d digits not contained in L. Denote an empty child/parent by ϵ. Define a representation $r_i(v)$ of a vertex v of a DBAG with children $ch_1(v)$ and $ch_2(v)$ and parents $pa_1(v)$ and $pa_2(v)$ recursively over i and the structure in the following way:

$$r_1(v) = f(l(v), r_1(ch_1(v)), r_1(ch_2(v))),$$

$$r_i(v) = f(l(v), r_i(ch_1(v)), r_i(ch_2(v)), r_{i-1}(pa_1(v)), r_{i-1}(pa_2(v)))$$

for $i > 1$. Note that $r_i(v)$ coincides with the functional form $term(x(v))$, x being the activation of a CRCC network with i hidden neurons and reduced dependence as specified in Theorem 7. Since this mirrors the functional dependence of CRCC, it yields unique values for large enough i. Denote by R_i the sequence of digits which is obtained by substituting the symbol f by l_f, '(' by $l_($, ',' by $l_,$, and ϵ by l_ϵ. Then we obtain unique numbers for every structure and vertex in the structure for large enough i. Thus, this proves step 1.

In the second step we show that the number $0.R_i(v)$ and $0.1^{|R_i(v)|}$, $|R_i(v)|$ being the length of $R_i(v)$, can be computed by a CRCC network. The general recurrence is of the form

$$0.1^{|R_i(v)|} = 0.1^{8d} \cdot 0.1^{|R_i(ch_1(v))|} \cdot 0.1^{|R_i(ch_2(v))|}$$
$$\cdot\, 0.1^{|R_{i-1}(pa_1(v))|} \cdot 0.1^{|R_{i-1}(pa_2(v))|}$$

for the length and

$0.R_i(v) =$
$0.1^d \cdot l_f + 0.1^{2d} \cdot l_(+ 0.1^{3d} \cdot l(v) + 0.1^{4d} \cdot l,$
$+ 0.1^{4d} \cdot 0.R_i(ch_1(v))$
$+ 0.1^{5d} \cdot 0.1^{|R_i(ch_1(v))|} \cdot n,$
$+ 0.1^{5d} \cdot 0.1^{|R_i(ch_1(v))|} \cdot 0.R_i(ch_2(v))$
$+ 0.1^{6d} \cdot 0.1^{|R_i(ch_1(v))|} \cdot 0.1^{|R_i(ch_2(v))|} \cdot l,$
$+ 0.1^{6d} \cdot 0.1^{|R_i(ch_1(v))|} \cdot 0.1^{|R_i(ch_2(v))|} \cdot 0.R_{i-1}(pa_1(v))$
$+ 0.1^{7d} \cdot 0.1^{|R_i(ch_1(v))|} \cdot 0.1^{|R_i(ch_2(v))|} \cdot 0.1^{|R_{i-1}(pa_1(v))|} \cdot l,$
$+ 0.1^{7d} \cdot 0.1^{|R_i(ch_1(v))|} \cdot 0.1^{|R_i(ch_2(v))|} \cdot 0.1^{|R_{i-1}(pa_1(v))|} \cdot 0.R_{i-1}(pa_2(v))$
$+ 0.1^{8d} \cdot 0.1^{|R_i(ch_1(v))|} \cdot 0.1^{|R_i(ch_2(v))|} \cdot 0.1^{|R_{i-1}(pa_1(v))|} \cdot 0.1^{|R_{i-1}(pa_2(v))|} \cdot l_)$

for the representation. This can be computed by a CRCC network with multiplicative neurons since the length does not depend on the representation, hence the restricted recurrence of CRCC is accounted for. As beforehand, the identity as activation function can be approximated arbitrarily well by σ.

Hence, CRCC extends causal models in the sense that it allows for general I/O-isomorphic transductions and supersource transductions on DBAGs, which extend trees to an important subclass of acyclic graphs. Note that this result also includes the fact, that non causal I/O-isomorphic transductions on trees can be approximated by CRCC models, thus, CRCC also extends the capacity of RCC on tree structures. Every (general) DPAG can be transformed to a DBAG by reenumeration of the children and extension of the fan-in/fan-out, as discussed in [22]. Hence, the overall picture as depicted in Figure 4.7 arises.

Actually, the capacity of CRCC is even larger: as discussed in Theorem 15, the functional form $term(\boldsymbol{x}(v))$ for CRCC neurons $\boldsymbol{x}$ and vertices v of DPAGs can uniquely be encoded by multiplicative neurons up to finite depth and for finite label alphabet. Therefore, all IO-isomorphic or supersource transductions of structures which are supported by CRCC networks, i.e. for which the expression $term(\boldsymbol{x}(v))$ yields unique representations, can be approximated by CRCC networks. We can identify general DPAGs with graphs with directed and enumerated connections from vertices pointing to their children and to their parents. As demonstrated in Theorem 7, all (labeled) paths within the graph starting from any vertex can be recovered from $term(\boldsymbol{x}(v))$ if the number of hidden neurons of the CRCC network is large enough. Thus, every two structures for which at least two different paths exist can be differentiated by CRCC networks, i.e. approximation completeness holds for this larger class of structures which are pairwise distinguishable by at least one path.

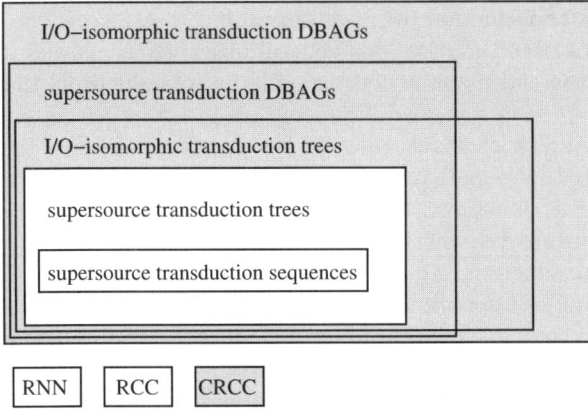

Fig. 4.7. Approximation capability of RCC and CRCC networks when considering approximation in probability

4.6 Conclusion

In this chapter we have analyzed, mainly from a computational point of view, Recursive Neural Networks models for the adaptive processing of structured data. These models are able to directly deal with the discrete nature of structured data, allowing to deal in a unified way both with discrete topologies and real valued information attached to each vertex of the structures. Specifically, RNNs allow for adaptive transduction on sequences and trees assuming a causal recursive processing system. However, causality affects computational capabilities and classes of data. Some of these capabilities can be recovered by contextual approaches, which provide an elegant approach for the extension of the classes of data that RNNs can deal with for learning.

We have reported and compared in a common framework the computational properties already presented in the literature for these models. Moreover, we have presented new results which allowed us to investigate the intimate relationships among models and classes of data. Specifically, we discussed the role of contextual information for adaptive processing of structured domains, evaluating the computational capability of the models with respect to the different classes of data. The use of contextual models allows to properly extend the class of transductions from sequences and trees, for which causal models are complete, to classes of DPAGs. Moreover, the characterization of the completeness introduced through this chapter allowed us to show that even if the causal models can treat DPAG structures, a proper extension to classes of DPAGs needs the use of contextual models. In fact, causal models induce collisions in the state representations of replicated substructures inserted into different contexts. This feature of causal models can be characterized through the definition of a constraint of invariance with respect to the unsharing construction we have defined for DPAGs. Contextual models, beside the extension to contextual transduction over sequences and trees, allow the relaxation of this constraint over DPAGs. This general observation allowed us to formally characterize the classes of structures and the transductions that the models can support.

This initial picture which relies on the functional form of the models, has been completed by results on the approximation capability using neural networks as transfer functions. Essentially, approximation completeness in probability can be proved wherever the functional representation supports the respective transduction. Note that, unlike [30], we discussed the practically relevant setting of approximation up to a set of small probability, such that the maximum recursive depth is limited in all concrete settings. For these situations, when considering multiplicative neurons, approximation completeness of RCC for supersource transductions on tree structures, and the approximation completeness of CRCC for IO-isomorphic and supersource transductions for tree structures and acyclic graphs, which are distinguishable by at least one path, results.

Interestingly, all cascade correlation models can be trained in a quite effective way with an extension of cascade correlation, determining in the same run the weights and structure of the neural networks. These models have already been successfully applied to QSAR and QSPR problems in chemistry [3, 4, 5, 6, 33, 7, 8, 9]. Although these tasks consist only of supersource transductions since the activity resp. relevant quantities of chemical molecules have to be predicted, the possibility to incorporate an expressive context proved beneficial for these tasks, as can be seen by an inspection of the context representation within the networks [34, 10].

References

1. Sperduti, A., Starita, A.: Supervised neural networks for the classification of structures. IEEE Trans. Neural Networks **8**(3) (1997) 714–735
2. Micheli, A., Portera, F., Sperduti, A.: A preliminary empirical comparison of recursive neural networks and tree kernel methods on regression tasks for tree structured domains. Neurocomputing **64** (2005) 73–92
3. Bianucci, A., Micheli, A., Sperduti, A., Starita, A.: Quantitative structure-activity relationships of benzodiazepines by Recursive Cascade Correlation. In IEEE, ed.: Proceedings of IJCNN '98 - IEEE World Congress on Computational Intelligence, Anchorage, Alaska (1998) 117–122
4. Bianucci, A.M., Micheli, A., Sperduti, A., Starita, A.: Application of cascade correlation networks for structures to chemistry. Journal of Applied Intelligence (Kluwer Academic Publishers) **12** (2000) 117–146
5. Micheli, A., Sperduti, A., Starita, A., Bianucci, A.M.: Analysis of the internal representations developed by neural networks for structures applied to quantitative structure-activity relationship studies of benzodiazepines. Journal of Chem. Inf. and Comp. Sci. **41**(1) (2001) 202–218
6. Micheli, A., Sperduti, A., Starita, A., Bianucci, A.M.: Design of new biologically active molecules by recursive neural networks. In: IJCNN'2001 - Proceedings of

the INNS-IEEE International Joint Conference on Neural Networks, Washington, DC (2001) 2732–2737
7. Bernazzani, L., Duce, C., Micheli, A., Mollica, V., Sperduti, A., Starita, A., Tiné, M.R.: Predicting physical-chemical properties of compounds from molecular structures by recursive ne ural networks. J. Chem. Inf. Model **46**(5) (2006) 2030–2042
8. Duce, C., Micheli, A., Solaro, R., Starita, A., Tiné, M.R.: Prediction of chemical-physical properties by neural networks for structures. Macromolecular Symposia **234**(1) (2006) 13–19
9. Duce, C., Micheli, A., Starita, A., Tiné, M.R., Solaro, R.: Prediction of polymer properties from their structure by recursive neural networks. Macromolecular Rapid Communications **27**(9) (2006) 711–715
10. Micheli, A., Sona, D., Sperduti, A.: Contextual processing of structured data by recursive cascade correlation. IEEE Trans. Neural Networks **15**(6) (2004) 1396–1410
11. Hammer, B.: Learning with Recurrent Neural Networks. Volume 254 of Springer Lecture Notes in Control and Information Sciences. Springer-Verlag (2000)
12. Elman, J.L.: Finding structure in time. Cognitive Science **14** (1990) 179–211
13. Fahlman, S.E., Lebiere, C.: The cascade-correlation learning architecture. In Touretzky, D., ed.: Advances in Neural Information Processing Systems 2, San Mateo, CA: Morgan Kaufmann (1990) 524–532
14. Fahlman, S.E.: The recurrent cascade-correlation architecture. In Lippmann, R., Moody, J., Touretzky, D., eds.: Advances in Neural Information Processing Systems 3, San Mateo, CA, Morgan Kaufmann Publishers (1991) 190–196
15. Sperduti, A.: On the computational power of recurrent neural networks for structures. Neural Networks **10**(3) (1997) 395–400
16. Carrasco, R., Forcada, M.: Simple strategies to encode tree automata in sigmoid recursive neural networks. IEEE TKDE **13**(2) (2001) 148–156
17. Gori, M., Kuchler, A., Sperduti, A.: On the implementation of frontier-to-root tree automata in recursive neural networks. IEEE Transactions on Neural Networks **10**(6) (1999) 1305–1314
18. Hammer, B.: On the learnability of recursive data. Mathematics of Control Signals and Systems **12** (1999) 62–79
19. Hammer, B.: Generalization ability of folding networks. IEEE TKDE **13**(2) (2001) 196–206
20. Micheli, A., Sona, D., Sperduti, A.: Bi-causal recurrent cascade correlation. In: IJCNN'2000 - Proceedings of the IEEE-INNS-ENNS International Joint Conference on Neural Networks. Volume 3. (2000) 3–8
21. Micheli, A., Sona, D., Sperduti, A.: Recursive cascade correlation for contextual processing of structured data. In: Proc. of the Int. Joint Conf. on Neural Networks - WCCI-IJCNN'2002. Volume 1. (2002) 268–273
22. Hammer, B., Micheli, A., Sperduti, A.: Universal approximation capability of cascade correlation for structures. Neural Computation **17**(5) (2005) 1109–1159
23. Kilian, J., Siegelmann, H.T.: The dynamic universality of sigmoidal neural networks. Information and Computation **128** (1996)
24. Siegelmann, H.T.: The simple dynamics of super Turing theories. Theoretical Computer Science **168** (1996)
25. Siegelmann, H.T., Sontag, E.D.: Analog computation, neural networks, and circuits. Theoretical Computer Science **131** (1994)

26. Siegelmann, H.T., Sontag, E.D.: On the computational power of neural networks. Journal of Computer and System Sciences **50** (1995)
27. Omlin, C., Giles, C.: Stable encoding of large finite-state automata in recurrent neural networks with sigmoid discriminants. Neural Computation **8** (1996)
28. Frasconi, P., Gori, M., Kuechler, A., Sperduti, A.: From sequences to data structures: Theory and applications. In Kolen, J., Kremer, S., eds.: A Field Guide to Dynamic Recurrent Networks. IEEE Press (2001) 351–374
29. Frasconi, P., Gori, M.: Computational capabilities of local-feedback recurrent networks acting as finite-state machines. IEEE Transactions on Neural Networks **7**(6) (1996) 1521–1524
30. Giles, C., Chen, D., Sun, G., Chen, H., Lee, Y., Goudreau, M.: Constructive learning of recurrent neural networks: limitations of recurrent cascade correlation and a simple solution. IEEE Transactions on Neural Networks **6**(4) (1995) 829–836
31. K., F., Nakamura, Y.: Approximation of dynamical systems by continuous time recurrent neural networks. Neural Networks **6**(6) (1993) 801–806
32. Hornik, K., Stinchcombe, M., White, H.: Multilayer feedforward networks are universal approximators. Neural Networks (1989) 359–366
33. Bianucci, A.M., Micheli, A., Sperduti, A., Starita, A.: A novel approach to QSPR/QSAR based on neural networks for structures. In Sztandera, L.M., Cartwright, H.M., eds.: Soft Computing Approaches in Chemistry. Springer-Verlag, Heidelberg (2003) 265–296
34. Micheli, A.: Recursive Processing of Structured Domains in Machine Learning. Ph.d. thesis: TD-13/03, Department of Computer Science, University of Pisa, Pisa, Italy (2003)

5

Markovian Bias of Neural-based Architectures With Feedback Connections

Peter Tiño[1], Barbara Hammer[2], and Mikael Bodén[3]

[1] University of Birmingham, Birmingham, UK   p.tino@cs.bham.ac.uk
[2] Clausthal University of Technology, Germany   hammer@in.tu-clausthal.de
[3] University of Queensland, Brisbane, Australia   mikael@itee.uq.edu.au

Summary. Dynamic neural network architectures can deal naturally with sequential data through recursive processing enabled by feedback connections. We show how such architectures are predisposed for suffix-based Markovian input sequence representations in both supervised and unsupervised learning scenarios. In particular, in the context of such architectural predispositions, we study computational and learning capabilities of typical dynamic neural network architectures. We also show how efficient finite memory models can be readily extracted from *untrained* networks and argue that such models should be used as baselines when comparing dynamic network performances in a supervised learning task. Finally, potential applications of the Markovian architectural predisposition of dynamic neural networks in bioinformatics are presented.

5.1 Introduction

There has been a considerable research activity in connectionist processing of sequential symbolic structures. For example, researchers have been interested in formulating models of human performance in processing linguistic patterns of various complexity (e.g. [1]).

Of special importance are dynamic neural network architectures capable of naturally dealing with sequential data through recursive processing. Such architectures are endowed with feedback delay connections that enable the models to operate with a "neural" memory in the form of past activations of a selected subset of neurons, sometimes referred to as recurrent neurons. It is expected that activations of such recurrent neurons will code all the "important" information from the past that is needed to solve a given task. In other words, the recurrent activations can be thought of as representing, in some "relevant" way, the temporal context in which the current input item is observed. The "relevance" is given by the nature of the task, be it next symbol prediction, or mapping sequences in a topographically ordered manner.

It was a bit surprising then, when researchers reported that when training dynamic neural networks to process language structures, activations of recurrent neurons displayed a considerable amount of structural differentiation even *prior to learning* [2, 3, 1]. Following [1], we refer to this phenomenon as the *architectural bias* of dynamic neural network architectures.

It has been recently shown, both theoretically and empirically, that the structural differentiation of recurrent activations before the training has much deeper implications [4, 5]. When initialized with small weights (standard initialization procedure), typical dynamic neural network architectures will organize recurrent activations in a suffix-based Markovian manner and it is possible to extract from such *untrained* networks predictive models comparable to efficient finite memory source models called variable memory length Markov models [6]. In addition, in such networks, recurrent activations tend to organize along a self-similar fractal substrate the fractal and multifractal properties of which can be studied in a rigorous manner [7]. Also, it is possible to rigorously characterize learning capabilities of dynamic neural network architectures in the early stages of learning [5].

Analogously, the well-known Self-Organizing Map (SOM) [8] for topographic low-dimensional organization of high-dimensional vectorial data has been reformulated to enable processing of sequential data. Typically, standard SOM is equipped with additional feed-back connections [9, 10, 11, 12, 13, 14]. Again, various forms of inherent predispositions of such models to Markovian organization of processed data have been discovered [15, 13, 16].

The aim of this paper is to present major developments in the phenomenon of architectural bias of dynamic neural network architectures in a unified framework. The paper has the following organization: First, a general state-space model formulation used throughout the paper is introduced in section 5.2. Then, still on the general high description level, we study in section 5.3 the link between Markovian suffix-based state space organization and contractive state transition maps. Basic tools for measuring geometric complexity of fractal sets are introduced as well. Detailed studies of several aspects of the architectural bias phenomenon in the supervised and unsupervised learning scenarios are presented in sections 5.4 and 5.5, respectively. Section 5.6 brings a flavor of potential applications of the architectural bias in bioinformatics. Finally, key findings are summarized in section 5.7.

5.2 General model formulation

We consider models that recursively process inputs $\boldsymbol{x}(t)$ from a set $\mathcal{X} \subseteq \mathbb{R}^{N_I}$ by updating their state $\boldsymbol{r}(t)$ in a bounded set $\mathcal{R} \subseteq \mathbb{R}^N$, and producing outputs $\mathbf{y}(t) \in \mathcal{Y}$, $\mathcal{Y} \subseteq \mathbb{R}^{N_O}$. State space model representation takes the form:

$$\mathbf{y}(t) = \boldsymbol{h}(\boldsymbol{r}(t)) \tag{5.1}$$

$$\boldsymbol{r}(t) = \boldsymbol{f}(\boldsymbol{x}(t), \boldsymbol{r}(t-1)). \tag{5.2}$$

Processing of an input time series $\boldsymbol{x}(t) \in \mathcal{X}$, $t = 1, 2, ...$, starts by initializing the model with some $\boldsymbol{r}(0) \in \mathcal{R}$ and then recursively updating the state

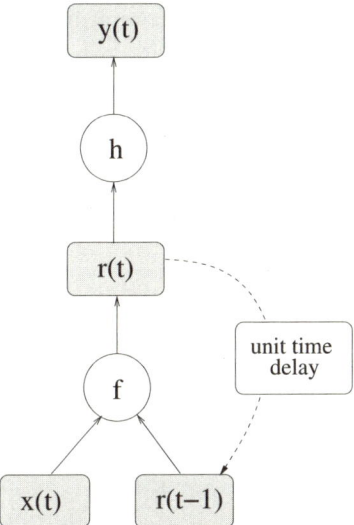

Fig. 5.1. The basic state space model used throughout the paper. Processing of an input time series $x(t) \in \mathcal{X}$, $t = 1, 2, ...$, is done by recursively updating the state $r(t) \in \mathcal{R}$ and output $y(t) \in \mathcal{Y}$ via functions $f : \mathcal{X} \times \mathcal{R} \to \mathcal{R}$ and $h : \mathcal{R} \to \mathcal{Y}$, respectively.

$r(t) \in \mathcal{R}$ and output $y(t) \in \mathcal{Y}$ via functions $f : \mathcal{X} \times \mathcal{R} \to \mathcal{R}$ and $h : \mathcal{R} \to \mathcal{Y}$, respectively. This is illustrated in figure 5.1.

We consider inputs coming from a finite alphabet $\mathcal{A}$ of A symbols. The set of all finite sequences over $\mathcal{A}$ is denoted by $\mathcal{A}^*$. The set $\mathcal{A}^*$ without the empty string ϵ is denoted by $\mathcal{A}^+$. The set of all sequences over $\mathcal{A}$ with exactly n symbols (n-blocks) is denoted by $\mathcal{A}^n$. Each symbol $s \in \mathcal{A}$ is mapped to its real-vector representation $c(s)$ by an injective coding function $c : \mathcal{A} \to \mathcal{X}$. The state transition process (5.2) can be viewed as a composition of fixed-input maps

$$f_s(r) = f(c(s), r), \quad s \in \mathcal{A}. \tag{5.3}$$

In particular, for a sequence $s_{1:n} = s_1...s_{n-2}s_{n-1}s_n$ over $\mathcal{A}$ and $r \in \mathcal{R}$, we write

$$\begin{aligned} f_{s_{1:n}}(r) &= f_{s_n}(f_{s_{n-1}}(...(f_{s_2}(f_{s_1}(r)))...)) \\ &= (f_{s_n} \circ f_{s_{n-1}} \circ ... \circ f_{s_2} \circ f_{s_1})(r). \end{aligned} \tag{5.4}$$

5.3 Contractive state transitions

In this section we will show that by constraining the system to contractive state transitions we obtain state organizations with Markovian flavor.

Moreover, one can quantify geometric complexity of such state space organizations using tools of fractal geometry.

5.3.1 Markovian state-space organization

Denote a Euclidean norm by $\|\cdot\|$. Recall that a mapping $\boldsymbol{F}: \mathcal{R} \to \mathcal{R}$ is said to be a contraction with contraction coefficient $\rho \in [0,1)$, if for any $\boldsymbol{r}, \boldsymbol{r}' \in \mathcal{R}$, it holds
$$\|\boldsymbol{F}(\boldsymbol{r}) - \boldsymbol{F}(\boldsymbol{r}')\| \leq \rho \cdot \|\boldsymbol{r} - \boldsymbol{r}'\|. \tag{5.5}$$

$\boldsymbol{F}$ is a contraction if there exists $\rho \in [0,1)$ so that $\boldsymbol{F}$ is a contraction with contraction coefficient ρ.

Assume now that each member of the family $\{\boldsymbol{f}_s\}_{s \in \mathcal{A}}$ is a contraction with contraction coefficient ρ_s and denote the weakest contraction rate in the family by
$$\rho_{max} = \max_{s \in \mathcal{A}} \rho_s.$$

Consider a sequence $s_{1:n} = s_1...s_{n-2}s_{n-1}s_n \in \mathcal{A}^n$, $n \geq 1$. Then for any two prefixes u and v and for any state $\boldsymbol{r} \in \mathcal{R}$, we have

$$\|\boldsymbol{f}_{us_{1:n}}(\boldsymbol{r}) - \boldsymbol{f}_{vs_{1:n}}(\boldsymbol{r})\| = \|\boldsymbol{f}_{s_{1:n-1}s_n}(\boldsymbol{f}_u(\boldsymbol{r})) - \boldsymbol{f}_{s_{1:n-1}s_n}(\boldsymbol{f}_v(\boldsymbol{r}))\| \tag{5.6}$$
$$= \|\boldsymbol{f}_{s_n}(\boldsymbol{f}_{s_{1:n-1}}(\boldsymbol{f}_u(\boldsymbol{r}))) - \boldsymbol{f}_{s_n}(\boldsymbol{f}_{s_{1:n-1}}(\boldsymbol{f}_v(\boldsymbol{r})))\| \tag{5.7}$$
$$\leq \rho_{max} \cdot \|\boldsymbol{f}_{s_{1:n-1}}(\boldsymbol{f}_u(\boldsymbol{r})) - \boldsymbol{f}_{s_{1:n-1}}(\boldsymbol{f}_v(\boldsymbol{r}))\|, \tag{5.8}$$

and consequently

$$\|\boldsymbol{f}_{us_{1:n}}(\boldsymbol{r}) - \boldsymbol{f}_{vs_{1:n}}(\boldsymbol{r})\| \leq \rho_{max}^n \cdot \|\boldsymbol{f}_u(\boldsymbol{r}) - \boldsymbol{f}_v(\boldsymbol{r})\| \tag{5.9}$$
$$\leq \rho_{max}^n \cdot diam(\mathcal{R}), \tag{5.10}$$

where $diam(\mathcal{R})$ is the diameter of the set $\mathcal{R}$, i.e. $diam(\mathcal{R}) = \sup_{x,y \in \mathcal{R}} \|x - y\|$. By similar arguments, for $L < n$,

$$\|\boldsymbol{f}_{s_{1:n}}(\boldsymbol{r}) - \boldsymbol{f}_{s_{n-L+1:n}}(\boldsymbol{r})\| \leq \rho^L \cdot \|\boldsymbol{f}_{s_{1:n-L}}(\boldsymbol{r}) - \boldsymbol{r}\| \tag{5.11}$$
$$\leq \rho^L \cdot diam(\mathcal{R}). \tag{5.12}$$

There are two related lessons to be learned from this exercise:

1. No matter what state $\boldsymbol{r} \in \mathcal{R}$ the model is in, the final states $\boldsymbol{f}_{us_{1:n}}(\boldsymbol{r})$ and $\boldsymbol{f}_{vs_{1:n}}(\boldsymbol{r})$ after processing two sequences with a long common suffix $s_{1:n}$ will lie close to each other. Moreover, the greater the length n of the common suffix, the closer lie the final states $\boldsymbol{f}_{us_{1:n}}(\boldsymbol{r})$ and $\boldsymbol{f}_{vs_{1:n}}(\boldsymbol{r})$.
2. If we could only operate with finite input memory of depth L, then all reachable states from an initial state $\boldsymbol{r}_0 \in \mathcal{R}$ could be collected in a finite set
$$\mathcal{C}_L(\boldsymbol{r}_0) = \{\boldsymbol{f}_w(\boldsymbol{r}_0) \mid w \in \mathcal{A}^L\}.$$

Consider now a long sequence $s_{1:n}$ over $\mathcal{A}$. Disregarding the initial transients for $1 \leq t \leq L-1$, the states $\boldsymbol{f}_{s_{1:t}}(\boldsymbol{r}_0)$ of the original model (5.2) can be approximated by the states $\boldsymbol{f}_{s_{t-L+1:t}}(\boldsymbol{r}_0) \in \mathcal{C}_L(\boldsymbol{r}_0)$ of the finite memory model to arbitrarily small precision, as long as the memory depth L is long enough.

What we have just described can be termed *Markovian organization of the model state space*. Information processing states that result from processing sequences sharing a common suffix naturally cluster together. In addition, the spatial extent of the cluster reflects the length of the common suffix - longer suffixes lead to tighter cluster structure.

If, in addition, the readout function $\boldsymbol{h}$ is Lipschitz with coefficient $\kappa > 0$, i.e. for any $\boldsymbol{r}, \boldsymbol{r}' \in \mathcal{R}$,

$$\|\boldsymbol{h}(\boldsymbol{r}) - \boldsymbol{h}(\boldsymbol{r}')\| \leq \kappa \cdot \|\boldsymbol{r} - \boldsymbol{r}'\|,$$

then

$$\|\boldsymbol{h}(\boldsymbol{f}_{us_{1:n}}(\boldsymbol{r})) - \boldsymbol{h}(\boldsymbol{f}_{vs_{1:n}}(\boldsymbol{r}))\| \leq \kappa \cdot \rho^n \cdot \|\boldsymbol{f}_u(\boldsymbol{r}) - \boldsymbol{f}_v(\boldsymbol{r})\|.$$

Hence, for arbitrary prefixes u, v, we have that for sufficiently long common suffix length n, the model outputs resulting from driving the system with $us_{1:n}$ and $vs_{1:n}$ can be made arbitrarily close. In particular, on the same long input sequence (after initial transients of length $L-1$), the model outputs (5.1) will be closely shadowed by those of the finite memory model operating only on the most recent L symbols (and hence on states from $\mathcal{C}_L(\boldsymbol{r})$), as long as L is large enough.

5.3.2 Measuring complexity of state space activations

Let us first introduce measures of size of geometrically complicated objects called fractals (see e.g. [17]). Let $K \subseteq \mathcal{R}$. For $\delta > 0$, a δ-*fine cover of* K is a collection of sets of diameter $\leq \delta$ that cover K. Denote by $N_\delta(K)$ the smallest possible cardinality of a δ-fine cover of K.

Definition 1. *The* upper *and* lower box-counting dimensions *of K are defined as*

$$\dim_B^+ K = \limsup_{\delta \to 0} \frac{\log N_\delta(K)}{-\log \delta} \quad \text{and} \quad \dim_B^- K = \liminf_{\delta \to 0} \frac{\log N_\delta(K)}{-\log \delta}, \quad (5.13)$$

respectively.

Let $\gamma > 0$. For $\delta > 0$, define

$$\mathcal{H}_\delta^\gamma(K) = \inf_{\Gamma_\delta(K)} \sum_{B \in \Gamma_\delta(K)} (diam(B))^\gamma, \quad (5.14)$$

where the infimum is taken over the set $\Gamma_\delta(K)$ of all countable δ-fine covers of K. Define
$$\mathcal{H}^\gamma(K) = \lim_{\delta \to 0} \mathcal{H}^\gamma_\delta(K).$$

Definition 2. *The* Hausdorff dimension *of the set K is*
$$\dim_H K = \inf\{\gamma|\ \mathcal{H}^\gamma(K) = 0\}. \tag{5.15}$$

It is well known that
$$\dim_H K \leq \dim_B^- K \leq \dim_B^+ K. \tag{5.16}$$

The Hausdorff dimension is more "subtle" than the box-counting dimensions: the former can capture details not detectable by the latter. For a more detailed discussion see e.g. [17].

Consider now a Bernoulli source $\mathcal{S}$ over the alphabet $\mathcal{A}$ and (without loss of generality) assume that all symbol probabilities are nonzero.

The system (5.2) can be considered an *Iterative Function System* (IFS) [18] $\{\boldsymbol{f}_s\}_{s \in \mathcal{A}}$. If all the fixed input maps $\boldsymbol{f}_s$ are contractions (contractive IFS), there exists a unique set $K \subseteq \mathcal{R}$, called the IFS attractor, that is invariant under the action of the IFS:
$$K = \bigcap_{s \in \mathcal{A}} \boldsymbol{f}_s(K).$$

As the system (5.2) is driven by an input stream generated by $\mathcal{S}$, the states $\{\boldsymbol{r}(t)\}$ converge to K. In other words, after some initial transients, state trajectory of the system (5.2) "samples" the invariant set K (chaos game algorithm).

It is possible to upper-bound fractal dimensions of K [17]:
$$\dim_B^+ K \leq \gamma, \quad \text{where} \quad \sum_{s \in \mathcal{A}} \rho_s^\gamma = 1.$$

If the IFS obeys the open set condition, i.e. all $\boldsymbol{f}_s(K)$ are disjoint, we can get a lower bound on the dimension of K as well: assume there exist $0 < \kappa_s < 1$, $s \in \mathcal{A}$, such that
$$\|\boldsymbol{f}_s(\boldsymbol{r}) - \boldsymbol{f}_s(\boldsymbol{r}')\| \geq \kappa_s \cdot \|\boldsymbol{r} - \boldsymbol{r}'\|.$$

Then [17],
$$\dim_H K \geq \gamma, \quad \text{where} \quad \sum_{s \in \mathcal{A}} \kappa_s^\gamma = 1.$$

5.4 Supervised learning setting

In this section, we will consider the implications of Markovian state-space organization for recurrent neural networks (RNN) formulated and trained in a supervised setting. We present the results for a simple 3-layer topology of RNN. The results can be easily generalized to more complicated architectures.

5.4.1 Recurrent Neural Networks and Neural Prediction Machines

Let $\boldsymbol{W}^{r,x}$, $\boldsymbol{W}^{r,r}$ and $\boldsymbol{W}^{y,r}$ denote real $N \times N_I$, $N \times N$ and $N_O \times N$ weight matrices, respectively. For a neuron activation function g, we denote its element-wise application to a real vector $\boldsymbol{a} = (a_1, a_2, ..., a_k)^T$ by $\mathbf{g}(\boldsymbol{a})$, i.e. $\mathbf{g}(\boldsymbol{a}) = (g(a_1), g(a_2), ..., g(a_k))^T$. Then, the state transition function $\boldsymbol{f}$ in (5.2) takes the form

$$\boldsymbol{f}(\boldsymbol{x}, \boldsymbol{r}) = \mathbf{g}(\boldsymbol{W}^{r,x}\boldsymbol{x} + \boldsymbol{W}^{r,r}\boldsymbol{r} + \boldsymbol{t}_f), \tag{5.17}$$

where $\boldsymbol{t}_f \in \mathbb{R}^N$ is a bias term.

Often, the output function $\boldsymbol{h}$ reads:

$$\boldsymbol{h}(\boldsymbol{r}) = \mathbf{g}(\boldsymbol{W}^{y,r}\boldsymbol{r} + \boldsymbol{t}_h), \tag{5.18}$$

with the bias term $\boldsymbol{t}_h \in \mathbb{R}^{N_O}$. However, for the purposes of symbol stream processing, the output function $\boldsymbol{h}$ is sometimes realized as a piece-wise constant function consisting of a collection of multinomial next-symbol distributions, one for each compartment of the partitioned state space $\mathcal{R}$. RNN with such an output function were termed *Neural Prediction Machines* (NPM) in [4]. More precisely, assuming the symbol alphabet $\mathcal{A}$ contains A symbols, the output space $\mathcal{Y}$ is the simplex

$$\mathcal{Y} = \{\mathbf{y} = (y_1, y_2, ..., y_A)^T \in \mathbb{R}^A \mid \sum_{i=1}^{A} y_i = 1 \text{ and } y_i \geq 0\}.$$

The next-symbol probabilities $\boldsymbol{h}(\boldsymbol{r}) \in \mathcal{Y}$ are estimates of the relative frequencies of symbols, conditional on RNN state $\boldsymbol{r}$. Regions of constant probabilities are determined by vector quantizing the set of recurrent activations that result from driving the network with the training sequence:

1. Given a training sequence $s_{1:n} = s_1 s_2 ... s_n$ over the alphabet $\mathcal{A} = \{1, 2, ..., A\}$, first re-set the network to the initial state $\boldsymbol{r}(0)$, and then, while driving the network with the sequence $s_{1:n}$, collect the recurrent layer activations in the set $\Gamma = \{\boldsymbol{r}(t) \mid 1 \leq t \leq n\}$.
2. Run your favorite vector quantizer with M codebook vectors $\boldsymbol{b}_1, ..., \boldsymbol{b}_M \in \mathcal{R}$, on the set Γ. Vector quantization partitions the state space $\mathcal{R}$ into M Voronoi compartments $V_1, ..., V_M$, of the codebook vectors[4] $\boldsymbol{b}_1, ..., \boldsymbol{b}_M$:

---
[4] ties are broken according to index order

$$V_i = \{r \mid \|r - b_i\| = \min_j \|r - b_j\|\}. \tag{5.19}$$

All points in V_i are allocated to the codebook vector b_i via a projection $\pi : \mathcal{R} \to \{1, 2, ..., M\}$,

$$\pi(r) = i, \quad \text{provided} \quad r \in V_i. \tag{5.20}$$

3. Re-set the network again with the initial state $r(0)$.
4. Set the [Voronoi compartment,next-symbol] counters $N(i, a)$, $i = 1, ..., M$, $a \in \mathcal{A}$, to zero.
5. Drive the network again with the training sequence $s_{1:n}$.
   For $1 \leq t < n$, if $\pi(r(t)) = i$, and the next symbol s_{t+1} is a, increment the counter $N(i, a)$ by one.
6. With each codebook vector associate the next symbol probabilities[5]

$$P(a \mid i) = \frac{N(i, a)}{\sum_{b \in \mathcal{A}} N(i, b)}, \quad a \in \mathcal{A}, \quad i = 1, 2, ..., M. \tag{5.21}$$

7. The output map is then defined as

$$h(r) = P(\cdot \mid \pi(r)), \tag{5.22}$$

where the a-th component of $h(r)$, $h_a(r)$, is the probability of next symbol being $a \in \mathcal{A}$, provided the current RNN state is r.

Once the NPM is constructed, it can make predictions on a test sequence (a continuation of the training sequence) as follows: Let $r(n)$ be the vector of recurrent activations after observing the training sequence $s_{1:n}$. Given a prefix $u_1 u_2, ... u_\ell$ of the test sequence, the NPM makes a prediction about the next symbol $u_{\ell+1}$ by:

1. Re-setting the network with $r(n)$.
2. Recursively updating states $r(n+t)$, $1 \leq t \leq \ell$, while driving the network with $u_1 u_2 ... u_\ell$.

---

[5] For bigger codebooks, we may encounter data sparseness problems. In such cases, it is advisable to perform smoothing of the empirical symbol frequencies by applying Laplace correction with parameter $\gamma > 0$:

$$P(a \mid i) = \frac{\gamma + N(i, a)}{\gamma \cdot A + \sum_{b \in \mathcal{A}} N(i, b)}.$$

The parameter γ can be viewed in the Dirichlet prior interpretation for the multinomial distribution $P(a \mid i)$ as the effective number of times each symbol $a \in \mathcal{A}$ was observed in the context of state compartment i, *prior* to counting the conditional symbol occurrences $N(i, a)$ in the training sequence. Typically, $\gamma = 1$, or $\gamma = A^{-1}$.

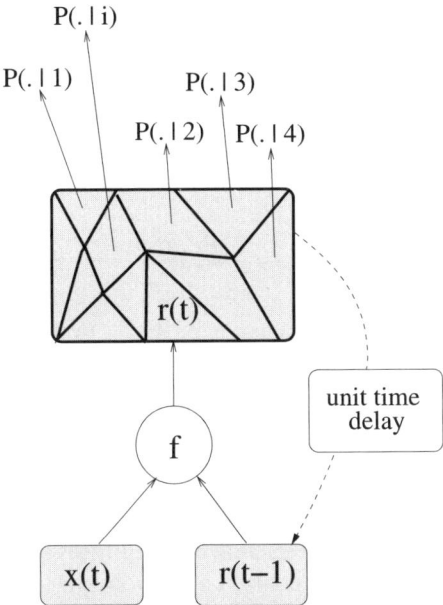

Fig. 5.2. Schematic illustration of Neural Prediction Machine. The output function is realized as a piece-wise constant function consisting of a collection of multinomial next-symbol distributions $P(\cdot|i)$, one for each compartment i of the state space $\mathcal{R}$.

3. The probability of $u_{\ell+1}$ occurring is

$$h_{u_{\ell+1}}(\boldsymbol{r}) = P(u_{\ell+1} \mid \pi(\boldsymbol{r}(n+\ell))). \tag{5.23}$$

A schematic illustration of NPM is presented in figure 5.2.

5.4.2 Variable memory length Markov models and NPM built on untrained RNN

In this section we will present intuitive arguments for a strong link between Neural Prediction Machines (NPM) constructed on untrained RNN and a class of efficient implementations of Markov models called *Variable memory length Markov models* (VLMM) [6]. The arguments will be made more extensive and formal in section 5.4.5.

In *Markov models* (MMs) of (fixed) order L, the conditional next-symbol distribution over alphabet $\mathcal{A}$, given the history of symbols $s_{1:t} = s_1 s_2 ... s_t \in \mathcal{A}^t$ observed so far, is written in terms of the last L symbols ($L \leq t$),

$$P(s_{t+1} \mid s_{1:t}) = P(s_{t+1} \mid s_{t-L+1:t}). \tag{5.24}$$

For large memory lengths L, the estimation of prediction probabilities $P(s|w)$, $w \in \mathcal{A}^L$, can become infeasible. To overcome this problem, VLMM were introduced. The key feature of VLMMs is that they permit prediction contexts of variable length, depending on the particular history of observed symbols.

Suppose we are given a long training sequence on which we calculate empirical frequencies $\hat{P}_\ell$ of ℓ-blocks over $\mathcal{A}$. Let $w \in \mathcal{A}^n$ be a potential prediction context of length n used to predict the next symbol $s \in \mathcal{A}$ according to the empirical estimates

$$\hat{P}(s|w) = \frac{\hat{P}_{n+1}(ws)}{\hat{P}_n(w)}.$$

If for a symbol $a \in \mathcal{A}$, such that $aw \in \mathcal{A}^{n+1}$, the empirical probability of the next symbol s,

$$\hat{P}(s|aw) = \frac{\hat{P}_{n+2}(aws)}{\hat{P}_{n+1}(aw)},$$

with respect to the extended context aw differs "significantly" from $\hat{P}(s|w)$, then extending the prediction context w with $a \in \mathcal{A}$ helps in the next-symbol predictions. Several decision criteria have been suggested in the literature. For example, extend the prediction context w with $a \in \mathcal{A}$, if the Kullback-Leibler divergence between the next-symbol distributions for the candidate prediction contexts w and aw, weighted by the prior distribution of the extended context aw, exceeds a given threshold [19] [20]. For other variants of decision criteria see [21] [22].

A natural representation of the set of prediction contexts, together with the associated next-symbol probabilities, has the form of a prediction suffix tree (PST) [19] [23]. The edges of PST are labeled by symbols from $\mathcal{A}$. From every internal node there is at most one outgoing edge labeled by each symbol. The nodes of PST are labeled by pairs $(s, \hat{P}(s|w))$, $s \in \mathcal{A}$, $w \in \mathcal{A}^+$, where w is a string associated with the walk starting from that node and ending in the root of the tree. Given a history $s_{1:t} = s_1 s_2 ... s_t$ of observed symbols up to time t, the corresponding prediction context is the deepest node in the PST reached by taking a walk labeled by the reversed string, $s_t ... s_2 s_1$, starting in the root.

It is a common practice to initialize the RNN weights with small numbers. In such cases, *prior to RNN training*, the state transition function (5.17) is initialized as a contraction. There is a good reason for this practice: unless one has a strong prior knowledge about the network dynamics [24], the sequence of desired bifurcations during the training process may be hard to achieve when starting from arbitrarily complicated network dynamics [25]. But, as argued in section 5.3.1, in *contractive NPMs*, histories of seen symbols sharing long common suffixes are mapped close to each other. Under finite input memory assumptions, such histories are likely to produce similar continuations and in contractive NPMs they are likely to appear in the same quantization region. Dense areas in the RNN state space correspond to symbol histories with long common suffixes and are given more attention by the vector quantizer. Hence, the prediction contexts are analogous to those of VLMM in that deep memory is used only when there are many different symbol histories sharing a long common suffix. In such situations it is natural to consider deeper past contexts when predicting the future. This is done automatically by the vector quantizer in NPM construction as it devotes more codebook vectors to the dense regions than to the sparsely inhabited areas of the RNN state space. More codebook vectors in dense regions imply tiny quantization compartments V_i that in turn group symbol histories with long shared suffixes.

Note, however, that while prediction contexts of VLMMs are built in a supervised manner, i.e. deep memory is considered only if it is dictated by the conditional distributions over the next symbols, in contractive NPMs prediction contexts of variable length are constructed in an unsupervised manner: prediction contexts with deep memory are accepted if there is a *suspicion* that shallow memory may not be enough, i.e. when there are many different symbol histories in the training sequence that share a long common suffix.

5.4.3 Fractal Prediction Machines

A special class of affine contractive NPMs operating with finite input memory has been introduced in [26] under the name *Fractal Prediction Machines* (FPM). Typically, the state space $\mathcal{R}$ is an N-dimensional unit hypercube[6] $\mathcal{R} = [0,1]^N$, $N = \lceil \log_2 A \rceil$. The coding function $\boldsymbol{c} : \mathcal{A} \to \mathcal{X} = \mathcal{R}$ maps each of the A symbols of the alphabet $\mathcal{A}$ to a unique vertex of the hypercube $\mathcal{R}$. The state transition function (5.17) is of particularly simple form, as the activation function g is identity and the connection weight matrices are set to $\boldsymbol{W}^{r,x} = \rho \boldsymbol{I}$ and $\boldsymbol{W}^{r,x} = (1-\rho)\boldsymbol{I}$, where $0 < \rho < 1$ is a contraction coefficient and $\boldsymbol{I}$ is the $N \times N$ identity matrix. Hence, the state space evolution (5.2) reads:

$$\boldsymbol{r}(t) = \boldsymbol{f}(\boldsymbol{x}(t), \boldsymbol{r}(t-1))$$
$$= \rho \boldsymbol{r}(t-1) + (1-\rho)\boldsymbol{x}(t). \quad (5.25)$$

The reset state is usually set to the center of $\mathcal{R}$, $\boldsymbol{r}_0 = \{\frac{1}{2}\}^N$. After fixing the memory depth L, the FPM states are confined to the set

$$\mathcal{C}_L(\boldsymbol{r}_0) = \{\boldsymbol{f}_w(\boldsymbol{r}_0) \mid w \in \mathcal{A}^L\}.$$

FPM construction proceeds in the same manner as that of NPM, except for the states $\boldsymbol{f}_{s_{1:t}}(\boldsymbol{r}_0)$ coding history of inputs up to current time t are approximated by their memory-L counterparts $\boldsymbol{f}_{s_{t-L+1:t}}(\boldsymbol{r}_0)$ from $\mathcal{C}_L(\boldsymbol{r}_0)$, as discussed in section 5.3.1.

---

[6] for $x \in \Re$, $\lceil x \rceil$ is the smallest integer y, such that $y \geq x$

5.4.4 Echo and Liquid State Machines

A similar architecture has been proposed by Jaeger in [27, 28]. So-called *Echo State Machines* (ESM) combine an untrained recurrent reservoir with a trainable linear output layer. Thereby, the state space model has the following form

$$\mathbf{y}(t) = \mathbf{h}(\mathbf{x}(t), \mathbf{r}(t), \mathbf{y}(t-1)) \tag{5.26}$$
$$\mathbf{r}(t) = \mathbf{f}(\mathbf{x}(t), \mathbf{r}(t-1), \mathbf{y}(t-1)). \tag{5.27}$$

For training, the function $\mathbf{f}$ is fixed and chosen in such a way that it has the echo state property, i.e. the network state $\mathbf{r}(t)$ is uniquely determined by left infinite sequences. In practice, the dependence of $\mathbf{r}$ on $\mathbf{y}$ is often dropped such that this property is equivalent to the fading property of recurrent systems and it can be tested by determining the Lipschitz constant of $\mathbf{f}$ (which must be smaller than 1). The readout $\mathbf{h}$ is trainable. It is often chosen as a linear function such that training can be done efficiently using linear regression. Unlike FPMs, ESMs usually work with very high dimensional state representations and a transition function which is obtained as the composition of a nonlinear activation function and a linear mapping with (sufficiently small) *random* weights. This combination should guarantee that the transition function has sufficiently rich dynamics to represent "important" properties of input series within its reservoir.

A similar idea is the base for so-called *Liquid State Machines* (LSM) as proposed by Maass [29]. These are based on a fixed recurrent mechanism in combination with a trainable readout. Unlike FPM and ESM, LSMs consider continuous time, and they are usually based on biologically plausible circuits of spiking neurons with sufficiently rich dynamics. It has been shown in [29] that LSMs are approximation complete for operators with fading memory under mild conditions on the models. Possibilities to estimate the capacity of such models and their generalization ability have recently been presented in [30]. However, the fading property is not required for LSMs.

5.4.5 On the computational power of recursive models

Here we will rigorously analyze the computational power of recursive models with contractive transition function. The arguments are partially taken from [5]. Given a sequence $s_{1:t} = s_1 \ldots s_t$, the *L-truncation* is given by

$$T_L(s_{1:t}) = \begin{cases} s_{t-L+1:t} & \text{if } t \leq L \\ s & \text{otherwise} \end{cases}$$

Assume $\mathcal{A}$ is a finite input alphabet and $\mathcal{Y}$ is an output set. A *Definite Memory Machine* (DMM) computes a function $f : \mathcal{A}^* \to \mathcal{Y}$ such that $f(s) = f(T_L(s))$ for some fixed memory length L and every sequence s. A *Markov model* (MM) fulfills $P(s_{t+1}|s_{1:t}) = P(s_{t+1}|T_L(s))$ for some L. Obviously, the function which maps a sequence to the next symbol probability given by a Markov model defines a definite memory machine. Variable length Markov models are subsumed by standard Markov models taking L as the maximum length. Therefore, we will only consider definite memory machines in the following.

State space models as defined in eqs. (5.1-5.2) induce a function $\boldsymbol{h} \circ \boldsymbol{f}_\bullet(\boldsymbol{r})$: $\mathcal{A}^* \to \mathcal{Y}, s_{1:t} \mapsto (\boldsymbol{h} \circ \boldsymbol{f}_{s_t} \circ \ldots \circ \boldsymbol{f}_{s_1})(\boldsymbol{r})$ as defined in (5.4) given a fixed initial value $\boldsymbol{r}$. As beforehand, we assume that the set of internal states $\mathcal{R}$ is bounded. Consider the case that $\boldsymbol{f}$ is a contraction w.r.t. $\boldsymbol{x}$ with parameter ρ and $\boldsymbol{h}$ is Lipschitz continuous with parameter κ. According to (5.12), $|\boldsymbol{f}_s(\boldsymbol{r}) - \boldsymbol{f}_{T_L(s)}(\boldsymbol{r})| \leq \rho^L \cdot diam(\mathcal{R})$. Note that the function $\boldsymbol{h} \circ \boldsymbol{f}_{T_L(\bullet)}(\boldsymbol{r})$ can be given by a definite memory machine. Therefore, if we choose $L = \log_\rho(\epsilon/(\kappa \cdot diam(\mathcal{R})))$, we find $|\boldsymbol{h} \circ \boldsymbol{f}_s(\boldsymbol{r}) - \boldsymbol{h} \circ \boldsymbol{f}_{T_L(s)}(\boldsymbol{r})| \leq \epsilon$. Hence, we can conclude that every recurrent system with contractive transition function can be approximated arbitrarily well by a finite memory model. This argument proves the Markovian property of RNNs with small weights:

Theorem 1. *Assume $\boldsymbol{f}$ is a contraction w.r.t. $\boldsymbol{x}$, and $\boldsymbol{h}$ is Lipschitz continuous. Assume $\boldsymbol{r}$ is a fixed initial state. Assume $\epsilon > 0$. Then there exists a definite memory machine with function $g : \mathcal{A}^* \to \mathcal{Y}$ such that $|g(s) - \boldsymbol{h} \circ \boldsymbol{f}_s(\boldsymbol{r})| \leq \epsilon$ for every input sequence s.*

As shown in [5], this argument can be extended to the situation where the input alphabet is not finite, but given by an arbitrary subset of a real-vector space.

It can easily be seen that standard recurrent neural networks with small weights implement contractions, since standard activation functions such as tanh are Lipschitz continuous and linear maps with sufficiently small weights are contractive. RNNs are often initialized by small random weights. Therefore, RNNs have a Markovian bias in early stages of training, first pushing solutions towards simple dynamics before going towards more complex models such as finite state automata and beyond.

It has been shown in [27] that ESMs without output feedback possess the so-called shadowing property, i.e. for every ϵ a memory length L can be found such that input sequences the last L entries of which are similar or identical lead to images which are at most ϵ apart. As a consequence, such mechanisms can also be simulated by Markovian models. The same applies to FPMs (the transition is a contraction by construction) and NPMs built on untrained RNNs randomly initialized with small weights.

The converse has also been proved in [5]: every definite memory machine can be simulated by a RNN with small weights. Therefore, an equivalence of small weight RNNs (and similar mechanisms) and definite memory machines holds. One can go a step further and consider the question in which cases *randomly* initialized RNNs with small weights can simulate a given definite memory machine provided a suitable readout is chosen.

The question boils down to the question whether the images of input sequences s of a fixed memory length L under the recursion $(\boldsymbol{f}_{s_L} \circ \ldots \circ \boldsymbol{f}_{s_1})(\boldsymbol{r})$ are of a form such that they can be mapped to desired output values by a trainable readout. We can assume that the codes for the symbols s_i are linearly independent. Further, we assume that the transfer function of the network is a nonlinear and non algebraic function such as tanh. Then the terms $(\boldsymbol{f}_{s_L} \circ \ldots \circ \boldsymbol{f}_{s_1})(\boldsymbol{r})$ constitute functions of the weights which are algebraically independent since they consist of the composition of linearly independent terms and the activation function. Thus, the images of sequences of length L are almost surely disjoint such that they can be mapped to arbitrary output values by a trainable readout provided the readout function $\boldsymbol{h}$ fulfills the universal approximation capability as stated e.g. in [31] (e.g. FNNs with one hidden layer). If the dimensionality of the output is large enough (dimensionality L^A), the outputs are almost surely linearly independent such that even a linear readout is sufficient. Thus, NPMs and ESMs with *random* initialization constitute universal approximators for finite memory models provided sufficient dimensionality of the reservoir.

5.4.6 On the generalization ability of recursive models

Another important issue of supervised adaptive models is their generalization ability, i.e. the possibility to infer from learned data to new examples. It is well known that the situation is complex for recursive models due to the relatively complex inputs: sequences of priorly unrestricted length can, in principle, encode behavior of priorly unlimited complexity. This observation can be mathematically substantiated by the fact that generalization bounds for recurrent networks as derived e.g. in [32, 33] depend on the input distribution. The situation is different for finite memory models for which generalization bounds independent of the input distribution can be derived. A formal argument which establishes generalization bounds of RNNs with Markovian bias has been provided in [5], whereby the bounds have been derived also for the setting of continuous input values stemming from a real-vector space. Here, we only consider the case of a finite input alphabet $\mathcal{A}$.

Assume $\mathcal{F}$ is the class of functions which can be implemented by a recurrent model with Markovian bias. For simplicity, we restrict to the case of binary classification, i.e. outputs stem from $\{0, 1\}$, e.g. adding an appropriate discretization to the readout $\boldsymbol{h}$. Assume P is an unknown underlying input-output distribution which has to be learned. Assume a set of m input-output pairs, $\mathcal{T}_m = \{(s^1, y^1), \ldots, (s^m, y^m)\}$, is given; the pairs are sampled from P in an i.i.d. manner[7]. Assume a learning algorithm outputs a function g when trained on the samples from $\mathcal{T}_m$. The *empirical error* of the algorithm

---
[7] independent identically distributed

is defined as
$$\hat{E}_{\mathcal{T}_m}(g) = \frac{1}{m} \sum_{i=1}^{m} |g(s^i) - y^i|.$$

This is usually minimized during training. The *real error* is given by the quantity
$$E_P(g) = \int |g(s) - y| \, dp(s,y).$$

A learning algorithm is said to generalize to unseen data if the real error is minimized by the learning scheme.

The capability of a learning algorithm of generalizing from the training set to new examples depends on the capacity of the function class used for training. One measure of the complexity is offered by the Vapnik Chervonenkis (VC) dimension of the function class $\mathcal{F}$. The VC dimension of $\mathcal{F}$ is the size of the largest set such that every binary mapping on this set can be realized by a function in $\mathcal{F}$. It has been shown by Vapnik [34] that the following bound holds for sample size m, $m \geq \mathcal{VC}(\mathcal{F})$ and $m > 2/\epsilon$:

$$P^m\{\mathcal{T}_m \mid \exists g \in \mathcal{F} \text{ s.t. } |E_P(g) - \hat{E}_{\mathcal{T}_m}(g)| > \epsilon\}$$
$$\leq 4 \cdot \left(\frac{2 \cdot e \cdot m}{\mathcal{VC}(\mathcal{F})}\right) \cdot \exp(-\epsilon^2 \cdot m/8)$$

Obviously, the number of different binary Markovian classifications with finite memory length L and alphabet $\mathcal{A}$ of A symbols is restricted by 2^{L^A}, thus, the VC dimension of the class of Markovian classifications is limited by L^A. Hence its generalization ability is guaranteed. Recurrent networks with Markovian bias can be simulated by finite memory models. Thereby, the necessary input memory length L can be estimated from the contraction coefficient ρ of the recursive function, the Lipschitz constant κ of the readout, and the diameter of the state space $\mathcal{R}$, by the term $-\log_\rho(2 \cdot \kappa \cdot diam(\mathcal{R}))$. This holds since we can approximate every binary classification up to $\epsilon = 0.5$ by a finite memory machine using this length as shown in Theorem 1. This argument proves the generalization ability of ESMs, FPMs, or NPMs with fixed recursive transition function and finite input alphabet $\mathcal{A}$ for which a maximum memory length can be determined a priori. The argumentation can be generalized to the case of arbitrary (continuous) inputs as shown in [5].

If we train a model (e.g. a RNN with small weights) in such a way that a Markovian model with finite memory length arises, whereby the contraction constant of the recursive part (and hence the corresponding maximum memory length L) is not fixed, these bounds do not apply. In this case, one can use arguments from the theory of structural risk minimization as provided e.g. in [35]: denote by $\mathcal{F}_i$ the set of recursive functions which can be simulated using memory length i (this corresponds to a contraction coefficient $(2 \cdot \kappa \cdot diam(\mathcal{R})^{-1/i})$. Obviously, the VC dimension of $\mathcal{F}_i$ is i^A. Assume a classifier is trained on a sample $\mathcal{T}_m$ of m i.i.d. input-output pairs (s^j, y^j), $j = 1, 2, ...m$, generated from P. Assume the number of errors on $\mathcal{T}_m$ of the trained classifier g is k, and the recursive model g is contained in $\mathcal{F}_i$. Then, according to [35] (Theorem 2.3), the generalization error $E_P(g)$ can be upper bounded by the following term with probability $1 - \delta$:

$$\frac{1}{m}\left((2+4\ln 2)k + 4\ln 2i + 4\ln\frac{4}{\delta} + 4i^A \cdot \ln\left(\frac{2em}{i^A}\right)\right).$$

(Here, we chose the prior probabilities used in [35] (Theorem 2.3) as $1/2^i$ for the probability that g is in $\mathcal{F}_i$ and $1/2^k$ that at most k errors occur.) Unlike the bounds derived e.g. in [15], this bound holds for any *posterior* finite memory length L obtained during training. In practice, the memory length can be estimated from the contraction and Lipschitz constants as shown above. Thus, unlike for general RNNs, generalization bounds which do not depend on the input distribution, but rather on the memory length, can be derived for RNNs with Markovian bias.

5.4.7 Verifying the architectural bias

The fact that one can extract from untrained recurrent networks Neural Prediction Machines (NPM, section 5.4.1) that yield comparable performance to Variable Memory Length Markov Models (VLMM, section 5.4.2) has been demonstrated e.g. in [4]. This is of vital importance for model building and evaluation, since in the light of these findings, the base line models in neural-based symbolic sequence processing tasks should in fact be VLMMs. If a neural network model after the training phase cannot beat a NPM extracted from an untrained model, there is obviously no point in training the model first place. This may sound obvious, but there are studies e.g. in the field of cognitive aspects of natural language processing, where trained RNN are compared with fixed order Markov models [1]. Note that such Markov models are often inferior to VLMM.

In [4], five model classes were considered:

- RNN trained via Real Time Recurrent Learning [36] coupled with Extended Kalman Filtering [37, 38]. There were A input and output units, one for each symbol in the alphabet $\mathcal{A}$, i.e. $N_I = N_O = A$. The inputs and desired outputs were encoded through binary one-of-A encoding. At each time step, the output activations were normalized to obtain next-symbol probabilities.
- NPM extracted from trained RNN.
- NPM extracted from RNN *prior to training*.
- Fixed order Markov models.
- VLMM.

The models were tested on two data sets:

- A series of quantized activations of a laser in a chaotic regime. This sequence can be modeled quite successfully with finite memory predictors, although, because of the limited finite sequence length, predictive contexts of variable memory depth are necessary.
- An artificial language exhibiting deep recursive structures. Such structures cannot be grasped by finite memory models and fully trained RNN should be able to outperform models operating with finite input memory and NPM extracted from untrained networks.

Model performance was measured by calculating the average (per symbol) negative log-likelihood on the hold-out set (not used during the training). As expected, given fine enough partition of the state space, the performances of RNNs and NPMs extracted from trained RNNs were almost identical. Because of finite sample size effects, fixed-order Markov models were beaten by VLMMs. Interestingly enough, when considering models with comparable number of free parameters, the performance of NPMs extracted prior to training mimicked that of VLMMs. Markovian bias of untrained RNN was clearly visible.

On the laser sequence, negative log-likelihood on the hold out set improved slightly through an expensive RNN training, but overall, almost the same performance could be achieved by a cheap construction of NPMs from randomly initialized *untrained* networks.

On the deep-recursion set, trained RNNs could achieve much better performance. However, it is essential to quantify how much useful information has really been induced during the training. As argued in this study, this can only be achieved by consulting VLMMs and NPMs extracted *before* training.

A large-scale comparison in [26] of Fractal Prediction Machines (FPM, section 5.4.3) with both VLMM an fixed order Markov models revealed an almost equivalent performance (modulo model size) of FPM and VLMM. Again, fixed order Markov models were inferior to both FPM and VLMM.

5.4.8 Geometric complexity of RNN state activations

It has been well known that RNN state space activations r often live on a fractal support of a self-similar nature [39]. As outlined in section 5.3.2, in case of contractive RNN, one can actually estimate the size and geometric complexity of the RNN state space evolution.

Consider a Bernoulli source $\mathcal{S}$ over the alphabet $\mathcal{A}$. When we drive a RNN with symbolic sequences generated by $\mathcal{S}$, we get recurrent activations $\{r(t)\}$ that tend to group in clusters sampling the invariant set K of the RNN iterative function system $\{f_s\}_{s \in \mathcal{A}}$.

Note that for each input symbol $s \in \mathcal{A}$, we have an affine mapping acting on $\mathcal{R}$:
$$Q_s(r) = W^{r,r} r + W^{r,x} c(s) + t_f.$$

The range of possible net-in activations of recurrent units for input s is then

$$C_s = \bigcup_{1 \leq i \leq N} [Q_s(\mathcal{R})]_i, \tag{5.28}$$

where $[B]_i$ is the slice of the set B, i.e. $[B]_i = \{r_i | \ \boldsymbol{r} = (r_1, ..., r_N)^T \in B\}$, $i = 1, 2, ..., N$.

By finding upper bounds on the values of derivatives of activation function,

$$g'_s = \sup_{v \in C_s} |g'(v)|, \quad g'_{max} = \max_{s \in \mathcal{A}} g'_s,$$

and setting

$$\rho_s^{max} = \alpha_+(\boldsymbol{W}^{r,r}) \cdot g'_s, \tag{5.29}$$

where $\alpha_+(\boldsymbol{W}^{r,r})$ is the largest singular value of $\boldsymbol{W}^{r,r}$, we can bound the upper box-counting dimension of RNN state activations [40]: Suppose $\alpha_+(\boldsymbol{W}^{r,r}) \cdot g'_{max} < 1$ and let γ_{max} be the unique solution of $\sum_{s \in \mathcal{A}} (\rho_s^{max})^\gamma = 1$. Then, $\dim_B^+ K \leq \gamma_{max}$.

Analogously, Hausdorff dimension of RNN sates can be lower-bounded as follows: Define

$$q = \min_{s, s' \in \mathcal{A}, s \neq s'} \|\boldsymbol{W}^{r,x}(\boldsymbol{c}(s) - \boldsymbol{c}(s'))\|$$

and assume $\alpha_+(\boldsymbol{W}^{r,r}) < \min\{(g'_{max})^{-1}, q \cdot diam(\mathcal{R})\}$. Let γ_{min} be the unique solution of $\sum_{s \in \mathcal{A}} (\rho_s^{min})^\gamma = 1$, where $\rho_s^{min} = \alpha_-(\boldsymbol{W}^{r,r}) \cdot \inf_{v \in C_s} |g'(v)|$ and $\alpha_-(\boldsymbol{W}^{r,r})$ is the smallest singular value of $\boldsymbol{W}^{r,r}$. Then, $\dim_H K \geq \gamma_{min}$.

Closed-form (less tight) bounds are also possible [40]:

$$\dim_B^+ K \leq -\frac{\log A}{\log N + \log W_{max}^{r,r} + \log g'_{max}},$$

where $W_{max}^{r,r} = \max_{1 \leq i,j \leq N} |W_{ij}^{r,r}|$; and

$$\dim_H K \geq -\frac{\log A}{\log \rho^{min}},$$

where $\rho^{min} = \min_{s \in \mathcal{A}} \rho_s^{min}$.

A more involved multifractal analysis of RNN state activations under more general stochastic sources driving the RNN input can be found in [7].

5.5 Unsupervised learning setting

In this section, we will study model formulations for constructing topographic maps of sequential data. While most approaches to topographic map formation assume that the data points are members of a finite-dimensional vector space of a fixed dimension, there has been a considerable interest in extending topographic maps to more general data structures, such as sequences or trees. Several modifications of standard Self-Organizing Map (SOM) [8] to sequences and/or tree structures have been proposed in the literature [41, 15]. Several modifications of SOM equip standard SOM with additional feed-back connections that allow for natural processing of recursive data types. Typical examples of such models are Temporal Kohonen Map [9], recurrent SOM [10], feedback SOM [11], recursive SOM [12], merge SOM [13] and SOM for structured data [14].

In analogy to well-known capacity results derived for supervised recursive models e.g. in [42, 43], some approaches to investigate the principled capacity of these models have been presented in the last few years. Thereby, depending on the model, the situation is diverse: whereas SOM for structured data and merge SOM (both with arbitrary weights) are equivalent to finite automata [44, 13], recursive SOM equipped with a somewhat simplified context model can simulate at least pushdown automata [45]. The Temporal Kohonen Map and recurrent SOM are restricted to finite memory models due to their restricted recurrence even with arbitrary weights [45, 44]. However, the exact capacity of most unsupervised recursive models as well as biases which arise during training have hardly been understood so far.

5.5.1 Recursive SOM

In this section we will analyze recursive SOM (RecSOM) [12]. There are two main reasons for concentrating on RecSOM. First, RecSOM transcends simple local recurrence of leaky integrators of earlier SOM-motivated models for processing sequential data and consequently can represent much richer dynamical behavior [15]. Second, the underlying continuous state-space dynamics of RecSOM makes this model particularly suitable for analysis along the lines of section 5.3.1.

The map is formed at the recurrent layer. Each unit $i \in \{1, 2, ..., N\}$ in the map has two weight vectors associated with it:

- $w_i^{r,x} \in \mathcal{X}$ – linked with an N_I-dimensional input $x(t)$ feeding the network at time t
- $w_i^{r,r} \in \mathcal{R}$ – linked with the context

$$r(t-1) = (r_1(t-1), r_2(t-1), ..., r_N(t-1))^T$$

containing map activations $r_i(t-1)$ from the previous time step.

Hence, the weight matrices $W^{r,x}$ and $W^{r,r}$ can be written as $W^{r,x} = ((w_1^{r,x})^T, (w_2^{r,x})^T, ..., (w_N^{r,x})^T)^T$ and $W^{r,r} = ((w_1^{r,r})^T, (w_2^{r,r})^T, ..., (w_N^{r,r})^T)^T$, respectively.

The activation of a unit i at time t is computed as

$$r_i(t) = \exp(-d_i(t)), \tag{5.30}$$

where

$$d_i(t) = \alpha \cdot \|\mathbf{x}(t) - \mathbf{w}_i^{r,x}\|^2 + \beta \cdot \|\mathbf{r}(t-1) - \mathbf{w}_i^{r,r}\|^2. \tag{5.31}$$

In eq. (5.31), $\alpha > 0$ and $\beta > 0$ are parameters that respectively quantify the influence of the input and the context on the map formation. The output of RecSOM is the index of the unit with maximum activation (best-matching unit),

$$y(t) = h(\mathbf{r}(t)) = \operatorname*{argmax}_{i \in \{1,2,\ldots,N\}} r_i(t). \tag{5.32}$$

Both weight vectors can be updated using a SOM-type learning [12]:

$$\Delta \mathbf{w}_i^{r,x} = \gamma \cdot \nu(i, y(t)) \cdot (\mathbf{x}(t) - \mathbf{w}_i^{r,x}), \tag{5.33}$$
$$\Delta \mathbf{w}_i^{r,r} = \gamma \cdot \nu(i, y(t)) \cdot (\mathbf{r}(t-1) - \mathbf{w}_i^{r,r}), \tag{5.34}$$

where $0 < \gamma < 1$ is the learning rate. Neighborhood function $\nu(i,k)$ is a Gaussian (of width σ) on the distance $d(i,k)$ of units i and k in the map:

$$\nu(i,k) = e^{-\frac{d(i,k)^2}{\sigma^2}}. \tag{5.35}$$

The "neighborhood width", σ, linearly decreases in time to allow for forming topographic representation of input sequences. A schematic illustration of RecSOM is presented in figure 5.3.

The time evolution (5.31) becomes

$$r_i(t) = \exp(-\alpha \|\mathbf{x}(t) - \mathbf{w}_i^{r,x}\|^2) \cdot \exp(-\beta \|\mathbf{r}(t-1) - \mathbf{w}_i^{r,r}\|^2). \tag{5.36}$$

We denote the Gaussian kernel of inverse variance $\eta > 0$, acting on $\mathbb{R}^N$, by $G_\eta(\cdot,\cdot)$, i.e. for any $\mathbf{u}, \mathbf{v} \in \mathbb{R}^N$,

$$G_\eta(\mathbf{u}, \mathbf{v}) = e^{-\eta \|\mathbf{u}-\mathbf{v}\|^2}. \tag{5.37}$$

Finally, the state space evolution (5.2) can be written for RecSOM as follows:

$$\mathbf{r}(t) = \mathbf{f}(\mathbf{x}(t), \mathbf{r}(t-1)) = (f_1(\mathbf{x}(t), \mathbf{r}(t-1)), \ldots, f_N(\mathbf{x}(t), \mathbf{r}(t-1))), \tag{5.38}$$

where
$$f_i(\mathbf{x}, \mathbf{r}) = G_\alpha(\mathbf{x}, \mathbf{w}_i^{r,x}) \cdot G_\beta(\mathbf{r}, \mathbf{w}_i^{r,r}), \quad i = 1, 2, \ldots, N. \tag{5.39}$$

The output response to an input time series $\{\mathbf{x}(t)\}$ is the time series $\{y(t)\}$ of best-matching unit indices given by (5.32).

5.5.2 Markovian organization of receptive fields in contractive RecSOM

As usual in topographic maps, each unit on the map can be naturally represented by its *receptive field* (RF). Receptive field of a unit i is the common suffix of all sequences for which that unit becomes the best-matching unit.

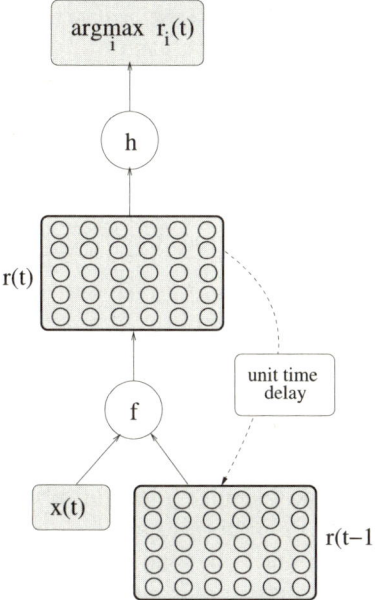

Fig. 5.3. Schematic illustration of RecSOM. Dimensions (neural units) of the state space $\mathcal{R}$ are topologically organized on a grid structure. The degree of "closeness" of units i and k on the grid is given by the neighborhood function $\nu(i,k)$. The output of RecSOM is the index of maximally activated unit (dimension of the state space).

It is common to visually represent trained topographic maps by printing out RFs on the map grid.

Let us discuss how *contractive* fixed-input maps $\boldsymbol{f}_s$ (5.3) shape the overall organization of RFs in the map. For each input symbol $s \in \mathcal{A}$, the autonomous dynamics

$$\boldsymbol{r}(t) = \boldsymbol{f}_s(\boldsymbol{r}(t-1)) \tag{5.40}$$

induces an output dynamics $y(t) = h(\boldsymbol{r}(t))$ of best-matching (winner) units on the map. Since for each input symbol $s \in \mathcal{A}$, the fixed-input dynamics (5.40) is a contraction, it will be dominated by a unique attractive fixed point $\boldsymbol{r}_s$. It follows that the output dynamics $\{y(t)\}$ on the map settles down in unit i_s, corresponding to the mode of $\boldsymbol{r}_s$. The unit i_s will be most responsive to input subsequences ending with long blocks of symbols s. Receptive fields of other units on the map will be organized with respect to the closeness of the units to the fixed-input winners i_s, $s \in \mathcal{A}$. When symbol s is seen at time t, the mode of the map activation profile $\boldsymbol{r}(t)$ starts drifting towards the unit i_s. The more consecutive symbols s we see, the more dominant the attractive fixed point of $\boldsymbol{f}_s$ becomes and the closer the winner position is to i_s.

It has indeed been reported that maps of sequential data obtained by RecSOM often seem to have a Markovian flavor: The units on the map become sensitive to recently observed symbols. Suffix-based RFs of the neurons are topographically organized in connected regions according to last seen symbols. If two prefixes $s_{1:p}$ and $s_{1:q}$ of a long sequence $s_1...s_{p-2}s_{p-1}s_p...s_{q-2}s_{q-1}s_q...$ share a common suffix of length L, it follows from (5.10) that for any $r \in \mathcal{R}$,

$$\|\boldsymbol{f}_{s_{1:p}}(\boldsymbol{r}) - \boldsymbol{f}_{s_{1:q}}(\boldsymbol{r})\| \leq \rho_{max}^L \cdot diam(\mathcal{R}).$$

For sufficiently large L, the two activations $\boldsymbol{r}^1 = \boldsymbol{f}_{s_{1:p}}(\boldsymbol{r})$ and $\boldsymbol{r}^2 = \boldsymbol{f}_{s_{1:q}}(\boldsymbol{r})$ will be close enough to have the same location of the mode,[8]

$$i_* = h(\boldsymbol{r}^1) = h(\boldsymbol{r}^2),$$

and the two subsequences $s_{1:p}$ and $s_{1:q}$ yield the same best matching unit i_* on the map, *irrespective of the position of the subsequences in the input stream*. All that matters is that the prefixes share a sufficiently long common suffix. We say that such an *organization of RFs on the map has a Markovian flavor*, because it is shaped solely by the suffix structure of the processed subsequences, and it does not depend on the temporal context in which they occur in the input stream[9].

RecSOM parameter β weighs the significance of importing information about possibly distant past into processing of sequential data. Intuitively, when β is sufficiently small, e.g. when information about the very recent inputs dominates processing in RecSOM, the resulting maps should have Markovian flavor. This intuition was given a more rigorous form in [16]. In particular, theoretical bounds on parameter β that guarantee contractiveness of the fixed input maps $\boldsymbol{f}_s$ were derived. As argued above, contractive fixed input mappings are likely to produce Markovian organizations of RFs on the RecSOM map.

Denote by $\boldsymbol{G}_\alpha(\boldsymbol{x})$ the collection of activations coming from the feed-forward part of RecSOM,

$$\boldsymbol{G}_\alpha(\boldsymbol{x}) = (G_\alpha(\boldsymbol{x}, \boldsymbol{w}_1^{r,x}), G_\alpha(\boldsymbol{x}, \boldsymbol{w}_2^{r,x}), ..., G_\alpha(\boldsymbol{x}, \boldsymbol{w}_N^{r,x})). \quad (5.41)$$

Then we have [16]:

Theorem 2. *Consider an input $\boldsymbol{x} \in \mathcal{X}$. If for some $\rho \in [0,1)$,*

$$\beta \leq \rho^2 \, \frac{e}{2} \, \|\boldsymbol{G}_\alpha(\boldsymbol{x})\|^{-2}, \quad (5.42)$$

then the mapping $\boldsymbol{f}_{\boldsymbol{x}}(\boldsymbol{r}) = \boldsymbol{f}(\boldsymbol{x}, \boldsymbol{r})$ is a contraction with contraction coefficient ρ.

---

[8] or at least mode locations on neighboring grid points of the map

[9] Theoretically, there can be situations where **(1)** locations of the modes of $\boldsymbol{r}^1$ and $\boldsymbol{r}^2$ are distinct, but the distance between $\boldsymbol{r}^1$ and $\boldsymbol{r}^2$ is small; or where **(2)** the modes of $\boldsymbol{r}^1$ and $\boldsymbol{r}^2$ coincide, while their distance is quite large. This follows from discontinuity of the output map h. However, in our extensive experimental studies, we have registered only a negligible number of such cases.

Corollary 1. *Provided*

$$\beta < \frac{e}{2N}, \qquad (5.43)$$

irrespective of the input symbol $s \in \mathcal{A}$, the fixed input map $\boldsymbol{f}_s$ of a RecSOM with N units will be a contraction.

The bound $e/(2N)$ may seem restrictive, but as argued in [15], the context influence has to be small to avoid instabilities in the model. Indeed, the RecSOM experiments of [15] used $N = 10 \times 10 = 100$ units and the map was trained with $\beta = 0.06$, which is only slightly higher than the bound $e/(2N) = 0.0136$. Obviously the bound $e/(2N)$ can be improved by considering other model parameters, as in Theorem 2.

These results complement Voegtlin's stability analysis of the parameter adaptation process during RecSOM training [12]: for $\beta < e/(2N)$, stability of weight updates with respect to small perturbations of the map activity $\boldsymbol{r}$ is ensured. Voegtlin also shows that if $\beta < e/(2N)$, small perturbations of the activities will decay (fixed input maps are locally contractive). The work in [16] extends this result to perturbations of arbitrary size. It follows that for each RecSOM model satisfying Voegtlin's stability bound on β, the fixed input dynamics for *any* input will be dominated by a unique attractive fixed point. This renders the map both Markovian quality and training stability.

5.5.3 Verifying Markovian organization of receptive fields in RecSOM

In [16] we posed the following questions:

- Is the architecture of RecSOM naturally biased towards Markovian representations of input streams? If so, under what conditions will Markovian organization of receptive fields (RF) occur? How natural are such conditions, i.e. can Markovian topographic maps be expected under widely-used architectures and (hyper)parameter settings in RecSOM?
- What can be gained by having a trainable recurrent part in RecSOM? In particular, how does RecSOM compare with a much simpler setting of *standard* SOM (no feedback connections) operating on a simple *non-trainable* FPM-based iterative function system (5.25)?

The experiments were performed on three data sets:

- A binary symbolic stream generated by a two-state first-order Markov chain used in the original RecSOM paper [12].
- The series of quantized activations of a laser in a chaotic regime from section 5.4.7.
- A corpus of written English - the novel "Brave New World" by Aldous Huxley. This is the second data set used to demonstrate RecSOM in [12].

Three methods of assessment of the topographic maps were used. The first two were suggested in [12].

- For each unit in the map its RF is calculated. The RFs are then visually presented on the map grid.
- As a measure of the amount of memory captured by the map, the overall quantizer depth of the map was calculated as

$$\mathrm{QD} = \sum_{i=1}^{N} p_i \ell_i, \tag{5.44}$$

where p_i is the probability of the RF of neuron i and ℓ_i is its length.
- Quantifying topography preservation in recursive extensions of SOM is not as straightforward as in traditional SOM. To quantify the maps' topographic order we first calculated the length of the longest common suffix shared by RFs of that unit and its immediate topological neighbors. The topography preservation measure (TP) is the average of such shared RF suffix lengths over all units in the map.

For the Markov chain series, when we chose values of parameters α and β from the region of stable RecSOM behavior reported in [12], after the training we almost always got a trivial single-attractive-fixed-point behavior of fixed input maps (5.40). As explained in section 5.5.2, this resulted in Markovian organization of RFs. Note that such a RF organization can be obtained "for free", i.e. without having a set of trainable feed-back connections, simply by constructing a standard SOM on a FPM-based iterative function system (5.25). We call such models IFS+SOM. This indeed turned out to be the case. Maps of RFs obtained by RecSOM and IFS+SOM of comparable sizes looked visually similar, with a clear Markovian organization. However, quantitative measures of quantizer depth (QD) and topography preservation (TP) revealed a significant advantage of RecSOM maps. The same applied to the chaotic laser data set. In most cases, the β parameter was above the upper bound of Theorem 2,

$$\Upsilon(s) = \frac{e}{2} \|\boldsymbol{G}_\alpha(\boldsymbol{c}(s))\|^{-2}, \tag{5.45}$$

guaranteeing contractive fixed input maps (5.40). Nevertheless, the organization of RFs was still Markovian. We performed experiments with β-values below $\epsilon/(2N)$ (see Corollary 1). In such cases, *irrespective of other training hyperparameters*, the fixed input maps (5.40) were always contractive, leading automatically to Markovian RF organizations.

When training RecSOM on the corpus of written English, for a wide variety of values of α and β that lead to stable map formation, non-Markovian organizations of RFs were observed. As a consequence, performance of IFS+SOM model, as measured by the QD and TP measures, was better, but it is important to realize that this does not necessarily imply that RecSOM maps were of inferior quality. The QD and TP are simply biased towards Markovian RF organizations. But there may well be cases where non-Markovian topographic maps are more desirable, as discussed in the next section.

5.5.4 Beyond Markovian organization of receptive fields

Periodic (beyond period 1), or aperiodic attractive dynamics (5.40) yield potentially complicated non-Markovian map organizations with "broken topography" of RFs. Two sequences with the same suffix can be mapped into distinct positions on the map, separated by a region of different suffix structure. Unlike in contractive RecSOM or IFS+SOM models, such context-dependent RecSOM maps embody a potentially unbounded input memory structure: the current position of the winner neuron is determined by the whole series of processed inputs, and not only by a history of recently seen symbols. To fully appreciate the meaning of the RF structure, we must understand the driving mechanism behind such context-sensitive suffix representations.

In what follows we will try to suggest one possible mechanism of creating non-Markovian RF organizations. We noticed that often Markovian topography was broken for RFs ending with symbols of high frequency of occurrence. It seems natural for such RFs to separate on the map into distinct islands with a more refined structure. In other words, it seems natural to distinguish on the map between temporal contexts in which RFs with high frequency suffixes occur.

Now, the frequency of a symbol $s \in \mathcal{A}$, input-coded as $c(s) \in \mathcal{X}$, in the input data stream will be reflected by the "feed-forward" part of the map consisting of weights $\boldsymbol{W}^{r,x}$. The more frequent the symbol is, the better coverage by the weights $\boldsymbol{W}^{r,x}$ it will get[10]. But good match with more feedforward weights $\boldsymbol{w}_i^{r,x}$ means higher kernel values of (see (5.37))

$$G_\alpha(\boldsymbol{c}(s), \boldsymbol{w}_i^{r,x}) = \exp\{-\alpha \|\boldsymbol{c}(s) - \boldsymbol{w}_i^{r,x}\|^2\},$$

and consequently higher L_2 norms of $\boldsymbol{G}_\alpha(\boldsymbol{c}(s))$ (5.41). It follows that the upper bound $\Upsilon(s)$ (5.45) on parameter β guaranteeing contractive fixed input map (5.40) for symbol s will be smaller. Hence, for more frequent symbols s, the interval $(0, \Upsilon(s))$ of β values for guaranteed Markovian organization of RFs ending with s shrinks and so a greater variety of stable maps can be formed with topography of RFs broken for suffixes containing s.

5.5.5 SOM for structured data

RecSOM constitutes a powerful model which has the capacity of at least pushdown automata as shown (for a simplified form) in [45]. It uses a very

---

[10] This is, strictly speaking, not necessarily true. It would be true in the standard SOM model. However, weight updates of $\boldsymbol{W}^{r,x}$ in RecSOM are influenced by the match in both the feedforward and recurrent parts of the model, represented by weights $\boldsymbol{W}^{r,x}$ and $\boldsymbol{W}^{r,x}$, respectively. Nevertheless, in all our experiments we found this correspondence between the frequency of an input symbol and its coverage by the feedforward part of RecSOM to hold.

complex context representation: the activation of all neurons of the map of the previous time step, usually a large number. This has the consequence that memory scales with N^2, N being the number of neurons, since the expected context is stored in each cell. SOM for structured data (SOMSD) and merge SOM (MSOM) constitute alternatives which try to compress the relevant information of the context such that the memory load is reduced.

SOMSD has been proposed in [14] for unsupervised processing of tree structured data. Here we restrict to sequences, i.e. we neglect branching. SOMSD compresses the temporal context by the *location of the winner* in the neural map. The map is given by units $i \in \{1, \ldots, N\}$ which are located on a lattice structure embedded in a real-vector space $\mathcal{R}^I \subset \mathbb{R}^d$, typically a two-dimensional rectangular lattice in $\mathbb{R}^2$. The location of neuron i in this lattice is referred to as I_i.

Each neuron has two vectors associated with it:

- $\boldsymbol{w}_i^{r,x} \in \mathcal{X}$ – linked with an N_I-dimensional input $\boldsymbol{x}(t)$ feeding the network at time t as in RecSOM
- $\boldsymbol{w}_i^{r,r} \in \mathcal{R}^I$ – linked with the context

$$\boldsymbol{r}^I(t-1) = I_{i_{t-1}} \text{ with } i_{t-1} = \operatorname*{argmin}_i \{r_i(t-1) \mid i = 1, \ldots, N\},$$

i.e. it is linked to the winner location depending on map activations $r_i(t-1)$ from the previous time step.

The activation of a unit i at time t is computed as

$$r_i(t) = \alpha \cdot \|\mathbf{x}(t) - \boldsymbol{w}_i^{r,x}\| + \beta \cdot \|\mathbf{r}^I(t-1) - \boldsymbol{w}_i^{r,r}\|. \tag{5.46}$$

where, as beforehand,

$$\boldsymbol{r}^I(t-1) = I_{i_{t-1}} \text{ with } i_{t-1} = \operatorname*{argmin}_i \{r_i(t-1) \mid i = 1, \ldots, N\}.$$

The readout extracts the winner

$$y(t) = h(\boldsymbol{r}(t)) = \operatorname*{argmin}_{i \in \{1, \ldots, N\}} r_i(t).$$

Note that this recursive dynamics is very similar to the dynamics of RecSOM as given in (5.31). The information stored in the context is compressed by choosing the winner location as internal representation.

Since $\mathcal{R}^I$ is a real-vector space, SOMSD can be trained in the standard Hebb style for both parts of the weights, moving $\boldsymbol{w}_i^{r,x}$ closer to $\boldsymbol{x}(t)$ and $\boldsymbol{w}_i^{r,r}$ closer to $\boldsymbol{r}^I(t-1)$ after the representation of a stimulus. The principled capacity of SOMSD can be characterized exactly. It has been shown in [15] that SOMSD can simulate finite automata in the following sense: assume $\mathcal{A}$ is a finite alphabet. Assume a finite automaton accepts the language $\mathcal{L} \subset \mathcal{A}^*$. Then there exists a constant delay τ, an encoding $C : \mathcal{A} \to \mathcal{X}^\tau$, and a SOMSD network such that $s \in \mathcal{L}$ if and only if the sequence $C(s_1)\ldots C(s_t)$ is mapped to neuron 1. Because of the finite context representation within the lattice in (5.46) (capacity at most the capacity of finite state automata), the equivalence of SOMSD for sequences and finite state automata results.

Obviously, unlike RecSOM, the dynamics of SOMSD is discontinuous since it involves the determination of the winner. Therefore, the arguments of section 5.5.2 which show a Markovian bias as provided for RecSOM do not apply to SOMSD. For SOMSD, an alternative argumentation is possible, assuming small β and two further properties which we refer to as *sufficient granularity* and as *distance preservation*. More precisely:

1. SOMSD has *sufficient granularity* given $\epsilon_1 > 0$ if for every $a \in \mathcal{A}$ and $\boldsymbol{r}^I \in \mathcal{R}^I$ a neuron i can be found such that $\|a - \boldsymbol{w}_i^{r,x}\| \leq \epsilon_1$ and $\|\boldsymbol{r}^I - \boldsymbol{w}_i^{r,r}\| \leq \epsilon_1$.
2. SOMSD is *distance preserving* given $\epsilon_2 > 0$ if for every two neurons i and j the following inequality holds $|\|I_i - I_j\| - \alpha \cdot \|\boldsymbol{w}_i^{r,x} - \boldsymbol{w}_j^{r,x}\| - \beta \cdot \|\boldsymbol{w}_i^{r,r} - \boldsymbol{w}_j^{r,r}\|| \leq \epsilon_2$. This inequality relates the distances of the locations of neurons on the lattice to the distances of their content.

Assume a SOMSD is given with initial context $\boldsymbol{r}_0$. We denote the output response to the empty sequence ϵ by $i_\epsilon = h(\boldsymbol{r}_0)$, the winner unit for an input sequence u by $i_u := h \circ \boldsymbol{f}_u(\boldsymbol{r}_0)$ and the corresponding winner location on the map grid by I_u. We can define an explicit Markovian metric on sequences in the following way: $d_{\mathcal{A}^*} : \mathcal{A}^* \times \mathcal{A}^* \to \mathbb{R}$, $d_{\mathcal{A}^*}(\epsilon, \epsilon) = 0$, $d_{\mathcal{A}^*}(u, \epsilon) = d_{\mathcal{A}^*}(\epsilon, u) = \|I_\epsilon - I_u\|$, and for two sequences $u = u_1 \ldots u_{p-1} u_p$, $v = v_1 \ldots v_{q-1} v_q$ over $\mathcal{A}$,

$$d_{\mathcal{A}^*}(u, v) = \alpha \cdot \|\boldsymbol{c}(u_p) - \boldsymbol{c}(v_q)\| + \beta \cdot d_{\mathcal{A}^*}(u_1 \ldots u_{p-1}, v_1 \ldots v_{q-1}). \quad (5.47)$$

Assume $\beta < 1$, then, the output of SOMSD is Markovian in the sense that the metric as defined in (5.47) approximates the metric induced by the winner location of sequences. More precisely, the following inequality is valid:

$$| d_{\mathcal{A}^*}(u, v) - \|I_u - I_v\| | \leq (\epsilon_2 + 4 \cdot (\alpha + \beta)\epsilon_1)/(1 - \beta).$$

This equation is immediate if u or v are empty. For nonempty sequences $u = u_{1:p-1} u_p$ and $v = v_{1:q-1} v_q$, we find

$$\begin{aligned}
| d_{\mathcal{A}^*}(u, v) - \|I_u - I_v\| | &= | \alpha \cdot \|\boldsymbol{c}(u_p) - \boldsymbol{c}(v_q)\| \\
&\quad + \beta \cdot d_{\mathcal{A}^*}(u_{1:p-1}, v_{1:q-1}) - \|I_u - I_v\| | \\
&\leq \epsilon_2 + \alpha \cdot | \|\boldsymbol{w}_{i_u}^{r,x} - \boldsymbol{w}_{i_v}^{r,x}\| - \|\boldsymbol{c}(u_p) - \boldsymbol{c}(v_q)\| | \\
&\quad + \beta \cdot | \|\boldsymbol{w}_{i_u}^{r,r} - \boldsymbol{w}_{i_v}^{r,r}\| - d_{\mathcal{A}^*}(u_{1:p-1}, v_{1:q-1}) | \\
&= (*)
\end{aligned}$$

Therefore, $\alpha \cdot \|\boldsymbol{w}_{i_u}^{r,x} - \boldsymbol{c}(u_p)\| + \beta \|\boldsymbol{w}_{i_u}^{r,r} - I_{u_{1:p-1}}\|$ is minimum. We have assumed sufficient granularity, i.e. there exists a neuron with weights at most ϵ_1 away from $\boldsymbol{c}(u_p)$ and $I_{u_{1:p-1}}$. Therefore,

$$\|\boldsymbol{w}_{i_u}^{r,x} - u_p\| \leq (\alpha + \beta)\epsilon_1/\alpha$$

and
$$\|w_{i_u}^{r,r} - I_{u_{1:p-1}}\| \leq (\alpha + \beta)\epsilon_1/\beta.$$

Using the triangle inequality, we obtain the estimation

$$(*) \leq \epsilon_2 + 4 \cdot (\alpha + \beta)\epsilon_1 + \beta \cdot |\ d_{\mathcal{A}^*}(u_{1:p-1}, v_{1:q-1}) - \|I_{u_{1:p-1}} - I_{v_{1:q-1}}\|\ |$$
$$\leq \cdots \leq (\epsilon_2 + 4 \cdot (\alpha + \beta)\epsilon_1)(1 + \beta + \beta^2 + \ldots)$$
$$\leq (\epsilon_2 + 4 \cdot (\alpha + \beta)\epsilon_1)/(1 - \beta),$$

since $\beta < 1$.

5.5.6 Merge SOM

SOMSD uses a compact context representation. However, it has the drawback that it relies on a fixed lattice structure. Merge SOM (MSOM) uses a different compression scheme which does not rely on the lattice, rather, it uses the *content* of the winner. Here context representations are located in the same space as inputs $\mathcal{X}$. A map is given by neurons which are associated with two vectors

- $w_i^{r,x} \in \mathcal{X}$ – linked with an N_I-dimensional input $x(t)$ feeding the network at time t as in RecSOM
- $w_i^{r,r} \in \mathcal{X}$ – linked with the context

$$r^M(t-1) = \gamma \cdot w_{i_{t-1}}^{r,x} + (1-\gamma) \cdot w_{i_{t-1}}^{r,r}$$

where $i_{t-1} = \mathrm{argmin}_i\{r_i(t-1) \mid i = 1, \ldots, N\}$ is the index of the best matching unit at time $t-1$. The feed-forward and recurrent weight vectors of the winner from the previous time step are merged using a fixed parameter $\gamma > 0$.

The activation of a unit i at time t is computed as

$$r_i(t) = \alpha \cdot \|\mathbf{x}(t) - w_i^{r,x}\| + \beta \cdot \|r^M(t-1) - w_i^{r,r}\|, \tag{5.48}$$

where $r^M(t-1) = \gamma \cdot w_{i_{t-1}}^{r,x} + (1-\gamma) \cdot w_{i_{t-1}}^{r,r}$ is the merged winner content of the previous time step. The readout is identical to SOMSD.

As for SOMSD, training in the standard Hebb style is possible, moving $w_i^{r,x}$ towards $x(t)$ and $w_i^{r,r}$ towards the context $r^M(t-1)$, given a history of stimuli up to time t. It has been shown in [13] that MSOM can simulate every finite automaton whereby we use the notation of simulation as defined for SOMSD. Since for every finite map only a finite number of different contexts can occur, the capacity of MSOM coincides with finite automata.

As for SOMSD, the transition function of MSOM is discontinuous since it incorporates the computation of the winner neuron. However, one can show that standard Hebbian learning biases MSOM towards Markovian models in the sense that optimum context weights which have a strong Markovian flavor are obtained by Hebbian learning for MSOM as stable fixed points of the learning dynamics. Thereby, the limit of Hebbian learning for vanishing neighborhood size is considered which yields the update rules

$$\Delta w_{i_t}^{r,x} = \eta \cdot (x(t) - w_{i_t}^{r,x}) \qquad (5.49)$$

$$\Delta w_{i_t}^{r,r} = \eta \cdot (r^M(t-1) - w_{i_t}^{r,r}) \qquad (5.50)$$

after time step t, whereby $x(t)$ constitutes the input at time step t, $r^M(t-1) = \gamma \cdot w_{i_{t-1}}^{r,x} + (1-\gamma) \cdot w_{i_{t-1}}^{r,r}$ the context at time step t, and i_t the winner at time step t; $\eta > 0$ is the learning rate.

Assume an input series $\{x(t)\}$ is presented to MSOM and assume the initial context r_0 is the null vector. Then, if a sufficient number of neurons is available such that there exists a disjoint winner for every input $x(t)$, the following weights constitute a stable fixed point of Hebbian learning as described in eqs. (5.49-5.50):

$$w_{i_t}^{r,x} = x(t) \qquad (5.51)$$

$$w_{i_t}^{r,r} = \sum_{j=1}^{t-1} \gamma \cdot (1-\gamma)^j x(t-j). \qquad (5.52)$$

The proof of this fact can be found in [15, 13]. Note that, depending on the size of γ, the context representation yields sequence codes which are very similar to the recursive states provided by fractal prediction machines (FPM) of section 5.4.3. For $\gamma < 0.5$ and unary inputs $x(t)$ (one-of-A encoding), (5.52) yields a fractal whereby the global position in the state space is determined by the most recent entries of the input series. Hence, a Markovian bias is obtained by Hebbian learning.

This can clearly be observed in experiments, compare figure 5.4[11] which describes a result of an experiment conducted in [13]: a MSOM with 644 neurons is trained on DNA sequences. The data consists of windows around potential splice sites from C.elegans as described in [46]. The symbols TCGA are encoded as the edges of a tetrahedron in $\mathbb{R}^3$. The windows are concatenated and presented to a MSOM using merge parameter $\gamma = 0.5$. Thereby, the neighborhood function used for training is not a standard SOM lattice, but training is data driven by the ranks as described for neural gas networks [47]. Figure 5.4 depicts a projection of the learned contexts. Clearly, a fractal structure which embeds sequences in the whole context space can be observed, substantiating the theoretical finding by experimental evidence.

Note that, unlike FPM which use a fixed fractal encoding, the encoding of MSOM emerges from training. Therefore, it takes the internal probability distribution of symbols into account, automatically assigning a larger space to those inputs which are presented more frequently to the map.

---

[11] We would like to thank Marc Strickert for providing the picture.

Fig. 5.4. Projection of the three dimensional fractal contexts which are learned when training a MSOM on DNA-strings whereby the symbols ACTG are embedded in three dimensions.

5.6 Applications in bioinformatics

In this section we briefly describe applications of the architectural bias phenomenon in the supervised learning scenario. Natural Markovian organization of biological sequences within the RNN state space may be exploited when solving some sequence-related problems of bioinformatics.

5.6.1 Genomic data

Enormous amounts of genomic sequence data has been produced in recent years. Unfortunately, the emergence of experimental data for the *functional parts* of such large-scale genomic repositories is lagging behind. For example, several hundred thousands of proteins are known, but only about 35,000 have had their structure determined. One third of all proteins are believed to be membrane associated, but less than a thousand have been experimentally well-characterized to be so.

As commonly assumed, genomic sequence data specifies the structure and function of its products (proteins). Efforts are underway in the field of bioinformatics to link genomic data to experimental observations with the goal of *predicting* such characteristics for the yet unexplored data [48]. DNA, RNA and proteins are all chain molecules, made up of distinct monomers, namely *nucleotides* (DNA and RNA) and *amino acids* (protein). We choose to cast such chemical compounds as symbol sequences. There are four different nucleotides and twenty different amino acids, here referred to as alphabets $\mathcal{A}_4$ and $\mathcal{A}_{20}$, respectively.

A functional region of DNA typically includes thousands of nucleotides. The average number of amino acids in a protein is around the 300 mark but sequence lengths vary greatly. The combinatorial nature of genomic sequence data poses a challenge for machine learning approaches linking sequence data with observations. Even when thousands of samples are available, the sequence space is sparsely populated. The *architectural bias* of predictive models may thus influence the result greatly. This section demonstrates that a Markovian state-space organization–as imposed by the recursive model in this paper– discerns patterns with a biological flavor [49].

In bioinformatics *motifs* are perceived as recurring patterns in sequence data (DNA, RNA and proteins) with an observed or conjectured biological significance (the character varies widely). Problems like promoter and DNA-binding site discovery [50, 51], identification of alternatively spliced exons [52] and glycosylation sites are successfully approached using the concept of sequence motifs. Motifs are usually quite short and to be useful may require a more permissive description than a list of consecutive sequence symbols. Hence, we augment the motif alphabet $\mathcal{A}$ to include the so-called wildcard symbol '*': $\tilde{\mathcal{A}} = \mathcal{A} \cup \{*\}$. Let $\mathcal{M} = m_1 m_2 ... m_{|\mathcal{M}|}$ be a motif where $m_i \in \tilde{\mathcal{A}}$. As outlined below the wildcard symbol has special meaning in the motif sequence matching process.

5.6.2 Recurrent architectures promote motif discovery

We are specifically interested if the Markovian state space organization promotes the discovery of a motif within a sequence. Following [49], *position-dependent* motifs are patterns of symbols relative to a fixed sequence-position. A motif *match* is defined as

$$\text{match}(s_{1:n}, \mathcal{M}) = \begin{cases} \text{T}, & \text{if } \forall i \in \{1, ..., |\mathcal{M}|\} \ [m_i = * \vee m_i = s_{n-|\mathcal{M}|+i}] \\ \text{F} & \text{otherwise} \end{cases}$$

where $n \geq |\mathcal{M}|$. Notice that the motif must occur at the end of the sequence.

To inspect how well sequences with and without a given motif $\mathcal{M}$ (T and F, respectively) are separated in the state-space, Bodén and Hawkins determined a motif match entropy for each of the Voronoi compartments $V_1, ..., V_M$, introduced in the context of NPM construction in section 5.4.1. The compartments were formed by vector quantizing the *final* activations produced by a neural-based architecture on a set of sequences (see Section 5.4). With each codebook vector $\boldsymbol{b}_i \in \mathcal{R}$, $i = 1, 2, ..., M$, we associate the probability $P(\text{T}|i)$ of observing a state $\boldsymbol{r} \in V_i$ corresponding to a sequence motif match. This probability is estimated by tracking the matching and non-matching sequences while counting the number of times V_i is visited.

The entropy for each codebook is defined as

$$H_i = -P(\text{T}|i) \log_2 P(\text{T}|i) - (1 - P(\text{T}|i)) \log_2 (1 - P(\text{T}|i)) \qquad (5.53)$$

and describes the homogeneity within the Voronoi compartment by considering the proportion of motif *positives* and *negatives*. An entropy of 0 indicates perfect homogeneity (contained states are exclusively for matching or non-matching sequences). An entropy of 1 indicates random organization (a matching state is observed by chance). The average codebook entropy, $H = M^{-1} \sum_{i=1}^{M} H_i$, is used to characterize the overall state-space organization.

In our experiments, the weights $\boldsymbol{W}^{r,x}$ and $\boldsymbol{W}^{r,r}$ were randomly drawn from a uniform distribution $[-0.5, +0.5]$. This, together with activation function g, ensured contractiveness of the state transition map $\boldsymbol{f}$ and hence (as explained in section 5.3.1) Markovian organization of the RNN state space.

In an attempt to provide a simple baseline alternative to our recursive neural-based models, a feedforward neural network was devised accepting the whole sequence at its input without any recursion. Given the maximum input sequence length ℓ_{max}, the input layer has $N_I \cdot \ell_{max}$ units and the input code of a sequence $s_{1:n} = s_1...s_n$ takes the form

$$\tilde{c}(s_{1:n}) = c(s_1) \bullet ... \bullet c(s_n) \bullet \{0\}^{N_I \cdot (\ell_{max}-n)},$$

where $\bullet$ denotes vector concatenation. Hence, $\tilde{c}(s_{1:n})$ is a concatenation of input codes for symbols in $s_{1:n}$, followed by $N_I \cdot (\ell_{max} - n)$ zeros. The hidden layer response of the feedforward network to $s_{1:n}$ is then

$$\tilde{\boldsymbol{f}}_{s_{1:n}} = \tilde{\boldsymbol{f}}(\tilde{c}(s_{1:n})) = \mathbf{g}(\tilde{\boldsymbol{W}}^{r,x}\tilde{c}(s_{1:n}) + \boldsymbol{t}_f), \qquad (5.54)$$

where the input-to-hidden layer weights (randomly drawn from $[-0.5, +0.5]$) are collected in the $N \times (N_I \cdot \ell_{max})$ matrix $\tilde{\boldsymbol{W}}^{r,x}$.

In each trial, a random motif $\mathcal{M}$ was first selected and then 500 random sequences (250 with a match, 250 without) were generated. Ten differently initialized networks (both RNN and the feed-forward baseline) were then subjected to the sequences and the average entropy H was determined. By varying sequence and motif lengths, the proportion of wildcards in the motifs, state space dimensionality and number of codebook vectors extracted from sequence states, Bodén and Hawkins established that recurrent architectures were clearly superior in discerning random but pre-specified motifs in the state space (see Figure 5.5). However, there was no discernable difference between the two types of networks (RNN and feedforward) for shift-invariant motifs (an alternative form which allows matches to appear at any point in the sequence). To leverage the bias, a recurrent network should thus continually monitor its state space.

These synthetic trials (see [49]) provide evidence that a recurrent network naturally organizes its state space at each step to enable classification of extended and flexible sequential patterns that led up to the current time step. The results have been shown to hold also for higher-order recurrent architectures [53].

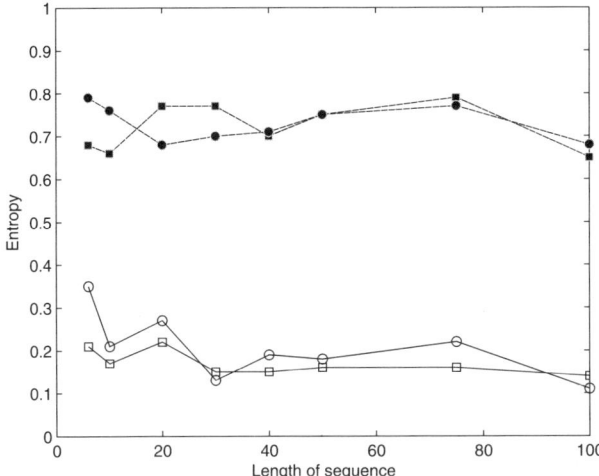

Fig. 5.5. The average entropy H (y-axis) for recurrent networks (unfilled markers) and baseline feedforward networks (filled markers) when $\mathcal{A}_4$ (squares) and $\mathcal{A}_{20}$ (circles) were used to generate sequences of varying length (x-axis) and position-dependent motifs (with length equal to half the sequence length and populated with 33% wildcards). The number of codebooks, M, was 10.

5.6.3 Discovering discriminative patterns in signal peptides

A long-standing problem of great biological significance is the detection of so-called signal peptides. Signal peptides are reasonably short protein segments that fulfill a transportation role in the cell. Basically, if a protein has a signal peptide it translocates to the exoplasmic space. The signal peptide is cleaved off in the process.

Viewed as symbol strings, signal peptides are rather diverse. However, there are some universal characteristics. For example, they always appear at the N-terminus of the protein, they are 15-30 residues long, consist of a stretch of seven to 15 hydrophobic (lipid-favoring) amino acids and a few well-conserved, small and neutral monomers close to the cleavage site. In Figure 5.6 nearly 400 mammalian signal peptides are aligned at their respective cleavage site to illustrate a subtle pattern of conservation.

In an attempt to simplify the sequence classification problem, each position of the sequence can first be tagged as belonging (T) or not belonging (F) to a signal peptide [54]. SignalP is a feedforward network that is designed to look at all positions of a sequence (within a carefully sized context window) and trained to classify each into the two groups (T and F) [54].

The Markovian character of the recursive model's state space suggests that replacing the feedforward network with a recurrent network may provide a better starting point for a training algorithm. However, since biological

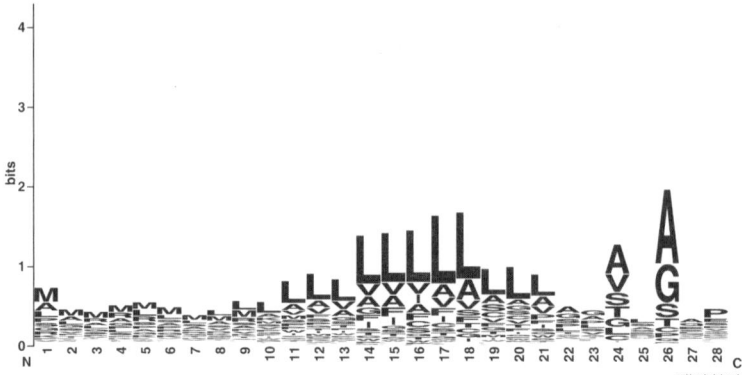

Fig. 5.6. Mammalian signal peptide sequences aligned at their cleavage site (between positions 26 and 27). Each amino acid is represented by its one-letter code. The relative frequency of each amino acid in each aligned position is indicated by the height of each letter (as measured in bits). The graph is produced using WebLogo.

sequences may exhibit dependencies in both directions (upstream as well as downstream) the simple recurrent network may be insufficient.

Hawkins and Bodén [53, 55] replaced the feedforward network with a *bi-directional* recurrent network [56]. The bi-directional variant is simply two conventional (uni-directional) recurrent networks coupled to merge two states each produced from traversing the sequence from one direction (either upstream or downstream relative the point of interest). In Hawkins and Bodén's study, the networks' ability was only examined after training.

After training, the average test error (as measured over known signal peptides and known negatives) dropped by approximately 25% compared to feedforward networks. The position-specific errors for both classes of architectures are presented in Figure 5.7. Several possible setups of the bi-directional recurrent network are possible, but the largest performance increase was achieved when the number of symbols presented at each step was 10 (for both the upstream and downstream networks).

Similar to signal peptides, mitochondrial targeting peptides and chloroplast transit peptides traffic proteins. In their case proteins are translocated to mitochondrion and chloroplast, respectively. Bi-directional recurrent networks applied to such sequence data show a similar pattern of improvement compared to feedforward architectures [53].

The combinatorial space of possible sequences and the comparatively small number of known samples make the application of machine learning to bioinformatics challenging. With significant data variance, it is essential to design models that guide the learning algorithm to meaningful portions of the model parameter space. By presenting exactly the same training data sets to two classes of models, we note that the architectural bias of recurrent networks seems to enhance the prospects of achieving a valid generalization. As noted

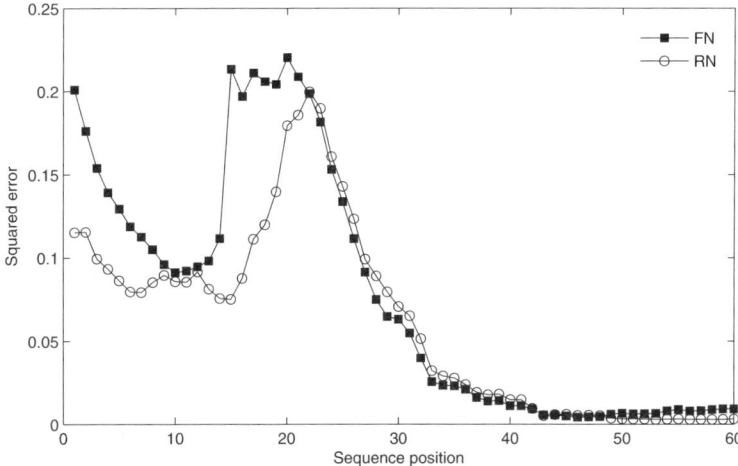

Fig. 5.7. The average error of recurrent and feedforward neural networks over the first 60 residues in a sequence test set discriminating between residues that belong to a signal peptide from those that do not. The signal peptide always appear at the N-terminus of a sequence (position 1 and onwards, average length is 23).

in [53] the locations of subtle sequential patterns evident in the data sets (cf. Figure 5.6) correspond well with those regions at which the recurrent networks excel. The observation lends empirical support to the idea that recursive models are well suited to this broad class of bioinformatics problems.

5.7 Conclusion

We have studied architectural bias of dynamic neural network architectures towards suffix-based Markovian input sequence representations. In the supervised learning scenario, the cluster structure that emerges in the recurrent layer *prior to training* is, due to the contractive nature of the state transition map, organized in a Markovian manner. In other words, the clusters emulate Markov prediction contexts in that histories of symbols are grouped according to the number of symbols they share in their suffix. Consequently, from such recurrent activation clusters it is possible to extract predictive models that correspond to variable memory length Markov models (VLMM). To appreciate how much information has been induced during the training, the dynamic neural network performance should always be compared with that of VLMM baseline models.

Apart from trained models (possibly with small weights, i.e. Markovian bias) various mechanisms which only rely on the dynamics of fixed or randomly initialized recurrence have recently been proposed, including fractal prediction machines, echo state machines, and liquid state machines. As discussed in the paper, these models almost surely allow an approximation of Markovian models even with random initialization provided a large dimensionality of the reservoir.

Interestingly, the restriction to Markovian models has benefits with respect to the generalization ability of the models. Whereas generalization bounds of general recurrent models necessarily rely on the input distribution due to the possibly complex input information, the generalization ability of Markovian models can be limited in terms of the relevant input length or, alternatively, the Lipschitz parameter of the models. This yields bounds which are independent of the input distribution and which are the better, the smaller the contraction coefficient. These bounds also hold for arbitrary real-valued inputs and for posterior bounds on the contraction coefficient achieved during training. The good generalization ability of RNNs with Markovian bias is particularly suited for typical problems in bioinformatics where dimensionality is high compared to the number of available training data. A couple of experiments which demonstrate this effect have been included in this article.

Unlike supervised RNNs, a variety of fundamentally different unsupervised recurrent networks has recently been proposed. Apart from comparably old biologically plausible leaky integrators such as the temporal Kohonen map or recurrent networks, which obviously rely on a Markovian representation of sequences, models which use a more elaborate temporal context such as the map activation, the winner location, or winner content have been proposed. These models have, in principle, the larger capacity of finite state automata or pushdown automata, respectively, depending on the context model. Interestingly, this capacity reduces to Markovian models under certain realistic conditions discussed in this paper. These include a small mixing parameter for recursive networks, a sufficient granularity and distance preservation for SOMSD, and Hebbian learning for MSOM. The field of appropriate training of complex unsupervised recursive models is, so far, widely unexplored. Therefore, the identification of biases of the architecture and training algorithm towards the Markovian property is a very interesting result which allows to gain more insight into the process of unsupervised sequence learning.

References

1. Christiansen, M., Chater, N.: Toward a connectionist model of recursion in human linguistic performance. Cognitive Science **23** (1999) 417–437
2. Kolen, J.: Recurrent networks: state machines or iterated function systems? In Mozer, M., Smolensky, P., Touretzky, D., Elman, J., Weigend, A., eds.: Proceedings of the 1993 Connectionist Models Summer School. Erlbaum Associates, Hillsdale, NJ (1994) 203–210
3. Kolen, J.: The origin of clusters in recurrent neural network state space. In: Proceedings from the Sixteenth Annual Conference of the Cognitive Science Society, Hillsdale, NJ: Lawrence Erlbaum Associates (1994) 508–513

4. Tiňo, P., Čerňanský, M., Beňušková, L.: Markovian architectural bias of recurrent neural networks. IEEE Transactions on Neural Networks **15** (2004) 6–15
5. Hammer, B., Tino, P.: Recurrent neural networks with small weights implement definite memory machines. Neural Computation **15** (2003) 1897–1929
6. Ron, D., Singer, Y., Tishby, N.: The power of amnesia. Machine Learning **25** (1996)
7. Tiňo, P., Hammer, B.: Architectural bias in recurrent neural networks: Fractal analysis. Neural Computation **15** (2004) 1931–1957
8. Kohonen, T.: The self-organizing map. Proceedings of the IEEE **78** (1990) 1464–1479
9. Chappell, G., Taylor, J.: The temporal kohonen map. Neural Networks **6** (1993) 441–445
10. Koskela, T., znd J. Heikkonen, M.V., Kaski, K.: Recurrent SOM with local linear models in time series prediction. In: 6th European Symposium on Artificial Neural Networks. (1998) 167–172
11. Horio, K., Yamakawa, T.: Feedback self-organizing map and its application to spatio-temporal pattern classification. International Journal of Computational Intelligence and Applications **1** (2001) 1–18
12. Voegtlin, T.: Recursive self-organizing maps. Neural Networks **15** (2002) 979–992
13. Strickert, M., Hammer, B.: Merge SOM for temporal data. Neurocomputing **64** (2005) 39–72
14. Hagenbuchner, M., Sperduti, A., Tsoi, A.: Self-organizing map for adaptive processing of structured data. IEEE Transactions on Neural Networks **14** (2003) 491–505
15. Hammer, B., Micheli, A., Strickert, M., Sperduti, A.: A general framework for unsupervised processing of structured data. Neurocomputing **57** (2004) 3–35
16. Tiňo, P., Farkaš, I., van Mourik, J.: Dynamics and topographic organization of recursive self-organizing maps. Neural Computation **18** (2006) 2529–2567
17. Falconer, K.: Fractal Geometry: Mathematical Foundations and Applications. John Wiley and Sons, New York (1990)
18. Barnsley, M.: Fractals everywhere. Academic Press, New York (1988)
19. Ron, D., Singer, Y., Tishby, N.: The power of amnesia. In: Advances in Neural Information Processing Systems 6, Morgan Kaufmann (1994) 176–183
20. Guyon, I., Pereira, F.: Design of a linguistic postprocessor using variable memory length markov models. In: International Conference on Document Analysis and Recognition, Monreal, Canada, IEEE Computer Society Press (1995) 454–457
21. Weinberger, M., Rissanen, J., Feder, M.: A universal finite memory source. IEEE Transactions on Information Theory **41** (1995) 643–652
22. Buhlmann, P., Wyner, A.: Variable length markov chains. Annals of Statistics **27** (1999) 480–513
23. Rissanen, J.: A universal data compression system. IEEE Trans. Inform. Theory **29** (1983) 656–664
24. Giles, C., Omlin, C.: Insertion and refinement of production rules in recurrent neural networks. Connection Science **5** (1993)
25. Doya, K.: Bifurcations in the learning of recurrent neural networks. In: Proc. of 1992 IEEE Int. Symposium on Circuits and Systems. (1992) 2777–2780
26. Tiňo, P., Dorffner, G.: Predicting the future of discrete sequences from fractal representations of the past. Machine Learning **45** (2001) 187–218

27. Jaeger, H.: The "echo state" approach to analysing and training recurrent neural networks. Technical Report GMD Report 148, German National Research Center for Information Technology (2001)
28. Jaeger, H., Haas, H.: Harnessing nonlinearity: Predicting chaotic systems and saving energy in wireless communication. Science **304** (2004) 78–80
29. Maass, W., Natschläger, T., Markram, H.: Real-time computing without stable states: A new framework for neural computation based on perturbations. Neural Computation **14** (2002) 2531–2560
30. Maass, W., Legenstein, R.A., Bertschinger, N.: Methods for estimating the computational power and generalization capability of neural microcircuits. In: Advances in Neural Information Processing Systems. Volume 17., MIT Press (2005) 865—872
31. Hornik, K., Stinchcombe, M., White, H.: Multilayer feedforward networks are universal approximators. Neural Networks **2** (1989) 359–366
32. Hammer, B.: Generalization ability of folding networks. IEEE Transactions on Knowledge and Data Engineering **13** (2001) 196–206
33. Koiran, P., Sontag, E.: Vapnik-chervonenkis dimension of recurrent neural networks. In: European Conference on Computational Learning Theory. (1997) 223–237
34. Vapnik, V.: Statistical Learning Theory. Wiley-Interscience (1998)
35. Shawe-Taylor, J., Bartlett, P.L.: Structural risk minimization over data-dependent hierarchies. IEEE Trans. on Information Theory **44** (1998) 1926–1940
36. Williams, R., Zipser, D.: A learning algorithm for continually running fully recurrent neural networks. Neural Computation **1** (1989) 270–280
37. Williams, R.: Training recurrent networks using the extended kalman filter. In: Proc. 1992 Int. Joint Conf. Neural Networks. Volume 4. (1992) 241–246
38. Patel, G., Becker, S., Racine, R.: 2d image modelling as a time-series prediction problem. In Haykin, S., ed.: Kalman filtering applied to neural networks. Wiley (2001)
39. Manolios, P., Fanelli, R.: First order recurrent neural networks and deterministic finite state automata. Neural Computation **6** (1994) 1155–1173
40. Tiño, P., Hammer, B.: Architectural bias in recurrent neural networks – fractal analysis. In Dorronsoro, J., ed.: Artificial Neural Networks - ICANN 2002. Lecture Notes in Computer Science, Springer-Verlag (2002) 1359–1364
41. de A. Barreto, G., Araújo, A., Kremer, S.: A taxanomy of spatiotemporal connectionist networks revisited: The unsupervised case. Neural Computation **15** (2003) 1255–1320
42. Kilian, J., Siegelmann, H.T.: On the power of sigmoid neural networks. Information and Computation **128** (1996) 48–56
43. Siegelmann, H.T., Sontag, E.D.: Analog computation via neural networks. Theoretical Computer Science **131** (1994) 331–360
44. Hammer, B., Micheli, A., Sperduti, A., Strickert, M.: Recursive self-organizing network models. Neural Networks **17** (2004) 1061–1086
45. Hammer, B., Neubauer, N.: On the capacity of unsupervised recursive neural networks for symbol processing. In d'Avila Garcez, A., Hitzler, P., Tamburrini, G., eds.: Workshop proceedings of NeSy'06. (2006)
46. Sonnenburg, S.: New methods for splice site recognition. Master's thesis, Diplom thesis, Institut für Informatik, Humboldt-Universität Berlin (2002)

47. Martinetz, T., Berkovich, S., Schulten, K.: 'neural-gas' networks for vector quantization and its application to time-series prediction. IEEE Transactions on Neural Networks **4** (1993) 558–569
48. Baldi, P., Brunak, S.: Bioinformatics: The machine learning approach. MIT Press, Cambridge, Mass (2001)
49. Bodén, M., Hawkins, J.: Improved access to sequential motifs: A note on the architectural bias of recurrent networks. IEEE Transactions on Neural Networks **16** (2005) 491–494
50. Bailey, T.L., Elkan, C.: Fitting a mixture model by expectation maximization to discover motifs in biopolymers. In: Proceedings of ISMB, Stanford, CA (1994) 28–36
51. Stormo, G.D.: Dna binding sites: representation and discovery. Bioinformatics **16** (2000) 16–23
52. Sorek, R., Shemesh, R., Cohen, Y., Basechess, O., Ast, G., Shamir, R.: A non-EST-based method for exon-skipping prediction. Genome Research **14** (2004) 1617–1623
53. Hawkins, J., Bodén, M.: The applicability of recurrent neural networks for biological sequence analysis. IEEE/ACM Transactions on Computational Biology and Bioinformatics **2** (2005) 243–253
54. Dyrlöv Bendtsen, J., Nielsen, H., von Heijne, G., Brunak, S.: Improved prediction of signal peptides: SignalP 3.0. Journal of Molecular Biology **340** (2004) 783–795
55. Hawkins, J., Bodén, M.: Detecting and sorting targeting papetides with neural networks and support vector machines. Journal of Bioinformatics and Computational Biology **4** (2006) 1–18
56. Baldi, P., Brunak, S., Frasconi, P., Soda, G., Pollastri, G.: Exploiting the past and the future in protein secondary structure prediction. Bioinformatics **15** (1999) 937–946

6

Time Series Prediction with the Self-Organizing Map: A Review

Guilherme A. Barreto

Department of Teleinformatics Engineering, Federal University of Ceará
Av. Mister Hull, S/N - C.P. 6005, CEP 60455-760, Center of Technology
Campus of Pici, Fortaleza, Ceará, Brazil
guilherme@deti.ufc.br

Summary. We provide a comprehensive and updated survey on applications of Kohonen's self-organizing map (SOM) to time series prediction (TSP). The main goal of the paper is to show that, despite being originally designed as an unsupervised learning algorithm, the SOM is flexible enough to give rise to a number of efficient supervised neural architectures devoted to TSP tasks. For each SOM-based architecture to be presented, we report its algorithm implementation in detail. Similarities and differences of such SOM-based TSP models with respect to standard linear and nonlinear TSP techniques are also highlighted. We conclude the paper with indications of possible directions for further research on this field.

6.1 Introduction

Time series prediction (TSP) (or forecasting) is a function approximation task whose goal is to estimate future values of a sequence of observations based on current and past values of the sequence. This is a rather general definition, which makes no distinction between the nature of the time series data samples; that is, whether they are deterministic (chaotic) or stochastic, linear or not, stationary or not, continuous or discrete, among other characterizations. Anyway, TSP is a mature field, with well-established linear and nonlinear techniques and well-defined data transformation methods contributed from many scientific domains, such as statistics, economy, engineering, hydrology, physics, mathematics and computer science.

From the last fifteen years on, a growing interest in TSP methods has been observed, particularly within the field of neural networks [1, 2]. To some extent, this interest can be explained by the fact that, in its simplest form, the TSP problem can be easily cast as a supervised learning problem, in which the input vectors are formed by the current and past values of a certain time series and the target values are the corresponding (one-step-ahead) future values. This fact has allowed users to explore the function approximation and generalization abilities of some well-known supervised neural architectures, such as the Multilayer Perceptron (MLP) and the Radial Basis Functions (RBF) networks. Not surprisingly, the vast majority of applications of neural network models to TSP tasks are based exactly on these two architectures and their variants.

The *Self-Organizing Map* (SOM) [3, 4, 5] is an important unsupervised neural architecture which, in contrast to the supervised ones, has been less applied to time series prediction. We believe that this occurs because TSP is usually understood as a function approximation problem, while the SOM is commonly viewed as a neural algorithm for data vector quantization (VQ), clustering and visualization [6, 7]. Despite this view, the use of the SOM as a stand-alone time series predictor is becoming more and more common in recent years, as we shall review in this paper.

In a global perspective, the main goal of the paper is then to show that, despite being originally designed as an unsupervised learning algorithm, the SOM is flexible enough to give rise to a number of efficient supervised neural architectures devoted to TSP tasks. To our knowledge, this paper is the first of its type, since former surveys on neural networks for time series prediction (see, for example, [2, 8, 9, 10]) only report just a few SOM-based applications (or even none at all!). In this regard, the contributions of this review are manifold, as we provide a comprehensive list of references, give detailed description of architectures, highlight relationships to standard linear and nonlinear TSP techniques and indicate possible directions for further research on this field. We do not follow a chronological order of presentation of the SOM-based models for TSP, but rather an order that favors understandability of the concepts involved[1].

The remainder of the paper is organized as follows. In Section 6.2 a brief description of the SOM algorithm is carried out. Then, in Section 6.3, we present the basic principles of time series prediction and discuss some reasons to using the SOM for this purpose. Several SOM-based models for TSP are presented from Section 6.4 to Section 6.8. In Section 6.9 possible directions for further research on the field are suggested. The paper is concluded in Section 6.10.

6.2 The Self-Organizing Map

The SOM is a well-known unsupervised neural learning algorithm. The SOM learns from examples a mapping from a high-dimensional continuous input space $\mathcal{X}$ onto a low-dimensional discrete space (lattice) $\mathcal{A}$ of q neurons which are arranged in fixed topological forms, e.g., as a rectangular

---

[1] Matlab codes of most SOM-based TSP models described in this paper can be downloaded from http://www.deti.ufc.br/~guilherme/competitive.html.

2-dimensional array[2]. The map $i^*(\mathbf{x}) : \mathcal{X} \to \mathcal{A}$, defined by the weight matrix $\mathbf{W} = (\mathbf{w}_1, \mathbf{w}_2, \ldots, \mathbf{w}_q)$, $\mathbf{w}_i \in \mathbb{R}^p \subset \mathcal{X}$, assigns to the current input vector $\mathbf{x}(t) \in \mathbb{R}^p \subset \mathcal{X}$ a neuron index

$$i^*(t) = \arg\min_{\forall i} \|\mathbf{x}(t) - \mathbf{w}_i(t)\|, \tag{6.1}$$

where $\|\cdot\|$ denotes the euclidean distance and t is the discrete time step associated with the iterations of the algorithm.

The weight vectors, $\mathbf{w}_i(t)$, also called prototypes or codebook vectors, are trained according to a competitive-cooperative learning rule in which the weight vectors of a winning neuron and its neighbors in the output array are updated after the presentation of the input vector:

$$\mathbf{w}_i(t+1) = \mathbf{w}_i(t) + \alpha(t) h(i^*, i; t)[\mathbf{x}(t) - \mathbf{w}_i(t)] \tag{6.2}$$

where $0 < \alpha(t) < 1$ is the learning rate and $h(i^*, i; t)$ is a weighting function which limits the neighborhood of the winning neuron. A usual choice for $h(i^*, i; t)$ is given by the Gaussian function:

$$h(i^*, i; t) = \exp\left(-\frac{\|\mathbf{r}_i(t) - \mathbf{r}_{i^*}(t)\|^2}{2\sigma^2(t)}\right) \tag{6.3}$$

where $\mathbf{r}_i(t)$ and $\mathbf{r}_{i^*}(t)$ are, respectively, the coordinates of the neurons i and i^* in the output array $\mathcal{A}$, and $\sigma(t) > 0$ defines the radius of the neighborhood function at time t. The variables $\alpha(t)$ and $\sigma(t)$ should both decay with time to guarantee convergence of the weight vectors to stable steady states. In this paper, we adopt for both an exponential decay, given by:

$$\alpha(t) = \alpha_0 \left(\frac{\alpha_T}{\alpha_0}\right)^{(t/T)} \quad \text{and} \quad \sigma(t) = \sigma_0 \left(\frac{\sigma_T}{\sigma_0}\right)^{(t/T)} \tag{6.4}$$

where α_0 (σ_0) and α_T (σ_T) are the initial and final values of $\alpha(t)$ ($\sigma(t)$), respectively.

Weight adjustment is performed until a steady state of global ordering of the weight vectors has been achieved. In this case, we say that the map has converged. The resulting map also preserves the topology of the input samples in the sense that adjacent patterns are mapped into adjacent regions on the map. Due to this topology-preserving property, the SOM is able to cluster input information and spatial relationships of the data on the map. Despite its simplicity, the SOM algorithm has been applied to a variety of complex problems [12, 13] and has become one of the most important ANN architectures.

---

[2] The number of neurons (q) is defined after experimention with the dataset. However, a useful rule-of-thumb is to set it to the highest power of two that is less or equal to $\sqrt{N}$, where N is the number of input vectors available [11].

6.3 Time Series Prediction in a Nutshell

Simply put, a (univariate) time series is an ordered sequence $\{x(t)\}_{t=1}^{N}$ of N observations of a scalar variable x, $x \in \mathbb{R}$. Since the dynamics of the underlying process generating the sequence is generally unknown, the main goal of the time series analyst is testing a number of hypothesis concerning the model that best fits to the time series at hand. Once a time series model has been defined, the next step consists of estimating its parameters from the avaliable data. Finally, the last step involves the evaluation of the model, which is usually performed through an analysis of its predictive ability.

A general modeling hypothesis assumes that the dynamics of the time series can be described by the *nonlinear regressive model* of order p, NAR(p):

$$x(t+1) = f\left[x(t), x(t-1), \ldots, x(t-p+1)\right] + \varepsilon(t) \quad (6.5)$$
$$= f\left[\mathbf{x}^{-}(t)\right] + \varepsilon(t) \quad (6.6)$$

where $\mathbf{x}^{-}(t) \in \mathbb{R}^p$ is the input or regressor vector, $p \in \mathbb{Z}^+$ is the order of the regression, $\mathbf{f}(\cdot)$ is a nonlinear mapping of its arguments and ε is a random variable which is included to take into account modeling uncertainties and stochastic noise. For a stationary time series, it is commonly assumed that $\varepsilon(t)$ is drawn from a gaussian white noise process. We assume that the order of the regression p is optimally estimated by a suitable technique, such as AIC, BIC or Cao's method.

As mentioned before, the structure of the nonlinear mapping $\mathbf{f}(\cdot)$ is unknown, and the only thing that we have available is a set of observables $x(1), x(2), \ldots, x(N)$, where N is the total length of the time series. Our task is to construct a physical model for this time series, given only the data. This model is usually designed to provide an estimate of the next value of the time series, $\hat{x}(t+1)$, an approach called one-step-ahead prediction:

$$\hat{x}(t+1) = \hat{f}\left[\mathbf{x}^{-}(t)|\boldsymbol{\Theta}\right] \quad (6.7)$$

where $\boldsymbol{\Theta}$ denotes the vector of adjustable parameters of the model. In nonlinear analysis of time series, Eq. (6.7) is commonly written in a slightly different way [14]:

$$x(t+1) = f\left[x(t), x(t-\tau), x(t-2\tau), \ldots, x(t-(p-1)\tau)|\boldsymbol{\Theta}\right] \quad (6.8)$$

where in this context the parameter p is usually called the embedding dimension and $\tau \in \mathbb{Z}^+$ is the embedding delay. Without loss of generality, in this paper we always assume $\tau = 1$. In this case, Eqs. (6.7) and (6.8) become equivalent.

The mapping $\hat{f}[\cdot|\boldsymbol{\Theta}]$ can be implemented through different techniques, such as linear models, polynomial models, neural network models, among others. For example, a linear version of Eq. (6.7), known as the autoregressive (AR) model [15], can be written as

$$\hat{x}(t+1) = \mathbf{a}^T \mathbf{x}^+(t) = a_0 + \sum_{l=1}^{p} a_l x(t-l+1), \qquad (6.9)$$

where $\mathbf{a} = [a_0 \; a_1 \; \cdots \; a_p]^T$ is the coefficient vector of the model, with the superscript T denoting the transpose of a vector, and $\mathbf{x}^+(t) = [1 \; x(t) \; \cdots \; x(t-p+1)]^T$ is an augmented regressor vector at time step t. This is one of the simplest time series models, since it assumes that the next value of the time series is a linear combination of the past p values. More sophisticated models, composed of combinations of linear, polynomial and/or neural network models, are also very common in the literature.

Independently of the chosen model, its parameters need to be computed during an estimation (or training) stage using only a portion of the available time series, e.g. the first half of the sequence. Once trained, the adequacy of the hypothesized model to the training data is usually carried out through the analysis of the generated prediction errors (or *residuals*),

$$e(t+1) = x(t+1) - \hat{x}(t+1), \qquad (6.10)$$

where $x(t+1)$ is the actually observed next value of the time series and $\hat{x}(t+1)$ is the predicted one. The model selection stage should be performed for different values of the parameter p. For each value of p, the model parameters should be estimated and the sequence of residuals computed. Usually, one chooses the lowest value of p that results in an approximately gaussian and uncorrelated sequence of residuals.

Once the time series model has been selected, its predictive ability can be assessed through an *out-of-sample prediction* scheme, carried out using the sequence of residuals computed over the remaining portion of the time series. As reviewed by Hyndman and Koehler [16], there exists a number of measures of predictive accuracy of a model, but in this paper we use only the *Normalized Root Mean Squared Error* (NRMSE):

$$NRMSE = \sqrt{\frac{(1/M)\sum_{t=1}^{M}(x(t) - \hat{x}(t))^2}{(1/M)\sum_{t=1}^{M}(x(t) - \bar{x})^2}} = \sqrt{\frac{\sum_{t=1}^{M}(x(t) - \hat{x}(t))^2(t)}{M \cdot s_x^2}} \qquad (6.11)$$

where $\bar{x}$ and s_x^2 are, respectively, the sample mean and sample variance of the testing time series, and M is the length of the sequence of residuals.

6.3.1 Why Should We Use the SOM for TSP?

This is the very first question we should try to answer before starting the description of SOM-based models for TSP. Since there are already available a huge number of models for this purpose, including neural network based ones, why do we need one more? Certainly, there is no definitive answers to this question, but among a multitude of reasons that could influence someone to apply the SOM to TSP problems, including curiosity about or familiarity with the algorithm, one can surely take the following two for granted.

Local Nature of SOM-based Models - Roughly speaking, SOM-based models for time series prediction belong to the class of models performing local function approximation. By local we denote those set of models whose output is determined by mathematical operations acting on localized regions of the input space. A global model, such as MLP-based predictors, makes use of highly distributed representations of the input space that difficult the interpretability of the results. Thus, SOM-based TSP models allow the user to better understand the dynamics of the underlying process that generates the time series, at least within the localized region used to compute the current model's output. This localized nature of SOM-based model combined with its topology-preserving property are useful, for example, if one is interested in time series segmentation or time series visualization problems.

Simple Growing Architectures - As one could expect, clustering of the input space is an essential step in designing efficient SOM-based TSP models. A critical issue in the process of partitioning the input space for the purpose of time series prediction is to obtain an appropriate estimation (guess?) of the number of prototype vectors. Over- or under-estimation of this quantity can cause, respectively, the appearance of clusters with very few pattern vectors that are insufficient to build a local model for them and clusters containing patterns from regions with different dynamics.

One of the main advantages of the SOM and related architectures over standard supervised ones, such as MLP and RBF networks, is that there is no need to specify in advance the number of neurons (and hence, prototypes) in the model. This becomes possible thanks to a number of growing competitive learning architectures currently available (see [17] for a detailed review). Such growing architectures are much easier to implement than their supervised counterparts, such as the cascade-correlation architecture [18]. They are useful even if one has already initialized a given SOM with a fixed number of neurons. In this case, if necessary, new knowledge can be easily incorporated into the network in the form of a new prototype vector.

6.4 The VQTAM Model

A quite intuitive and straightforward way of using the SOM for time series prediction purposes is by performing vector quantization simultaneously on the regressor vectors $\mathbf{x}^-(t)$, $t = p, \ldots, N-1$, and their associated one-step-ahead observations $x(t+1)$. This is exactly the idea behind the *Vector-Quantized Temporal Associative Memory* (VQTAM) model [19, 20]. This model can be understood as a generalization to the temporal domain of a SOM-based associative memory technique that has been widely used to learn static (memoryless) input-output mappings, specially within the domain of robotics [21].

6 Time Series Prediction with the Self-Organizing Map: A Review

In a general formulation of the VQTAM model, the input vector at time step t, $\mathbf{x}(t)$, is composed of two parts. The first part, denoted $\mathbf{x}^{in}(t) \in \mathbb{R}^p$, carries data about the input of the dynamic mapping to be learned. The second part, denoted $\mathbf{x}^{out}(t) \in \mathbb{R}^m$, contains data concerning the desired output of this mapping. The weight vector of neuron i, $\mathbf{w}_i(t)$, has its dimension increased accordingly. These changes are formulated as follows:

$$\mathbf{x}(t) = \begin{pmatrix} \mathbf{x}^{out}(t) \\ \mathbf{x}^{in}(t) \end{pmatrix} \quad \text{and} \quad \mathbf{w}_i(t) = \begin{pmatrix} \mathbf{w}_i^{out}(t) \\ \mathbf{w}_i^{in}(t) \end{pmatrix} \quad (6.12)$$

where $\mathbf{w}_i^{in}(t) \in \mathbb{R}^p$ and $\mathbf{w}_i^{out}(t) \in \mathbb{R}^m$ are, respectively, the portions of the weight (prototype) vector which store information about the inputs and the outputs of the desired mapping.

For the univariate time series prediction tasks we are interested in, the following definitions apply:

$$x^{out}(t) = x(t+1) \quad \text{and} \quad \mathbf{x}^{in}(t) = \mathbf{x}^{-}(t) \quad (6.13)$$
$$w_i^{out}(t) = w_i^{out}(t) \quad \text{and} \quad \mathbf{w}_i^{in}(t) = \mathbf{w}_i^{-}(t) \quad (6.14)$$

where $\mathbf{w}_i^{-}(t) \in \mathbb{R}^p$ is the portion of the weight vector associated with the regressor vector $\mathbf{x}^{-}(t)$. Note that the vectors $\mathbf{x}^{out}(t)$ and $\mathbf{w}_i^{out}(t)$ in Eq. (6.12) are now reduced to the scalar quantities $x(t+1)$ and $w_i^{out}(t)$, respectively.

The winning neuron at time step t is determined based only on $\mathbf{x}^{in}(t)$:

$$i^*(t) = \arg\min_{i \in \mathcal{A}} \{\|\mathbf{x}^{in}(t) - \mathbf{w}_i^{in}(t)\|\} = \arg\min_{i \in \mathcal{A}} \{\|\mathbf{x}^{-}(t) - \mathbf{w}_i^{-}(t)\|\} \quad (6.15)$$

However, for updating the weights, both $\mathbf{x}^{in}(t)$ and $x^{out}(t)$ are used:

$$\mathbf{w}_i^{in}(t+1) = \mathbf{w}_i^{in}(t) + \alpha(t)h(i^*, i; t)[\mathbf{x}^{in}(t) - \mathbf{w}_i^{in}(t)] \quad (6.16)$$
$$w_i^{out}(t+1) = w_i^{out}(t) + \alpha(t)h(i^*, i; t)[x^{out}(t) - w_i^{out}(t)] \quad (6.17)$$

where $0 < \alpha(t) < 1$ is the learning rate and $h(i^*, i; t)$ is a time-varying gaussian neighborhood function as defined in Eq. (6.3). In words, Eq. (6.16) performs topology-preserving vector quantization on the input space and Eq. (6.17) acts similarly on the output space of the mapping being learned. As training proceeds, the SOM learns to associate the input prototype vectors $\mathbf{w}_i^{in}$ with the corresponding output prototype vectors w_i^{out} (see Figure 6.1).

During the out-of-sample prediction phase, the one-step-ahead estimate at time step t is then given by

$$\hat{x}(t+1) = w_{i^*}^{out}(t), \quad (6.18)$$

where the winning neuron at time step t is found according to Eq. (6.15), for each new incoming regressor vector $\mathbf{x}^{in}(t)$. Instead of using a single output weight $w_{i^*}^{out}(t)$, one can compute an improved one-step-ahead prediction as the average value of the output weights of the K ($K > 1$) closest prototypes.

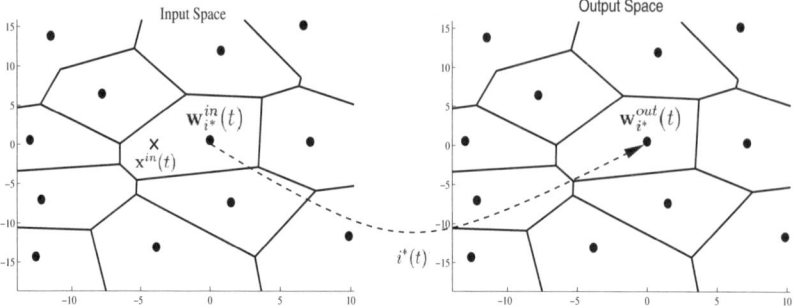

Fig. 6.1. Mapping between input and output Voronoi cells of a VQTAM model. The symbols '•' denote the prototype vectors of each input/output cell. The symbol '×' denotes an input regressor.

Related Architectures: VQTAM's philosophy is the same as that of the local model proposed in the seminal paper by Farmer and Sidorowich [22]. These authors used all the vectors $\mathbf{x}(t) = [x^{out}(t)\ \mathbf{x}^{in}(t)]$, $t = 1, \ldots, N - p$, to compute predictions for a given time series. The procedure is quite simple: find the closest vector to the current input vector on the basis of $\mathbf{x}^{in}(t)$ only and, then, use the prediction associated to the closest vector as an approximation of the desired one. Roughly speaking, the VQTAM approach to TSP can then be understood as a VQ-based version of the Farmer-Sidorowich method.

Baumann and Germond [23] and Lendasse et al. [24] developed independently SOM-based architectures which are in essence equivalent to the VQTAM and successfully applied them to the prediction of electricity consumption. More recently, Lendasse et al. [25] used standard unsupervised competitive learning to implement a number of vector quantization strategies for time series prediction purposes. One of them consists in assigning different weights to the components of the regressor vector $\mathbf{x}^{in}(t)$ and to the target output value $x^{out}(t) = x(t + 1)$. In this case, the vector $\mathbf{x}(t)$ is now written as

$$\mathbf{x}(t) = [x^{out}(t)\ |\ \mathbf{x}^{in}(t)]^T \qquad (6.19)$$
$$= [kx(t+1)\ |\ a_1 x(t)\ a_2 x(t-1)\ \cdots\ a_p x(t-p+1)]^T, \qquad (6.20)$$

where $k > 0$ is a constant weight determined through a cross-validation procedure, and a_j, $j = 1, \ldots, p$, are the coefficients of an AR(p) model (see Eq. (6.9)), previously fitted to the time series of interest. This simple strategy was able to improve the performance of VQ-based models in a benchmarking time series prediction task.

Despite its simplicity the VQTAM and variants can be used as a stand-alone time series predictor or as the building block for more sophisticated models as we see next.

6.4.1 Improving VQTAM Predictions Through Interpolation

Since the VQTAM is essentially a vector quantization algorithm, it may require too many neurons to provide small prediction errors when approximating continuous mappings. This limitation can be somewhat alleviated through the use of interpolation methods specially designed for the SOM architecture, such as geometric interpolation [26], topological interpolation [27] and the *Parameterized Self-Organizing Map* (PSOM) [28].

Smooth output values can also be obtained if we use the VQTAM model to design RBF-like networks. For instance, an RBF network with q gaussian basis functions and a single output neuron can be built directly from the learned prototypes $\mathbf{w}_i^{in}$ and w_i^{out}, without requiring additional training. In this case, the one-step-ahead prediction at time step t can be computed as follows:

$$\hat{x}(t+1) = \frac{\sum_{i=1}^{q} w_i^{out} G_i(\mathbf{x}^{in}(t))}{\sum_{i=1}^{q} G_i(\mathbf{x}^{in}(t))} \quad (6.21)$$

where $G_i(\mathbf{x}^{in}(t))$ is the response of this basis function to the current input vector $\mathbf{x}^{in}(t)$:

$$G_i(\mathbf{x}^{in}(t)) = \exp\left(-\frac{\|\mathbf{x}^{in}(t) - \mathbf{w}_i^{in}\|^2}{2\gamma^2}\right) \quad (6.22)$$

where $\mathbf{w}_i^{in}$ plays the role of the center of the i-th basis function and $\gamma > 0$ defines its radius (or *spread*). This method has been successfully applied to time series prediction [25] and related tasks, such as system identification and adaptive filtering [29].

Results obtained by the application of the plain VQTAM and an VQTAM-based RBF model to an one-step-ahead prediction task are shown in Figure 6.2. Only the first 500 predictions are shown in this figure. Details about the generation of the time series data are given in Appendix 1. For this simulation, the plain VQTAM model has only 5 neurons. The VQTAM-based RBF model has also 5 neurons and a common value for the spread of the gaussian basis functions ($\gamma = 0.40$). As expected, the VQTAM has a poor prediction performance ($NRMSE = 0.288$), while the RBF-like interpolation method performs quite well for the same number of neurons ($NRMSE = 0.202$). Obviously, even for the VQTAM-based RBF model, the predictions are poor for high curvature points.

6.4.2 Rule Extraction from a Trained VQTAM Model

As mentioned in the beginning of this section, better interpretability is a clear-cut advantage of local models over global ones. In particular, interpretation of the results can be carried out by means of rule extraction procedures and prototype-based self-organizing algorithms are well-suited to this task [30]. For

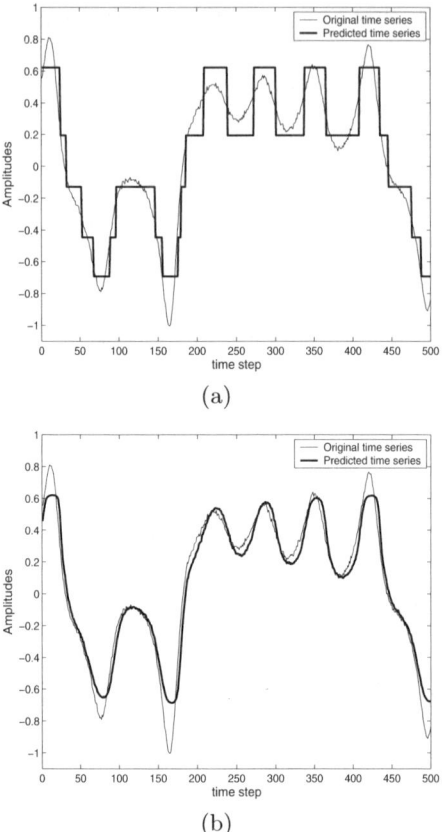

Fig. 6.2. One-step-ahead prediction results for the (a) VQTAM model and (b) for a VQTAM-based RBF model.

example, through a trained VQTAM approach one can easily build interval-based rules that helps explaining the predicted values. This procedure can be summarized in a few steps as follows.

Step 1 - Present the set of training data vectors once again to the VQTAM model and find the neuron each vector $\mathbf{x}^{in}(t)$ is mapped to.

Step 2 - Let $\mathbf{x}_j^i = [x_{j,0}^i \ x_{j,1}^i \ \cdots \ x_{j,p}^i]^T$ denote the j-th training data vector that has been mapped to neuron i. Then, find the subset $\mathcal{X}_i = \{\mathbf{x}_1^i, \mathbf{x}_2^i, \ldots, \mathbf{x}_{n_i}^i\}$ of n_i ($0 < n_i \ll N - p$) training data vectors which are mapped to neuron i. Note that $\sum_{i=1}^q n_i = N - p$.

Step 3 - Find the range of variation of the l-th component $x_{j,l}^i$ within $\mathcal{X}_i$:

$$[x_l^i]^- \leq x_{j,l}^i \leq [x_l^i]^+, \qquad j = 1, \ldots, n_i \qquad (6.23)$$

Table 6.1. Interval-based rules obtained for a VQTAM model with $q = 5$ neurons.

Neuron	$x(t+1)$	$x(t)$	$x(t-1)$	$x(t-2)$	$x(t-3)$	$x(t-4)$
$i=1$	[-0.94,-0.38]	[-0.94,-0.44]	[-0.94,-0.50]	[-0.94,-0.54]	[-0.95,-0.52]	[-0.93,-0.49]
$i=2$	[-0.67,-0.12]	[-0.63,-0.17]	[-0.60,-0.21]	[-0.58,-0.26]	[-0.61,-0.24]	[-0.67,-0.22]
$i=3$	[-0.36,0.10]	[-0.36,0.09]	[-0.32,0.07]	[-0.29,0.05]	[-0.32,0.09]	[-0.37,0.16]
$i=4$	[-0.14,0.55]	[-0.08,0.49]	[-0.03,0.47]	[0.03,0.43]	[0.02,0.48]	[0.00,0.58]
$i=5$	[0.15,1.00]	[0.24,1.00]	[0.34,1.00]	[0.41,1.00]	[0.38,1.00]	[0.33,1.00]

where $\left[x_l^i\right]^- = \min_{\forall j}\{x_{j,l}^i\}$ and $\left[x_l^i\right]^+ = \max_{\forall j}\{x_{j,l}^i\}$.

Step 4 - Then, associate the following interval-based prediction rule R_i to neuron i:

$$R_i : \begin{cases} \text{IF} \quad \left[x_1^i\right]^- \leq x(t) \leq \left[x_1^i\right]^+ \text{ and } \left[x_2^i\right]^- \leq x(t-1) \leq \left[x_2^i\right]^+ \text{ and } \cdots \\ \quad \cdots \text{ and } \left[x_p^i\right]^- \leq x(t-p+1) \leq \left[x_p^i\right]^+, \\ \text{THEN } \left[x_0^i\right]^- \leq x(t+1) \leq \left[x_0^i\right]^+ \end{cases}$$

(6.24)

Once all the rules are determined, a certain rule R_i is activated whenever $i^*(t) = i$, for a new incoming vector $\mathbf{x}^{in}(t)$. As an example, using the nonlinear time series data described in Appendix 1, the resulting five interval-based prediction rules obtained by a VQTAM model with $q = 5$ neurons are shown in Table 6.1. Figure 6.3 shows the time steps in which the fifth rule in Table 6.1 is activated during the out-of-sample (testing) prediction task.

It is worth emphasizing that fuzzy linguistic labels, such as very high, high, medium, small and very small, can be assigned by an expert to each component of the rule R_i to facilitate comprehension of the different clusters in the data. Such an approach has been already successfully tested in a number of real-world complex tasks, such as analysis of telecommunication networks [31], data mining [32, 33] and classification of multispectral satellite images [34].

6.5 The Double Vector Quantization Method

Simon et al. [35, 36] proposed the *Double Vector Quantization* (DVQ) method, specifically designed for long-term time series prediction tasks. The DVQ method has been successfully evaluated in a number of real-world time series, such as the far-infrared chaotic laser and electrical consumption data sets. Using VQTAM's notation, the DVQ method requires two SOM networks: one to cluster the regressors $\mathbf{x}^{in}(t)$, and a second one to cluster the associated *deformations* $\Delta\mathbf{x}^{in}(t) = \mathbf{x}^{in}(t+1) - \mathbf{x}^{in}(t)$. By definition, each deformation $\Delta\mathbf{x}^{in}(t)$ is associated to a single regressor $\mathbf{x}^{in}(t)$. The number of neurons in each SOM is not necessarily the same.

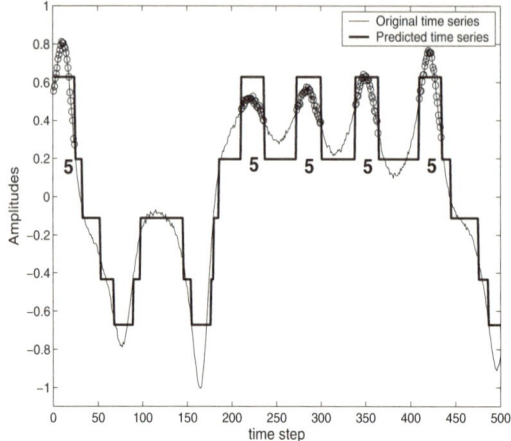

Fig. 6.3. Regions where the rule R_5 associated to neuron $i = 5$ is activated are highlighted with small open circles ('o').

Double quantization of regressors and deformations only gives a static characterization of the past evolution of the time series. It is worth noting, however, that the associations between the deformations $\Delta \mathbf{x}^{in}(t)$ and their corresponding regressors $\mathbf{x}^{in}(t)$, can contain useful dynamical information about how the series has evolved between a regressor and the next one. These associations can be stochastically modelled by a transition probability matrix $P = [p_{ij}]$, $1 \leq i \leq q_1$ and $1 \leq j \leq q_2$, where q_1 and q_2 are the number of prototypes in the first and second SOMs, respectively.

Let $\mathbf{w}_i^{(1)}$, $i = 1, \ldots, q_1$, be the i-th prototype of the first SOM. Similarly, let $\mathbf{w}_j^{(2)}$, $j = 1, \ldots, q_2$, be the j-th prototype of the second SOM. Then, each element p_{ij} is estimated as

$$\hat{p}_{ij} = \frac{\#\{\mathbf{x}^{in}(t) \in \mathbf{w}_i^{(1)} \text{ and } \Delta \mathbf{x}^{in}(t) \in \mathbf{w}_j^{(2)}\}}{\#\{\mathbf{x}^{in}(t) \in \mathbf{w}_i^{(1)}\}}, \quad (6.25)$$

where the symbol $\#$ denotes the cardinality of a set. In fact, $\hat{p}_{ij}$ is an estimate of the conditional probability that the deformation $\Delta \mathbf{x}^{in}(t)$ is mapped to the prototype $\mathbf{w}_i^{(2)}$ given that the regressor $\mathbf{x}^{in}(t)$ is already mapped to $\mathbf{w}_i^{(1)}$.

Once the two SOMs are trained and the transition matrix $\hat{P} = [\hat{p}_{ij}]$ is estimated, one-step-ahead predictions using the DVQ model can be computed as summarized in the steps.

1. **Step 1** - Find the closest prototype $\mathbf{w}_{i^*}^{(1)}(t)$ to the current regressor vector $\mathbf{x}^{in}(t)$ using Eq. (6.15).

2. **Step 2** - Choose at random a deformation prototype $\mathbf{w}_{j*}^{(2)}(t)$ from the set $\{\mathbf{w}_j^{(2)}\}$, $1 \leq j \leq q_2$, according to the conditional probability distribution defined by row $i^*(t)$ of the transition matrix, i.e. $P[i^*(t), 1 \leq j \leq q_2]$.
3. **Step 3** - Compute an estimate of the next regressor vector as follows:

$$\hat{\mathbf{x}}^{in}(t+1) = \mathbf{x}^{in}(t) + \mathbf{w}_{j*}^{(2)}(t). \quad (6.26)$$

4. **Step 4** - Extract the one-step-ahead prediction from $\hat{\mathbf{x}}^{in}(t+1)$, i.e.

$$\hat{x}(t+1) = \hat{x}_1^{in}(t+1), \quad (6.27)$$

where $\hat{x}_1^{in}(t+1)$ is the first component of $\hat{\mathbf{x}}^{in}(t+1)$.

To estimate $h > 1$ values into the future, the Steps 1-4 are then iterated, inserting each new one-step-ahead estimate $\hat{x}(t+1)$ into the regressor vector of the next iteration of the algorithm. Note that, for each time iteration t, the Steps 1-4 need to be repeated using a Monte-Carlo procedure, since the random choice of deformation according to the conditional probability distributions given by the rows of the transition matrix is stochastic.

6.6 Local AR Models from Clusters of Data Vectors

Lehtokangas et al. [37] and Vesanto [38] independently proposed a SOM-based model that associates a local linear AR(p) model to each neuron i. In this approach, there is a coefficient vector $\mathbf{a}_i \in \mathbb{R}^p$, defined as

$$\mathbf{a}_i = [a_{i,0} \; a_{i,1} \; \cdots \; a_{i,p}]^T, \quad (6.28)$$

associated to each prototype vector $\mathbf{w}_i$. Thus, the total number of adjustable parameters of this model is $2pq$ (i.e., q units of p-dimensional weight vectors plus q units of p-dimensional coefficient vectors).

The procedure for building local AR models requires basically four steps. The first two steps are equivalent to training the VQTAM model, being repeated here just for the sake of convenience.

1. **Data preparation** - From a training time series of length N, one can build $N - p$ input vectors $\mathbf{x}(t) = [x(t+1) \; \mathbf{x}^-(t)]^T$, $t = p, \ldots, N-1$, by sliding a time window of width $p+1$ over the time series.
2. **SOM training** - The training data vectors are presented to the SOM for a certain number of epochs. The order of presentation of the data vectors can be randomly changed at each training epoch. The winning neuron at time step t is found using $\mathbf{x}^-(t)$ only, as in Eq. (6.15). For weight updating, however, one should use the entire vector $\mathbf{x}(t)$.
3. **Data partitioning** - Once training is completed, the set of training data vectors is presented once again to the SOM in order to find the neuron each data vector is mapped to. Let $\mathcal{X}_i = \{\mathbf{x}_1^i, \mathbf{x}_2^i, \ldots, \mathbf{x}_{n_i}^i\}$ denote the

subset of n_i ($0 < n_i \ll N - p$) training data vectors which are mapped to neuron i, with $\mathbf{x}_j^i = [x_{j,0}^i \ x_{j,1}^i \ \cdots \ x_{j,p}^i]^T$ representing the j-th training data vector that has been mapped to neuron i.

4. **Parameter estimation** - The coefficient vector $\mathbf{a}_i$ of the i-th AR local model is computed through the pseudoinverse (least squares) technique using the data vectors in the set $\mathcal{X}_i$:

$$\mathbf{a}_i = \left(\mathbf{R}_i^T \mathbf{R}_i + \lambda \mathbf{I}\right)^{-1} \mathbf{R}_i^T \mathbf{p}_i \qquad (6.29)$$

where $0 < \lambda \ll 1$ is a small constant, $\mathbf{I}$ is an identity matrix of dimension $p+1$, and the vector $\mathbf{p}_i$ and the matrix $\mathbf{R}_i$ are defined, respectively, as

$$\mathbf{p}_i = \begin{pmatrix} x_{1,0}^i \\ x_{2,0}^i \\ \vdots \\ x_{n_i,0}^i \end{pmatrix} \quad \text{and} \quad \mathbf{R}_i = \begin{pmatrix} 1 & x_{1,1}^i & x_{1,2}^i & \cdots & x_{1,p}^i \\ 1 & x_{2,1}^i & x_{2,2}^i & \cdots & x_{2,p}^i \\ \vdots & \vdots & \vdots & \ddots & \vdots \\ 1 & x_{n_i,1}^i & x_{n_i,2}^i(t) & \cdots & x_{n_i,p}^i \end{pmatrix}. \qquad (6.30)$$

Numerical problems may arise when implementing Eq. (6.29), which can be caused by any of the following situations:

1. No data vector is assigned to neuron i at all, i.e. the cardinality of the corresponding set $\mathcal{X}_i$ is zero. In this case, the best thing to do is to eliminate this neuron from the set of neurons allowed to build the local models.
2. Some data vectors are indeed assigned to neuron i, but the cardinality of $\mathcal{X}_i$ is much less than p. In this case, it is recommended to join the subset $\mathcal{X}_i$ of this neuron with that of its closest neighbor.

Once all the (q) local AR models have been determined, one-step-ahead predictions can be computed for each new incoming vector $\mathbf{x}^+(t)$ as follows:

$$\hat{x}(t+1) = \mathbf{a}_{i^*}(t) \mathbf{x}^+(t) = a_{i^*,0}(t) + \sum_{l=1}^{p} a_{i^*,l}(t) x(t-l+1) \qquad (6.31)$$

where $\mathbf{a}_{i^*}(t)$ is the coefficient vector of the current winning neuron. It is worth noting that if the number of neurons is set to one ($q=1$), the network reduces to an ordinary AR(p) model.

Related Architectures: Koskela et al. [39] introduced a SOM-based approach that uses a temporal variant of the SOM, called the *Recurrent SOM*, to cluster the training data vectors. They argue that this type of dynamic SOM algorithm can better capture the temporal dependencies of consecutive samples. The approach introduced by Sfetsos and Siriopoulos [40] uses the well-known K-means clustering algorithm to partition data vectors for building local AR models. Poliker and Geva [41] used a fuzzy clustering algorithm for the same purpose. In [42], data partitioning is performed with the SOM, while local TSP models are built using functional networks. More recently, Pavlidis et al. [43] introduced a local nonlinear approach and applied it successfully to financial time series prediction tasks. They used the

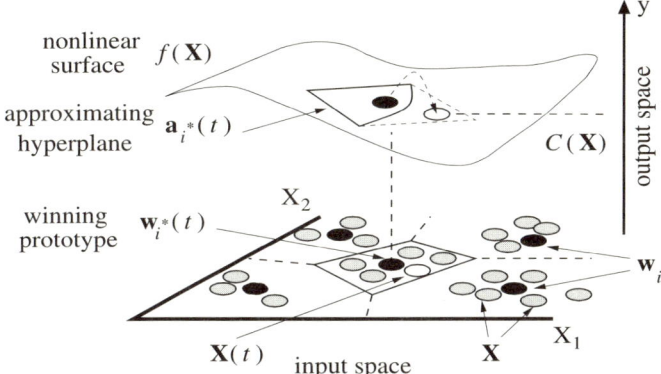

Fig. 6.4. Sketch of the local modeling approach implemented by the LLM architecture.

Growing Neural Gas algorithm [17] for time series clustering purposes and a feedforward supervised neural network (e.g. MLP) to build local nonlinear TSP models.

It is worth noting that each of the methods above uses a different clustering algorithm to partition the data. How sensitive are the local linear/nonlinear AR models with respect to the choice of the clustering algorithm? This is an open question. We return to this issue in Section 6.9.

6.7 Online Learning of Local Linear Models

One of the first applications of the SOM to time series prediction was proposed by Walter et al. [44] using the *Local Linear Map* (LLM) model. The underlying idea is the same of that presented in Section 6.6, i.e. to associate to each neuron a linear AR model. Thus, the total number of parameters of the LLM model is also $2pq$. Once trained, the one-step-ahead prediction is provided by the local AR model associated with the current winning neuron as defined by Eq. (6.31) (see Figure 6.4).

Unlike the model described in Section 6.6, the LLM model updates the coefficient vectors of the local models simultaneously with the clustering of the data vectors, thus reducing the time devoted to the model building process. Another important advantage is a drastic reduction of the computational efforts, since there is no need to compute the (pseudo)inverses of q matrices $\mathbf{R}_i$, $i = 1, \ldots, q$, as shown in Eq. (6.29).

For the LLM model, the input vector at time step t is defined as $\mathbf{x}^{in}(t) = \mathbf{x}^-(t)$. Hence, the weight vector of neuron i, $i = 1, \ldots, q$, is defined as $\mathbf{w}_i^{in}(t) = \mathbf{w}_i^-(t)$. Then, weight updating rule follows exactly the one given in Eq. (6.16). The learning rule of the coefficient vectors $\mathbf{a}_i(t)$ is an extension of the well-known LMS rule that takes into account the influence of the neighborhood function $h(i^*, i; t)$:

$$\begin{aligned}\mathbf{a}_i(t+1) &= \mathbf{a}_i(t) + \alpha h(i^*, i; t) e_i(t) \mathbf{x}^+(t) \\ &= \mathbf{a}_i(t) + \alpha h(i^*, i; t)[x(t+1) - \mathbf{a}_i^T(t)\mathbf{x}^+(t)]\mathbf{x}^+(t)\end{aligned} \quad (6.32)$$

where $e_i(t)$ is the prediction error of the i-th local model and $0 < \alpha < 1$ is the learning rate of the coefficient vectors. Once trained, one-step-ahead predictions are computed according to Eq. (6.31).

Related Architectures: Stokbro et al. [45], much the same way as implemented by the LLM architecture, associated a linear AR filter to each hidden unit of an RBF network, applying it to prediction of chaotic time series. The main difference between this approach and the LLM architecture is that the former allows the activation of more than one hidden unit, while the latter allows only the winning neuron to be activated. In principle, by combining the output of several local linear models instead of a single one, one can improve generalization performance. Martinetz et al. [46] introduced a variant of the LLM that is based on the Neural Gas algorithm and applied it successfully to TSP. It is important to point out the existence of a growing LLM algorithm introduced by Fritzke [47], which can be easily applied to TSP using the formulation just described.

It is worth pointing out the LLM model bears strong resemblance to the widely-known *Threshold AR* (TAR) model [48]. This is a nonlinear time-series model comprised of a number of linear AR sub-models (local linear AR models, in our jargon). Each sub-model is constructed for a specific amplitude range, which is classified by amplitude thresholds and a time-delay amplitude. In this basic formulation, the switching betwen sub-AR models is governed by a *scalar quantization* process built over the amplitudes of the observations. In this sense, the LLM can be understood as a VQ-based implementation of the TAR model.

6.8 Time-Varying Local AR Models from Prototypes

The last SOM-based model for time series prediction we describe was introduced by Barreto et al. [49], being called the *KSOM* model. The KSOM model combines the vector quantization approach of the VQTAM model with that of building local linear AR models.

However, instead of q time-invariant local linear models, the KSOM works with a single time-variant linear AR model whose coefficient vector is recomputed at every time step t, directly from a subset of K ($K \ll q$) weight vectors extracted from a trained VQTAM model. This subset contains the weight vectors of the current K first winning neurons, denoted by $\{i_1^*(t), i_2^*(t), \ldots, i_K^*(t)\}$:

$$i_1^*(t) = \arg\min_{\forall i} \left\{ \|\mathbf{x}^{in}(t) - \mathbf{w}_i^{in}(t)\| \right\} \tag{6.33}$$

$$i_2^*(t) = \arg\min_{\forall i \neq i_1^*} \left\{ \|\mathbf{x}^{in}(t) - \mathbf{w}_i^{in}(t)\| \right\}$$

$$\vdots \qquad \vdots \qquad \vdots$$

$$i_K^*(t) = \arg\min_{\forall i \neq \{i_1^*,\ldots,i_{K-1}^*\}} \left\{ \|\mathbf{x}^{in}(t) - \mathbf{w}_i^{in}(t)\| \right\}$$

The out-of-sample prediction at time step t is then computed as

$$\hat{x}(t+1) = \mathbf{a}^T(t)\mathbf{x}^+(t), \tag{6.34}$$

where the current coefficient vector $\mathbf{a}(t)$ is computed by means of the pseudoinverse method as:

$$\mathbf{a}(t) = \left(\mathbf{R}^T(t)\mathbf{R}(t) + \lambda \mathbf{I}\right)^{-1} \mathbf{R}^T(t)\mathbf{p}(t), \tag{6.35}$$

where $\mathbf{I}$ is a identity matrix of order p and $\lambda > 0$ (e.g. $\lambda = 0.01$) is a small constant added to the diagonal of $\mathbf{R}^T(t)\mathbf{R}(t)$ to make sure that this matrix is full rank. The prediction vector $\mathbf{p}(t)$ and the regression matrix $\mathbf{R}(t)$ at time t are defined as

$$\mathbf{p}(t) = [w_{i_1^*,1}^{out}(t) \ w_{i_2^*,1}^{out}(t) \ \cdots \ w_{i_K^*,1}^{out}(t)]^T, \tag{6.36}$$

$$\mathbf{R}(t) = \begin{pmatrix} w_{i_1^*,1}^{in}(t) & w_{i_1^*,2}^{in}(t) & \cdots & w_{i_1^*,p}^{in}(t) \\ w_{i_2^*,1}^{in}(t) & w_{i_2^*,2}^{in}(t) & \cdots & w_{i_2^*,p}^{in}(t) \\ \vdots & \vdots & \vdots & \vdots \\ w_{i_K^*,1}^{in}(t) & w_{i_K^*,2}^{in}(t) & \cdots & w_{i_K^*,p}^{in}(t) \end{pmatrix}. \tag{6.37}$$

It is worth emphasizing that the KSOM model should be used when the number of neurons is much larger than K (e.g. $q > 2K$). If this is not the case, the choice of another SOM-based model, such as the one described in Section 6.6 or the LLM model, is highly recommended. As an example, we compare the performance of the LLM and KSOM models using only $q = 5$ neurons. The results are shown in Figure 6.5. As expected, the one-step-ahead prediction performance of the LLM model ($NRMSE = 0.039$) is much better than that of the KSOM model ($NRMSE = 0.143$, $K = 4$), since the number of neurons is too small. It is easy to see that the results of the LLM model present also less variability.

Related Architecture: Principe and Wang [50] proposed a neural architecture for nonlinear time series prediction which is equivalent to KSOM in the sense that a time-varying coefficient vector $\mathbf{a}(t)$ is computed from K prototype vectors of a trained SOM using the pseudoinverse method. However, the required prototype vectors are not selected as the K nearest prototypes to the current input vector, but rather automatically selected as the winning prototype at time t and its $K-1$ topological neighbors. If a perfect topology

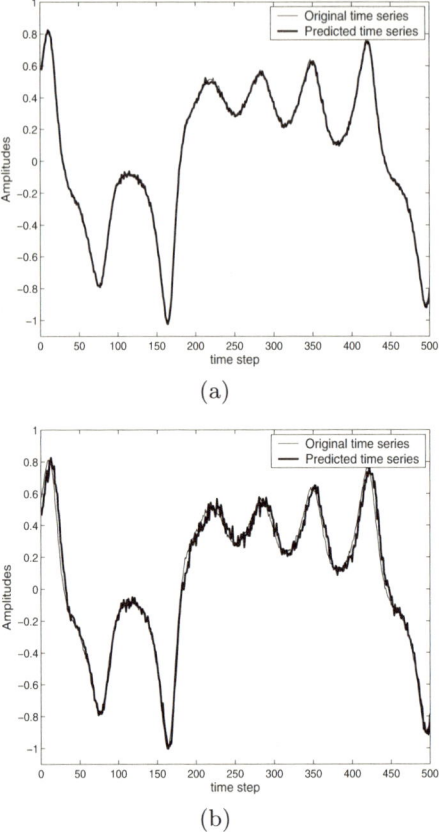

Fig. 6.5. One-step-ahead prediction results for the (a) LLM model and (b) for the KSOM model.

preservation is achieved during SOM training, the neurons in the topological neighborhood of the winner are also the closest ones to the current input vector. However, if an exact topology preserving map is not achieved, as usually occurs for multidimensional data, the KSOM provides more accurate results.

6.9 Directions for Further Work

The SOM-based TSP models described in this paper have shown, among other things, that the SOM is a highly flexible neural architecture, in the sense that it can be successfully applied to unsupervised learning tasks (e.g. data clustering and visualization), as well as to typical supervised learning tasks (e.g. function approximation).

Despite the relatively large number of TSP models developed based on the SOM, this field is still in its first infancy if compared to those based on the MLP or the RBF networks. In the remaining part of this paper we list a number of issues that still need to addressed carefully in the future to allow SOM-based applications to time series prediction to establish themselves as a rich and mature field of research.

Time Series Clustering - Clustering of the input vectors is an important step in designing SOM-based models for time series prediction. A recent survey by Liao [51] only reports a number of methods to cluster time series data, but fails in providing hints about which one should be preferred for handling a certain time series of interest (I wonder if it is really possible!).

Currently, most of SOM-based TSP models use conventional clustering methods for static (i.e. non-temporal) data (see [6] for a review of these methods). However, it is still not clear how useful temporal information present in time series data can be extracted using such methods[3]. More recently, several temporal vector quantization algorithms have been proposed [54, 39, 55, 56, 57, 58], specially within the field of neural networks, that aim at performing better than static clustering methods when dealing with time series data.

Independently of the type of clustering methods, if static or temporal ones, much work still needs to be developed in this regard, since there is no comprehensive empirical or theoretical study evaluating how sensitive is a certain local time series model to the vector quantization algorithm used to cluster the data.

Multi-Step-Ahead Time Series Prediction - The vast majority of the works described in this paper were concerned in computing one-step-ahead predictions. However, the computation of predictions for larger time horizons is much more challenging. Just a few papers have explored this application using SOM-based TSP models [50, 36, 19, 35], revealing that there is still room for applications and models.

Two approaches have been used for multi-step-ahead TSP using SOM-based models. In the first one, called *recursive prediction* [50], every new one-step-ahead estimate $\hat{x}(t+1)$ is fed back to the input regressor until the desired prediction horizon is reached. In the second one, called *vector forecasting* [36, 19, 35], a whole bunch of future values within a desired horizon is predicted at once. In this case, the one-step-ahead prediction model in Eq. (6.5) is generalized and written as

$$[x(t+1), x(t+2), \ldots, x(t+h)] = f[\mathbf{x}^-(t)] + \varepsilon(t), \qquad (6.38)$$

where $h > 1$ is the desired prediction horizon. To use the VQTAM model for vector forecasting purposes is straightforward as long as we apply the following definitions:

---

[3] There is even a recent controversy on this issue. See [52] and [53] for more detail.

154 Guilherme A. Barreto

$$\mathbf{x}^{out}(t) = [x(t+1), x(t+2), \ldots, x(t+h)] \tag{6.39}$$
$$\mathbf{x}^{in}(t) = [x(t), x(t-1), \ldots, x(t-p+1)] \tag{6.40}$$

Time Series Visualization and Knowledge Discovery - The SOM is particularly well-known in the field of data mining as a nonlinear projection and visualization tool. Thus, an interesting field of application would be temporal knowledge discovery, where the SOM can be used to capture regular behaviors within time series data. There are already some papers on this topic [59, 60, 61, 62, 63], but much more still can be done, specially for the analysis of multivariate time series [64].

Prediction of Multivariate Time Series - A challenging task arises when one is interested in predicting two or more variables at the same time. While gathering the bibliography for this survey, we were unable to find applications of SOM-based models for multivariate time series prediction. Hence, this seems to be an entirely open research field, starving for new models and applications.

6.10 Conclusion

In this paper we reviewed several applications of Kohonen's SOM-based models to time series prediction. Our main goal was to show that the SOM can perform efficiently in this task and can compete equally with well-known neural architectures, such as MLP and RBF networks, which are more commonly used. In this sense, the main advantages of SOM-based models over MLP- or RBF-based models are the inherent local modeling property, which favors the interpretability of the results, and the facility in developing growing architectures, which alleviates the burden of specifying an adequate number of neurons (prototype vectors).

Acknowledgment

The authors would like to thank CNPq (grant #506979/2004-0) and CAPES (PRODOC grant) for their financial support.

APPENDIX

The time series data set $\{x(t)\}_{t=1}^{N}$ used in this paper is generated by numerical integration of the Lorenz dynamical system of equations:

$$\begin{aligned}
\frac{dx}{dt} &= a[y(t) - x(t)] \\
\frac{dy}{dt} &= bx(t) - y(t) - x(t)z(t) \\
\frac{dz}{dt} &= x(t)y(t) - cz(t)
\end{aligned} \tag{6.41}$$

where a, b and c are constants. Using the Euler forward method, a discrete time version of the Lorenz system is given by

$$\begin{aligned} x(t+1) &= x(t) + ha[y(t) - x(t)] \\ y(t+1) &= y(t) + h[bx(t) - y(t) - x(t)z(t)] \\ z(t+1) &= z(t) + h[x(t)y(t) - cz(t)] \end{aligned} \quad (6.42)$$

where $h > 0$ is the step size.

A chaotic time series with $N = 5000$ sample points was generated using $h = 0.01$, $a = 10$, $b = 28$ and $c = 8/3$. The time series was then rescaled to the range $[-1, +1]$. Finally, white gaussian noise, with zero mean and standard-deviation $\sigma_\varepsilon = 0.01$, is added to each sample point, i.e.

$$\tilde{x}(t) = x(t) + \varepsilon(t), \qquad \varepsilon(t) \sim N(0, \sigma_\varepsilon^2). \quad (6.43)$$

The first 4000 sample points $\{\tilde{x}(t)\}_{t=1}^{4000}$ are used for training the SOM-based TSP models, while the remaining 1000 points are used for performance evaluation purposes.

For all the simulations performed in this paper we use a unidimensional SOM architecture. The order of regression is always $p = 5$. For a given number of neurons (q), we train a given SOM-based model for $10 \times q$ epochs. Initial and final neighborhood widths are set to $\sigma_0 = q/2$ and $\sigma_T = 10^{-3}$, respectively. Initial and final values of the learning rate are set to $\alpha_0 = 0.5$ and $\sigma_T = 10^{-3}$, respectively. Weights are randomly initialized.

References

1. Weigend, A., Gershefeld, N.: Time Series Prediction: Forecasting the Future and Understanding the Past. Addison-Wesley (1993)
2. Palit, A., Popovic, D.: Computational Intelligence in Time Series Forecasting: Theory and Engineering Applications. 1st edn. Springer (2005)
3. Kohonen, T.: The self-organizing map. Neurocomputing **21** (1998) 1–6
4. Kohonen, T.: Self-Organizing Maps. 2nd extended edn. Springer-Verlag, Berlin, Heidelberg (1997)
5. Kohonen, T.: The self-organizing map. Proceedings of the IEEE **78**(9) (1990) 1464–1480
6. Xu, R., Wunsch, D.: Survey of clustering algorithms. IEEE Transactions on Neural Networks **16**(3) (2005) 645–678
7. Flexer, A.: On the use of self-organizing maps for clustering and visualization. Intelligent Data Analysis **5**(5) (2001) 373–384
8. Frank, R., Davey, N., Hunt, S.: Time series prediction and neural networks. Journal of Intelligent and Robotic Systems **31**(1-3) (2001) 91–103
9. Zhang, G., Patuwo, B.E., Hu, M.Y.: Forecasting with artificial neural networks: The state of the art. International Journal of Forecasting **14**(1) (1998) 35–62
10. Dorffner, G.: Neural networks for time series processing. Neural Network World **6**(4) (1996) 447–468

11. Pedreira, C.E., Peres, R.: Preliminary results on noise detection and data selection for vector quantization. In: Proceedings of the IEEE World Congress on Computational Intelligence (WCCI'06). (2006) 3617–3621
12. Kohonen, T., Oja, E., Simula, O., Visa, A., Kangas, J.: Engineering applications of the self-organizing map. Proceedings of the IEEE **84**(10) (1996) 1358–1384
13. Oja, M., Kaski, S., Kohonen, T.: Bibliography of self-organizing map (SOM) papers: 1998-2001 addendum. Neural Computing Surveys **3** (2003) 1–156
14. Kantz, H., Schreiber, T.: Nonlinear Time Series Analysis. Cambridge University Press (1999)
15. Box, G., Jenkins, G.M., Reinsel, G.: Time Series Analysis: Forecasting & Control. 3rd edn. Prentice-Hall (1994)
16. Hyndman, R.J., Koehler, A.B.: Another look at measures of forecast accuracy. International Journal of Forecasting **22**(4) (2006) 679–688
17. Fritzke, B.: Unsupervised ontogenetic networks. In Beale, R., Fiesler, E., eds.: Handbook of Neural Computation. IOP Publishing/Oxford University Press (1996) 1–16
18. Fahlman, S., Lebiere, C.: The cascade-correlation learning architecture. Technical Report CMU-CS-90-100, School of Computer Science, Carnegie Mellon University (1991)
19. Barreto, G.A., Araújo, A.F.R.: Identification and control of dynamical systems using the self-organizing map. IEEE Transactions on Neural Networks **15**(5) (2004) 1244–1259
20. Barreto, G., Araújo, A.: A self-organizing NARX network and its application to prediction of chaotic time series. In: Proceedings of the IEEE-INNS Int. Joint Conf. on Neural Networks, (IJCNN'01). Volume 3., Washington, D.C. (2001) 2144–2149
21. Barreto, G.A., Araújo, A.F.R., Ritter, H.J.: Self-organizing feature maps for modeling and control of robotic manipulators. Journal of Intelligent and Robotic Systems **36**(4) (2003) 407–450
22. Farmer, J., Sidorowich, J.: Predicting chaotic time series. Physical Review Letters **59**(8) (1987) 845–848
23. Baumann, T., Germond, A.J.: Application of the Kohonen network to short-term load forecasting. In: Proceedings of the Second International Forum on Applications of Neural Networks to Power Systems (ANNPS'93). (1993) 407–412
24. Lendasse, A., Lee, J., Wertz, V., Verleysen, M.: Forecasting electricity consumption using nonlinear projection and self-organizing maps. Neurocomputing **48**(1-4) (2002) 299–311
25. Lendasse, A., Francois, D., Wertz, V., Verleysen, M.: Vector quantization: a weighted version for time-series forecasting. Future Generation Computer Systems **21**(7) (2005) 1056–1067
26. Göppert, J., Rosenstiel, W.: Topology preserving interpolation in self-organizing maps. In: Proceedings of the NeuroNIMES'93. (1993) 425–434
27. Göppert, J., Rosenstiel, W.: Topological interpolation in SOM by affine transformations. In: Proceedings of the European Symposium on Artificial Neural Networks (ESANN'95). (1995) 15–20
28. Walter, J., Ritter, H.: Rapid learning with parametrized self-organizing maps. Neurocomputing **12** (1996) 131–153
29. Barreto, G., Souza, L.: Adaptive filtering with the self-organizing map: a performance comparison. Neural Networks **19**(6-7) (2006) 785–798

30. Hammer, B., Rechtien, A., Strickert, M., Villmann, T.: Rule extraction from self-organizing networks. Lecture Notes in Computer Science **2415** (2002) 877–883
31. Lehtimäki, P., Raivio, K., Simula, O.: Self-organizing operator maps in complex system analysis. Lecture Notes in Computer Science **2714** (2003) 622–629
32. Malone, J., McGarry, K., Wermter, S., Bowerman, C.: Data mining using rule extraction from Kohonen self-organising maps. Neural Computing & Applications **15**(1) (2005) 9–17
33. Drobics, M., Bodenhofer, U., Winiwarter, W.: Mining clusters and corresponding interpretable descriptions - a three-stage approach. Expert Systems **19**(4) (2002) 224–234
34. Pal, N.R., Laha, A., Das, J.: Designing fuzzy rule based classifier using self-organizing feature map for analysis of multispectral satellite images. International Journal of Remote Sensing **26**(10) (2005) 2219–2240
35. Simon, G., Lendasse, A., Cottrell, M., Fort, J.C., Verleysen, M.: Time series forecasting: obtaining long term trends with self-organizing maps. Pattern Recognition Letters **26**(12) (2005) 1795–1808
36. Simon, G., Lendasse, A., Cottrell, M., Fort, J.C., Verleysen, M.: Double quantization of the regressor space for long-term time series prediction: method and proof of stability. Neural Networks **17**(8-9) (2004) 1169–1181
37. Lehtokangas, M., Saarinen, J., Kaski, K., Huuhtanen, P.: A network of autoregressive processing units for time series modeling. Applied Mathematics and Computation **75**(2-3) (1996) 151–165
38. Vesanto, J.: Using the SOM and local models in time-series prediction. In: Proceedings of the Workshop on Self-Organizing Maps (WSOM'97), Espoo, Finland (1997) 209–214
39. Koskela, T., Varsta, M., Heikkonen, J., Kaski, S.: Time series prediction using recurrent SOM with local linear models. International Journal of Knowledge-based Intelligent Engineering Systems **2**(1) (1998) 60–68
40. Sfetsos, A., Siriopoulos, C.: Time series forecasting with a hybrid clustering scheme and pattern recognition. IEEE Transactions on Systems, Man, and Cybernetics **A-34**(3) (2004) 399–405
41. Poliker, S., Geva, A.: Hierarchical-fuzzy clustering of temporal-patterns and its application for time-series prediction. Pattern Recognition Letters **20**(14) (1999) 1519–1532
42. Sanchez-Maroño, N., Fontela-Romero, O., Alonso-Betanzos, A., Guijarro-Berdiñas, B.: Self-organizing maps and functional networks for local dynamic modeling. In: Proceedings of the European Symposium on Artificial Neural Networks (ESANN'95). (2003) 39–44
43. Pavlidis, N.G., Tasoulis, D., Plagianakos, V., Vrahatis, M.: Computational intelligence methods for financial time series modeling. International Journal of Bifurcation and Chaos **16**(7) (2006) 2053–2062
44. Walter, J., Ritter, H., Schulten, K.: Non-linear prediction with self-organizing map. In: Proceedings of the IEEE International Joint Conference on Neural Networks (IJCNN'90). Volume 1. (1990) 587–592
45. Stokbro, K., Umberger, D.K., Hertz, J.A.: Exploiting neurons with localized receptive fields to learn chaos. Complex Systems **4**(3) (1990) 603–622
46. Martinetz, T.M., Berkovich, S.G., Schulten, K.J.: Neural-gas network for vector quantization and its application to time-series prediction. IEEE Transactions on Neural Networks **4**(4) (1993) 558–569

47. Fritzke, B.: Incremental learning of local linear mappings. In: Proceedings of the International Conference On Artificial Neural Networks (ICANN'95). (1995) 217–222
48. Tong, H.: Non-linear Time Series: A Dynamic System Approach. Oxford University Press (1993)
49. Barreto, G., Mota, J., Souza, L., Frota, R.: Nonstationary time series prediction using local models based on competitive neural networks. Lecture Notes in Computer Science **3029** (2004) 1146–1155
50. Principe, J.C., Wang, L.: Non-linear time series modeling with self-organizing feature maps. In: Proceedings of the IEEE Workshop on Neural Networks for Signal Processing (NNSP'95). (1995) 11–20
51. Liao, T.W.: Clustering of time series data - a survey. Pattern Recognition **38**(11) (2005) 1857–1874
52. Keogh, E., Lin, J., Truppel, W.: Clustering of time series subsequences is meaningless: implications for previous and future research. In: Proceedings of the 3rd IEEE International Conference on Data Mining. (2003) 115–122
53. Chen, J.: Making subsequence time series clustering meaningful. In: Proceedings of the 5th IEEE International Conference on Data Mining. (2005) 1–8
54. Chappell, G.J., Taylor, J.G.: The temporal Kohonen map. Neural Networks **6**(3) (1993) 441–445
55. Barreto, G.A., Araújo, A.F.R.: Time in self-organizing maps: An overview of models. International Journal of Computer Research **10**(2) (2001) 139–179
56. Principe, J., Euliano, N., Garani, S.: Principles and networks for self-organization in space-time. Neural Networks **15**(8-9) (2002) 1069–1083
57. Voegtlin, T.: Recursive self-organizing maps. Neural Networks **15**(8-9) (2002) 979–991
58. Baier, V.: Motion perception with recurrent self-organizing maps based models. In: Proceedings of the 2005 IEEE International Joint Conference on Neural Networks (IJCNN'05). (2005) 1182–1186
59. Van Wijk, J., Van Selow, E.: Cluster and calendar based visualization of time series data. In: Proceedings of the IEEE Symposium on Information Visualization (InfoVis'99). (1999) 4–9
60. Fu, T.C., Chung, F., Ng, V., Luk, R.: Pattern discovery from stock time series using self-organizing maps. In: Workshop Notes of KDD'2001 Workshop on Temporal Data. (2001) 27–37
61. Chicco, G., Napoli, R., Piglione, F.: Load pattern clustering for short-term load forecasting of anomalous days. In: Proceedings of the 2001 IEEE Porto Power Tech Conference (PPT'01). (2001) 1–6
62. Tsao, C.Y., Chen, S.H.: Self-organizing maps as a foundation for charting or geometric pattern recognition in financial time series. In: Proceedings of the IEEE International Conference on Computational Intelligence for Financial Engineering. (2003) 387–394
63. Lin, J., Keogh, E., Lonardi, S.: Visualizing and discovering non-trivial patterns in large time series databases. Information Visualization **4**(3) (2005) 61–82
64. Guimarães, G.: Temporal knowledge discovery for multivariate time series with enhanced self-organizing maps. In: Proceedings of the IEEE-INNS-ENNS International Joint Conference on Neural Networks (IJCNN'00). (2000) 165–170

7

A Dual Interaction Perspective for Robot Cognition: Grasping as a "Rosetta Stone"

Helge Ritter, Robert Haschke, and Jochen J. Steil

Neuroinformatics Group, Bielefeld University, Germany
{helge,rhaschke,jsteil}@techfak.uni-bielefeld.de

Summary. One of the major milestones to higher cognition may be the ability to shape movements that involve very complex interactions with the environment. Based on this hypothesis we argue that the study and technical replication of manual intelligence may serve as a "Rosetta Stone" for designing cognitive robot architectures. The development of such architectures will strongly benefit if improvements to their interaction capabilities in the task domain become paired with efficient modes of interaction in the domain of configuring and restructuring such systems. We find that this "dual interaction perspective" is closely connected with the challenge of integrating holistic and symbolic aspects of representations. In the case of grasping, this requires a very tight marriage between continuous control and more discrete, symbol-like representations of contact and object states. As a concrete implementation, we propose a two layered architecture, where the lower, subsymbolic layer offers a repository of elementary dynamical processes implemented as specialised controllers for sensori-motor primitives. These controllers are represented and coordinated in the upper, symbolic layer, which employs a hierarchy of state machines. Their states represent semantically related classes of dynamic sensori-motor interaction patterns. We report on the application of the proposed architecture to a robot system comprising a 7-DOF redundant arm and a five-fingered, 20-DOF anthropomorphous manipulator. Applying the dual interaction approach, we have endowed the robot with a variety of grasping behaviours, ranging from simple grasping reflexes over visually instructed "imitation grasping" to grasping actions initiated in response to spoken commands. We conclude with a brief sketch of cognitive abilities that we now feel within close reach for the described architecture.

7.1 Introduction

One of the most remarkable feats of nature is the evolution of the cognitive abilities of the human brain. Despite the use of neurons which are – compared to technical transistor devices – slow, of low accuracy and of high tolerance, evolution has found architectures that operate on highly complex perception and control tasks in real time. These by far outperform our most sophisticated technical solutions not only in speed, but also in robustness and adaptability.

What can we learn from these systems in order to replicate at least partially some of their enticing properties, bringing robots (or other artificial systems) closer to a level that might deserve the term "cognitive"?

Brains seem always intimately connected with action and change: as soon as evolution started to endow organisms with modestly rich capabilities of motion, we also see the advent of nerve cells that are invoked in the organisation of such motion. Ordering organisms according to the sophistication of their self-controlled movements suggests a strong correlation with the complexity of their brains: "simple" central pattern generators of a handful of cells for generating periodic movements adequate for a homogeneous environment such as water, added sensorimotor-complexity in the form of a cerebellum for coping with the more varying demands of walking on a solid ground, the advent of cortex for even higher flexibility, enabling exploration and context-specific assembly of complex motor programs that later can become automated by learning. Finally there is generalisation of manipulation of physical objects and tool use to the domain of "mental objects" in one's own brain or in the brain of a conspecific: communication and thinking.

We may look at this evolution from at least two perspectives: the first perspective is that of neurophysiology, trying to understand the evolution of the physical structure of the brain. Since single neurons are the elements that can be studied best, this approach faces a strong bottom-up bias, confronting us with a task that is at least as monumental as reverse-engineering the architecture of the operating system of a computer when we have access to a small subset of its transistors. The increasing availability of more global windows into the brain begins to enrich our picture also from a top-down perspective to the extent that we can devise clever experiments that tell us more than just where a brain function seems to be located.

A totally different approach results when we step back from the brain's hardware and instead take the perspective that our main concern is the *functional architecture* needed for a cognitive system. Then we look at the issue of implementation only to the extent that it may pose significant constraints of what architectures are realisable at all. If this route can be made productive, the outcome will be functional invariants that characterise cognitive systems per se. These invariants possibly have very different mappings onto actual hardware, depending on whether the target system employs neural tissue, silicon processors or even some still undiscovered technology of the future.

7.2 Grasping as a "Rosetta Stone" for Cognitive Robots

Following this route, an obvious question to ask is: can we find a *key problem* that is more manageable than the whole, but sufficiently rich to provide essential insights into the architectural principles enabling natural cognition? We believe that there is an affirmative answer to this question: *The study and technical replication of manual intelligence* can be regarded as a good "Rosetta Stone" for designing an architecture for cognitive robots.

Using our own hands, we hardly notice the complexity of the involved processes: depending on the presence of other objects and our intended action, we select an appropriate approach direction and pre-shaping of our hand, initiate finger contact and create new contacts while closing our fingers. We adjust our finger forces and positions according to the object properties perceived during our grasp, and we break existing and create new contacts for controlled re-grasping.

From an abstract perspective, manipulation involves the structuring of a complex physical interaction between a highly redundant, articulated manipulator and a physical object, which can be highly variable. Dextrous manipulation is a form of mechanical control that is pervaded by frequent discontinuous changes of the contact topology between the actuator and the object. At the same time, these contact switches are accompanied by significant uncertainties in geometrical and physical parameters of the object (and, in robot reality, of the actuator too!). As a result, dextrous manipulation calls for an unusually tight marriage between continuous control and more discrete, symbol-like representations that can deal with the issues of topology-switching and encapsulation of parameter uncertainty. Viewed from the perspective of architecture research, it is this tight coupling of continuous, sensorimotor control and discrete, symbol-like representations that seems to be a major preparatory step for realising system structures which can finally lead to cognition with its remarkable integration of holistic and symbolic processing.

Once the necessary control architecture capable of controlling complex *external physical states* was in place, it did not really matter whether the controlled state referred to physical objects or to more abstract entities. One may speculate, that this control machinery was first applied to the organisation of the brain itself, finally enabling self-awareness [1]. From an architectural point of view, this would have required the creation of some "internal interface", making a subset of the brain's subsymbolic activity available in a similar more "tangible" fashion as physical objects outside. Once such an interface had become available, the control capabilities could become "elevated" with relatively small effort from the physical to the *mental domain*, leading to thinking and reasoning, and a new, consciously accessible "symbolic" representation of oneself. From this point, it appears no longer as a distant step to extend this representation also to other individuals, ultimately refining the manipulation of mental objects to enable coordination of "mental state objects" across individuals: communication.

Although speculative, from a functional point of view the drawn picture offers a very natural road map for the generalisation of movement control sophistication. In this picture, hands and manipulation have the pivotal role of preparing the transition from the organisation of more rigid behaviour to mental operations and thinking. Therefore, the synthesis of dextrous manipulation

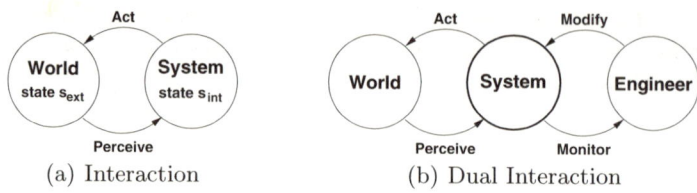

Fig. 7.1. Interaction perspective on architectures. Left (a): Classical interaction scheme, focusing on task-interaction between world and system only. Right (b): The dual interaction perspective takes a second branch of "config-interaction" between engineer and system into account.

skills for robots appears to have great promise for driving us towards "the right architecture" to finally master the step to higher levels of cognition.

7.3 A Dual Interaction Perspective on Artificial Systems

The grasping scenario focuses our efforts on a key task: to organise a usually complex *interaction pattern* with the environment. Here, the notion of interaction pattern refers to the *overall behaviour* of the system due to different situations in the world. This requires to select adequate sensors and motor actions from a broad repertoire of elementary behaviour patterns or skills in order to realise a highly structured, situation-dependent coupling between selected regions of the world and the sensors and actuators of the robot or organism.

From a formal point of view, any architecture for the interaction between a robot (or any other technical or biological system) and the world implements a system of the form depicted in fig. 7.1(a), where the state of the world and its change due to actions are at best partly known. The aim of an architecture then is two-fold: first, to implement the system as the result of interactions between more elementary subsystems, and second, to shape the *overall behaviour* of the system in such a manner that an adequate, task- and situation-dependent interaction pattern results. The subsystems of the architecture should be organised in a modular and hierarchical manner, and they should obey a parsimonious set of carefully selected design principles, requiring the recurring application of a restricted set of design patterns at different levels, leading to a partial self-similarity of the architecture design.

So far, our discussion was focused on the "task-interaction" between the system and the world in its application domain. However, as an artefact, any technical system inevitably is involved in a second sphere of interaction, namely the interaction with its human developer(s). We may distinguish this second type of interaction with the term *config-interaction*. Current architectures of cognitive systems are largely blueprinted by humans. Engineers have to interact with the architecture during development, tuning and maintenance which is heavily facilitated by a well-structured and easily accessible config-interaction.

As seen from fig. 7.1(b), perception and action flow is reversed for config-interaction: Actions are inwards directed and have the goal to alter the structure of the system, perception is outwards directed and has the aim to provide system monitoring. Both interaction types strive for contrary representations of the internal system state. While config-interaction has a strong bias to well-structured, symbolic representations which are easily comprehensible by humans, task-interaction is designed towards successful system behaviour in its application context and hence may prefer a distributed, less structured coding.

Despite these differences, task- and config-interaction share an important commonality: Both of them support a form of human-machine cooperation towards certain goals using a range of actions in which the system has to participate. (While we take this active view for granted in the case of task-interaction, it is not yet wide-spread when considering config-interaction. However, we anticipate that "cooperative" monitoring and reconfiguration interfaces will gain importance as systems become more and more complex.)

Embracing config-interaction as an important factor that may affect architecture design as much as task-interaction leads us to a *dual interaction perspective* on architectures for artificial cognitive systems. It emphasises the need to explicitly consider the development of a proper architecture as a coupled optimisation problem in which the constraints resulting from the dual goals of successful task-interaction and easy config-interaction compete and may exert partly opposing forces on potential solutions.

A first area of "opposing forces" concerns system interconnectivity. Experience teaches us that most architectures – due to their fixed modular structure – lack a flexible interconnection between the low- and high-level cognitive processes. A strong interconnection would allow to employ dynamic and context-specific representations of symbolic and subsymbolic information between different involved processes as well as a fine-grained top-down control of low-level information processing. Both capabilities are necessary for a cognitive system to flexibly adapt itself to changing task requirements. Contrary, a strong interconnection complicates the view for the engineer. The example of artificial neural networks shows, that good human readability and maintenance (serving good config-interaction) does not by itself follow from functional efficiency (serving good task-interaction), while some early approaches of classical AI provide examples that the opposite conclusion does not work either.

Another typical drawback of classical system approaches is their inflexible organisation of process flow which is mostly hand-coded statically into the system. This typically causes system deadlocks due to unforeseen system dynamics and at the same time prohibits dynamical extension through online acquisition of new skills and control patterns. Nevertheless, we think that this capability of online learning on a high system level is a major prerequisite for autonomous cognitive systems which interact with humans in their environments. On the other hand, we foresee that autonomous learning may serve "task-interaction", but erode "config-interaction" by changing the inner working of the system in a way that is not fully comprehensible for the human engineer.

As we see, with a significant number of specialised processing modules at hand, topics of communication and coordination between components, as well as integration and fusion of information become major issues in the design of complex cognitive systems. To date, use cases of highly integrated cognitive systems running in a real-world scenario and in close interaction with the human while integrating a large number of different system behaviours in a unified framework are only sparsely reported. This reflects the difficulty to simultaneously serve the demanding goals of task- and config-interaction. A few examples of such systems acting in a robotics or pure vision domain can be found in [2, 3, 4, 5]. Some of our own earlier work along these lines is summarised in [6, 7]. Experience from this work led to the dual interaction perspective described above and to ideas for its implementation in a compact and highly manageable architecture employing hierarchical state machines (HSMs). Before describing our approach, therefore, it seems appropriate to insert a short technical section motivating and explaining some HSM basics. Readers familiar with HSMs can directly proceed to the next section.

7.4 Hierarchical State Machines

Typically the correct action in response to an event depends not only on the event itself but also on the internal state of various submodules which in turn evolve depending on the input history of the whole system. It is one major task of large reactive systems to monitor this current state and finally choose the right decisions and actions if a new event has to be handled. Actually, this decision process establishes the behaviour of the system.

From a programming perspective, the dependence on the event history very often leads to deeply nested if-else or switch-case constructs. Most reactive programs start out fairly simple and well structured, but as more features and behaviours are added, more flags and variables have to be introduced to capture the relevant state and event history such that the logical expressions formed from the various variables become increasingly complex. Therefore it is much more elegant to express the behaviour using more abstract programming tools, e.g. finite state machines [8] or petri-nets [9]. Their deployment can drastically reduce the number of execution paths through the code, simplify the conditions tested at each branching point, and simplify transitions between different modes of execution. Most important, its states and transitions can be visualised concisely.

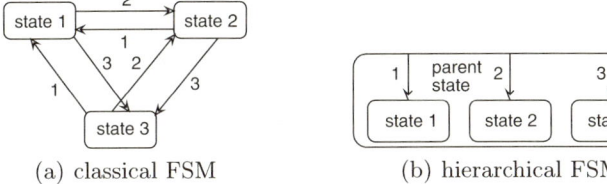

Fig. 7.2. Due to their hierarchically nested states and inheritance of state transitions from parent to child states, simple finite state machines (a) can be expressed much more concisely using hierarchical state machines (b).

One major drawback of both finite state machines and petri-nets is the quadratical increase of state transitions w.r.t. the number of states. Take for instance a default "error" state of a component, which is targeted from every possible state in case of an error. In this case the transition with its associated action(s) has to be duplicated for every *source*-state, although essentially all transitions are identical. Clearly, the structure and behaviour of such FSMs is poorly represented in their state transition diagrams and the large number of transitions tends to obscure any visualisation and makes their implementation an error-prone process.

Additionally, classical FSMs suffer from another major and closely related problem: To express an infinite, or at least large number of states, an appropriate number of states has to be introduced, e.g. a vending machine needs a particular state for every possible amount of money already inserted. Nevertheless, the logical state "waiting for money" is the same for all these states. To simplify the state diagram it is useful to augment the state machine with so called extended state variables, which can be used in arithmetic and conditional expressions just like the state flags before.

However, the borderline between real states and extended state variables is blurred, because any extended state variables can be expressed as (a possibly large) number of real states and vice versa. Hence, a tradeoff must be found keeping conditional expressions in state transitions simple and using extended state variables to combine similar behaviour. Practical experience shows, that in a complex system extended variables are not sufficient to avoid a "state explosion", e.g. the introduction of a new state for each semantically different hand movement in the robotics grasping scenario introduced later.

In practice, many states are similar and differ only by a few transitions. Therefore, what is needed is an *inheritance mechanism for transitions*, analogous to the inheritance of attributes and methods that is familiar from object oriented programming. The formalism of statecharts, proposed in [10] and extended in [8], provides exactly that: a way of capturing the common behaviour in order to share it among many states. Its most important innovation is the introduction of hierarchically nested states, which is why statecharts are also called hierarchical state machines (HSMs). As an example consider the hierarchical state diagram shown in fig. 7.2b. Here all states of the simple FSM of fig. 7.2a are embedded in a grouping parent state, which defines the common state transitions inherited by all sub-states. Consider for example state 1 getting event 3 which should trigger a transition to state 3 regardless of the current state. Because state 1 doesn't handle this event explicitly, the transition rules of the super-state apply, which in turn provoke the intended transition to state 3. This transition inheritance mechanism tremendously facilitates the proper structuring of transitions, since now sub-states need only define the differences from their super-states. By the same token, *state differentiation* (splitting an existing state into several slightly different states if additional information becomes available to the system) is facilitated, making incremental system development an easier process. In the context of robot grasping, for example all grasp postures can be grouped together and inherit from its super-state a transition to an "error" state, which also triggers the actuation of a relax posture.

7.5 A HSM-based Architecture for Grasping

In the following, we describe our approach to an architecture for grasping which allows both, a tight interconnection and coordination between subprocesses as well as a highly structured and modular design of the system and its state representation. We employ dynamically configurable hierarchical state machines, which reflect the high-level, symbolic system state and coordinate several behaviour controllers directly interfacing to the world on a subsymbolic level. Hence the approach tries to consider the opposing constraints of the task- and config-interaction.

Due to their symbolic nature, several HSMs can be flexibly connected by a suitable communication framework. Here, we make use of the excellent extensibility and human-readability of XML-based representations and employ a very flexible communication layer (AMI, cf. Sec. 7.7) that has been developed previously to support distributed processing for interactive systems. Continuing to take grasping as our "Rosetta stone", we will explain the main ideas in this context.

Once the external conditions (object location and shape, possible obstacles around) have been determined, e.g. by a suitable vision front-end, the core part of the grasping process involves a coordinated movement of arm and hand in order to grasp the object. Our architecture distinguishes two separate layers of processing (see fig. 7.3):

Controllers. Both the reaching movement of the arm and the actuation of the hand can be regarded as the evolution of dynamical systems constituted by controllers which implement a specific low-level behaviour. Each controller binds to a specific subset of sensory inputs and motor outputs. For example, a controller to actuate given joint angles would bind to angle sensors and joint motors. To actuate postures specified in Cartesian coordinates another

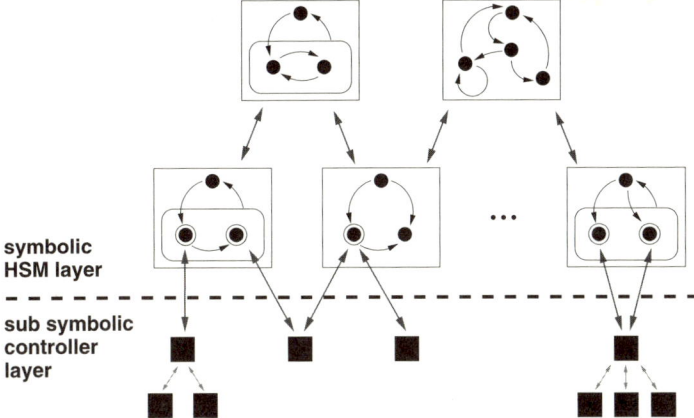

Fig. 7.3. Sketch of the system architecture comprising a high-level, symbolic HSM layer and a low-level, subsymbolic controller layer. Controller states within the HSM (indicated as double circles) are bound to a particular set of controllers. When entered, the state configures its associated controllers through parameterised events and then awaits termination of the process.

controller would bind to the end-effector position instead. To implement a visual servoing controller, one would bind the Cartesian target positions to the output of a vision-based tracker of the end-effector. Each controller might offer a parameterisation interface, e.g. to change the default execution velocity, smoothness or accuracy of a motion. However, the most typical parameterisation is the actually targeted posture of the robot.

HSMs. While controllers provide elementary behavioural skills on a subsymbolic level, hierarchical state machines provide a symbolic abstraction layer which allows a clearly structured, *semantic* view of the interaction realised by the controllers. States represent different behavioural modes of the system. In particular, states can be bound to low-level controllers, meaning that the associated interaction pattern is currently being executed and the HSM is waiting for its termination. Each of these *controller states* maintains a fixed selection of active controllers bound in a particular fashion to relevant sensory inputs and motor actuators.

Transitions between states are triggered by internal or external events which may carry optional parameters to transmit additional information, e.g. the object type or required hand posture for a GRASP event. Besides the state change itself, each transition can evoke *actions*, e.g. the generation of *parameterised events* to trigger transitions in concurrent HSM processes or to parameterise and (de)activate involved controller(s).

All actions associated to a transition are executed and finished before the next event is processed by an HSM. This avoids race conditions in a simple and efficient manner, because actions cannot be interrupted by concurrent process activity. To ensure high responsiveness of the system to human instructions or urgent events, all actions should have a short execution period. Anyway, to allow long-running remote procedure calls, we generate a parameterised event to transmit the procedure request and enter a state waiting for the termination event of the triggered activity. In the meantime we can also respond to other incoming events to react e.g. to emergency conditions.

The proposed two-layer architecture provides a clear separation between the *semantic structure* of the interaction dynamics and its subsymbolic parts encapsulated at the level of controller(s). On the one hand, this permits to easily upgrade controllers by more sophisticated ones, e.g. replacing an initial PID controller by a hierarchical controller employing adaptive neural networks.

On the other hand, state nesting and transition inheritance permit a flexible top-down and incremental shaping of the task-interaction at a semantic level: Whenever a task requires to refine an existing task-interaction pattern into a number of more specialised patterns, we can express this at the state level by introducing a corresponding number of (mutually exclusive) sub-states. As explained in the previous section, each sub-state can inherit all properties of its parent state, so that we only need to specify the "delta part" of its more specialised interaction dynamics.

Connecting a HSM in the described way to a layer of low-level controllers, the HSM performs the usually highly non-trivial task of "glueing together" a set of (possibly complex) dynamical systems. While the HSM representation alone gives no theoretical foundation of how to achieve a desired combination behaviour, *it does* provide a very powerful and structured interface for the config-interaction with the human designer. This largely facilitates to explore and tune different alternatives and, in this way, to arrive at a robust solution in an "evolutionary" manner.

7.6 Application in an Imitation Grasping Scenario

In the concrete example of the coupled hand-arm system, the hand subsystem is realised by the HSM depicted in fig. 7.4. The `handPosture` state corresponds to an actuation of the hand using a particular controller. It is differentiated into three sub-states for attaining a pre-grasp posture (`pre-grasping`), performing finger closure from a pre-grasp to a target grasp posture (`grasping`), and for opening the hand (`stretching`). Transitions into any of these sub-states share the emission of a parameterised `TARGET` event to the hand controller, which provides the targeted hand posture and suitable joint stiffness values.

The distinction between different hand postures within these sub-states is achieved by appropriately parameterised events, i.e. the `PRE-GRASP` and `GRASP` events carry information about the concrete grasp type to be used. This information is finally forwarded to the controller as the targeted posture. Instead of introducing new sub-states for different postures (which would complicate the state diagram) the three employed states only capture the *semantic state*

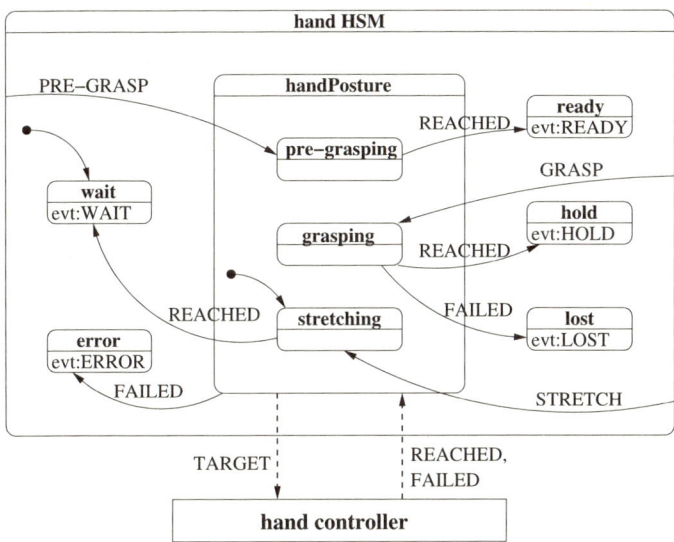

Fig. 7.4. HSM-implementation of the hand control layer: entry actions marked as "evt: ..." generate appropriate events within the overall system, which in this case trigger state transitions in the arm control layer (see fig. 7.5).

of the grasping process. Arcs emanating from a black dot, indicate initial transitions to a more detailed sub-state, if the parent state is entered directly.

A second, similar HSM is used to shape the high-level structure of the arm subsystem. Here, sub-states of an embracing moving state represent semantically different movement strategies of the arm, i.e. movements w.r.t. different coordinate systems, and a planned arm motion avoiding collisions. Again, different arm postures are specified through parameterised events to the HSM states as well as the associated controllers.

Once each dynamical subsystem has been packaged as a HSM box, we arrive at the task of coupling a number of such processes. Here again, we can use event communication between the HSM-subsystems in order to coordinate their concurrent activity at the next higher integration level. Inevitably, at this level we are no longer spared from the enormous complexity of coordinating complex concurrent processes. However, having cleverly deployed the power of HSMs to flexibly coordinate a small number of (controller) processes in each HSM box already and providing the resulting subsystems with judiciously designed interfaces, the task of their coordination at the next higher level becomes much more manageable.

As a concrete example, fig. 7.5 shows the organisation of the grasping sequence based on two cooperating HSMs for the hand and arm control system. Our implementation of the grasping sequence starts with an initial manipulator movement to a coarsely determined offset position above the object. The according parameterisation event, called TAKEOBJECT, is generated by a

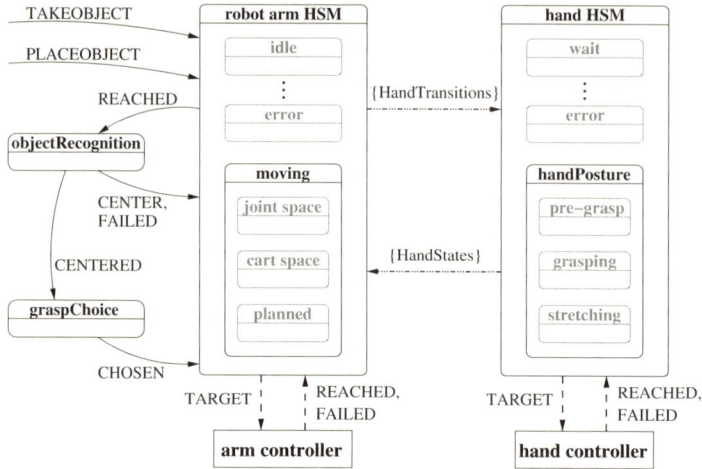

Fig. 7.5. Implementation of the grasping behaviour employing two HSMs running as separate processes and communicating via events.

coordination instance if a grasping command was articulated and the 3D visual object detection has recognised an appropriate object. To disambiguate between similar objects on the table, the grasping region might be restricted further by a pointing gesture. Due to the limited spatial resolution in object detection, a subsequent fine-positioning above the grasping target is required. Hence, the manipulator is centred above the object employing a local, visually driven feedback loop alternating between the states objectRecognition and armMovement until the visual centre of gravity assumes a predefined position.

Next an appropriate grasp prototype has to be selected. Following Cutkosky's taxonomy of human prehensile grasps [11] we implemented four basic grasp prototypes: "two finger pinch", "two finger precision", "all finger precision", and "power grasp". The appropriate prototype is selected either based on previously acquired knowledge "How to grasp a specific class of objects" or according to a visual observation of an instructing human hand posture, which will be associated to the given object for later reuse. Depending on the selected grasp prototype, the robot hand has to be aligned along some particular axis – typically the main axis of the object – and an initial hand posture is achieved.

When this pre-grasp posture has been reached, we initiate a purely haptically driven finger closure phase. Empirically we found, that by careful fine-tuning of the four employed grasp prototypes, the described grasping architecture can capture the relevant "reactive grasp knowledge" sufficiently well to allow successful grasps for a considerable range of everyday objects [6]. Many alternative approaches to grasping put a high emphasis on rather elaborate grasp planning [12, 13, 14], achieved with sophisticated optimisation methods to find optimal contact points [15, 16], and with very precise joint position control. In this view, we find it remarkable that our approach is quite reliable for a large variety of objects and parametrically robust against uncertainties and small errors in the localisation of the object, the actuation of a particular hand posture, and the controller parameters.

Due to its flexibility, the present grasping architecture could be enhanced in many ways. One promising direction will be the discrimination of a larger number of different contact states between fingers and object, and modifying the closure synergy accordingly. Such control switches, but also initiation and termination of the whole process, as well as diagnosing failure and error conditions, and, possibly, triggering of suitable response actions, can all be rather easily adopted by adding the required interaction patterns in the form of additional states to the HSM depicted in fig. 7.5.

7.7 Managing Interconnectivity

So far, we have focused only on the hand-arm component of our robot system in order to illustrate the basic ideas behind our architecture approach. Evidently, the core hand-arm behaviours have to be embedded into a network of additional skills in order to coordinate grasping actions with vision and spoken commands. At this point, it may have become clear how the described HSM-based interaction architecture can be extended to include additional skill modules, each consisting of a cluster of related dynamical low-level controllers encapsulated as a HSM box which offers a semantic interface to the engineer for tailoring the overall behaviour of the system.

However, when tying an increasing number of HSM subsystems together, a second key issue for a complex cognitive system comes into view: how to realise and manage *a high degree of interconnectivity* between a large number of functional elements that must operate concurrently and in a largely asynchronous manner. (This seems to have also been a significant challenge for the brain: The most part of its volume is taken by the *connections between* neurons, not the neurons themselves.)

Having encapsulated the subsymbolic layer of our robot system within HSM boxes with a flexible coarse-grained interface, we may hope to escape the necessity to implement a similarly fine-grained interconnectivity as encountered in the brain. Instead, we can hope to exploit the convenience of communication via parametrised events in order to manage the communication between even a large number of modules.

Working with a technical system, we again face "evolutionary pressures" arising from the config-interaction. A major issue is system maintainability, which has several facets in its own. First of all, the evolution of a large system progresses slowly and design decisions on data representations made in the early development process might turn out to be not powerful enough for subsequently added applications. As a consequence, data formats have to be extended, which in classical approaches using binary data formats requires

```
<OBJECT>
  <CLASS reliability="0.87">
    cup
  </CLASS>
  <CENTER x="32" y="44"/>
  <REGION>
    <RECT x="12" y="30"
          w="40" h="80"/>
  </REGION>
  <IMAGE uri="xcf://img/123"/>
</OBJECT>
```

```
<EVENT name="RobotCommands">
  <SELECT robot="LeftArm"/>
  <MAXSPEED xyz="200" ypr="50"/>
  <POSTURE> home </POSTURE>
  <MOVETO mode="relToTool">
    <POS> 10 0 0 </POS>
    <EULER> 90 35 0 </EULER>
  </MOVETO>
  <NOTIFY id="GraspModule"/>
</EVENT>
```

Fig. 7.6. Two examples of XML messages used within the robotics system. Their semantics is easily recognised. While the left message shows the output of an object classifier, the right message shows a commands script accepted by robots.

modifications of all affected processes in turn, thus causing a considerable amount of programming effort. Instead, a data format is desirable which can be easily extended on demand without the need to modify the overall system. Second, for debugging purposes system developers should be able to easily inspect the data flow within the system in an online fashion during system operation, or to record these data for an offline evaluation.

These requirements call for an embedding of the "atomic" event format into a coherent communication framework based on a human-readable data format which should be easily extensible and simultaneously can ensure type safety in a scalable manner. Additionally, we anticipate the need to accumulate and interact with significant databases for storing episodes or general semantic world knowledge about objects and actions.

Looking for a suitable framework fulfilling these needs, we found the *Active Memory Infrastructure*[1] (AMI) [17] ideally suited for the task. The AMI framework is based on the *XML based Communication Framework* (XCF) [18] which provides an easy-to-use middleware supporting synchronous and asynchronous remote method invocation (RMI) as well as publisher-subscriber communication semantics. Exposed methods of server processes are bound and invoked dynamically on client side, with XML schemas optionally providing runtime type safety of exchanged parameters. In the same manner messages published via streams can be checked for correctness.

On top of the communication framework, an *Active XML Memory* provides a flexible basis for coordination and shared data management. First of all, the Active XML Memory provides the memory needed by cognitive systems to track a complex interaction context in terms of perceived objects, commands and episodes as well as to persistently store long-term knowledge acquired during learning. Processes can insert XML documents together with attached binary data into the database and thus provide information

---

[1] available for download at http://xcf.sourceforge.net

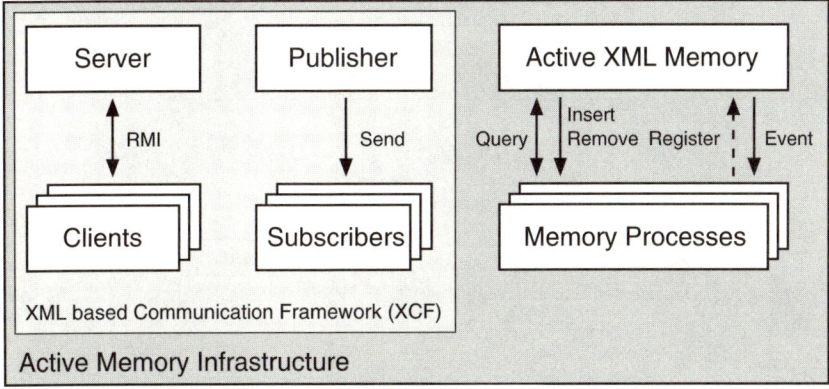

Fig. 7.7. The XML-based Communication Framework (XCF) provides distributed process communication including 1:n stream semantics as well as n:1 remote method invocations (RMI). The Active XML Memory build upon XCF supplements persistent (and transient) XML memory as well as event-triggered notification.

for subsequent use by other processes. To selectively retrieve documents by their semantic structure or content the framework allows to specify XPath expressions, e.g. of the form "/OBJECT[@reliability>0.8]".

Basic coordination between components is provided by a flexible event-notification mechanism extending the mere storage functionality of the database. Processes interested in particular documents inserted into (or removed from) the database can subscribe to an insertion (or removal) event. Again, interesting documents are selected by specifying a restricting XPath expression. If a matching XML document is inserted (or removed), the Memory *actively* transmits this document to the subscriber.

The framework supports all the necessary processes and primitives to run an *active blackboard* where system components can subscribe to XML messages inserted into the database, or pasted on the blackboard to keep this picture. If any other component inserts a matching XML message, the subscriber is automatically (and asynchronously) notified about this event. Coordination thus matches ideally with the event format of the HSM layer and is not bound to explicit links between a fixed set of components present in the system.

7.8 Towards Higher Cognition

The AMI-framework can be easily employed to address issues such as information fusion from different modalities or active forgetting of information that is no longer needed. For example, Wrede et al [17, 19] employ a dedicated information fusion process to integrate unreliable 3D position information over time and from different low-level recognition processes to arrive at more robust hypotheses about positions of visually recognised objects.

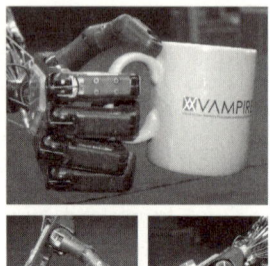

Fig. 7.8. Considered scenario for imitation grasping integrating a 7-DOF Mitsubishi robot arm, a 20-DOF dextrous robot hand and a 4-DOF stereo camera head.

Starting from a previous robotics setup developed in the course of the special research unit "SFB 360" and providing a large number of specialised processing modules [20], we have implemented a robot system (fig. 7.8) whose grasping abilities connect sensori-motor control with vision and language understanding. In reaction to a spoken command, the robot can grasp a variety of every-day objects from a table [21]. By integrating a special vision module tailored specifically to the classification of human hand postures, a human instructor can teach the robot to grasp new objects by showing an appropriate grasp posture with his own hand. This posture is recognised and classified as one of the four grasp prototypes by the visual system. Currently, we integrate other previously developed processing modules, which allow the robot to pay attention to objects on the table and to interpret pointing gestures to disambiguate between similar objects by further narrowing the region of interest.

The implemented system can be seen as a rather large-scale exercise in coordinating the modality- and context-specific interaction patterns of a significant number of low-level subsymbolic processes packaged into HSM boxes with a semantic interface at a symbolic level. To coordinate these boxes, as well as data from a number of additional processes directly publishing to the AMI blackboard, once again the technique of a coordinating HSM is adopted (see fig. 7.9).

In view of this coordinating HSM instance, all subprocesses – as well implemented as HSMs – can be treated like a "pseudo-controller" providing some specialised functionality: moving an arm, grasping, recognising objects, finding or tracking a hand, looking for salient points, listening to user commands. Hence, they can be represented in the coordinating state machine as a kind of "skill states", which have to be orchestrated to achieve a particular behaviour. From this point of view, an action sequence is represented as a

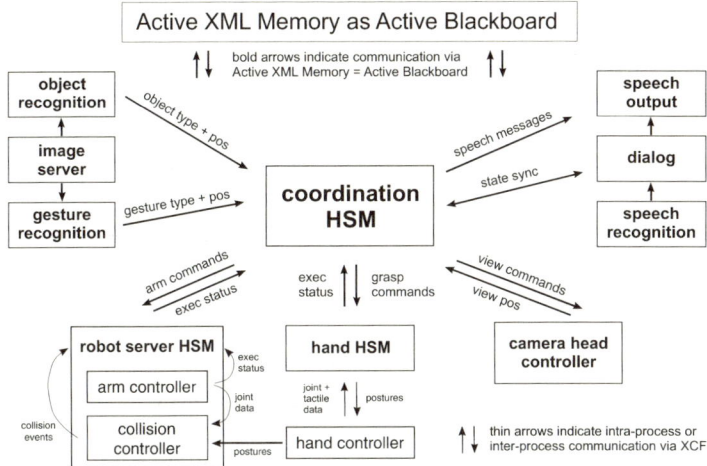

Fig. 7.9. Sketch of the overall system architecture of our cognitive robot.

sequence of state transitions within the coordinating HSM, each triggered by a set of predefined events from the subprocesses. Simultaneously each transition can generate events to trigger the activation or deactivation of some skill states within the subprocesses.

The flexible config-interaction achieved with the HSM layer offers us an interesting and potentially very powerful way of making system changes: Designing the coordinating state machine such that its structure can be loaded or extended on demand from a (textual) XML description, we can easily instantiate different (or partially modify existing) coordinating patterns from a "memory" of such patterns. From a functional perspective, this resembles a first step towards a capability of flexible and very high-level "self-reprogramming", utilising memories of previously successful coordination strategies. This makes it possible to implement very high-level learning techniques, e.g. based on imitation learning or heuristic forms of "introspection" and connect them with lower level approaches, such as the application of "datamining" techniques to analyse occuring data streams in an on-line fashion and synthesise new or modify existing behavioural skills.

To restrict the overall complexity of the coordinating state machine, one can also introduce several functional building blocks – each based on a HSM –, which coordinate specific aspects of the overall system only. Again, on the next higher level a state machine would coordinate these functional building blocks. Proceeding in this recursive manner, the system remains much easier configurable and provides a natural mechanism for encapsulation of increasingly complex behaviour at all levels while avoiding complex synchronisation and scheduling schemes.

7.9 Summary and Discussion

Taking an evolutionary perspective on cognitive architectures, it appears that the control of hands might be a good "Rosetta Stone" problem to master the step from automatically executed low-level behaviour patterns to highly flexible and adaptive behaviour, ultimately leading to mental cognition.

Arguing that the shaping of interaction is the core task for any cognitive system, we have presented a *dual interaction* view on the development of architectures for artificial cognitive systems. A novel element of this view is the emphasis that the shaping of interactions has to occur in two complementary spheres: the task-directed interaction with the world (*task-interaction*), and the interaction with the engineer that develops and maintains the system (*config-interaction*). As a suitable approach to meet the contrary objectives of this dual optimisation problem, we have proposed a two-layer architecture.

The first layer accommodates the necessary building blocks from which we can compose the required task-interaction patterns. Each building block supports a particular elementary dynamical process, implemented in the form of a suitable controller connecting (a subset of the) sensors with (a subset of the) actuators. To establish a desired task-interaction pattern requires to combine and suitably parametrise a subset of such controllers.

The aim of the second layer is to provide a good config-interaction interface for the engineer to achieve the necessary shaping of task-interaction primitives into a coherent task-interaction pattern. For this symbolic layer, we propose to use hierarchical state machines (HSMs) whose states represent semantically related classes of task-interaction patterns. Transitions between such states are coupled via an emission of parametrised events to the controller layer.

As an extensible technical framework to implement the required asynchronous communication as well as to maintain the required basic world knowledge and context-specific data, we employ a global message blackboard built on the XML-based Active Memory Infrastructure described in [17].

From a more general architecture perspective, this approach offers a principled way for the coordination of *continuous dynamical systems (controllers)* by *discrete dynamical systems* implemented as HSMs that act as "discrete dynamical glue". HSMs offer powerful structuring principles (state encapsulation, transition inheritance) to forge complex continuous sensorimotor processes at a level that is much better suited for semantic labelling than the to-be-coordinated continuous processes themselves.

Applying the described scheme to the case of robot grasping with a five-fingered anthropomorphous manipulator, we are able to realise surprisingly robust grasping capabilities for a wide range of common household objects. Based on these very encouraging results, we are currently porting the capabilities of a previously developed system with a more limited robot manipulator, but coupled with speech-understanding and binocular vision capabilities [2, 6, 7], to the new architecture. Developed over a time horizon of several years, this previous system had reached the limits of its extensibility and maintainability. We view the current porting process as a rather stringent test of our architecture concept, anticipating significant gains in the ability to restructure the system towards even higher cognitive skills, in particular imitation learning.

While our porting project has been triggered by the pressures of evolving a system in the technical domain over the years, it is tempting to speculate that somewhat similar pressures may have been acting on the development of our own "cognitive machinery": as the potentially available skills reached a certain level of complexity, it may have become necessary to introduce something like config-interaction also within the brain. Language in combination with explicit reasoning and planning seems to resemble a config-interaction *into our own brain* that allows us to configure our "pre-rational" skills in a very flexible and rapid way. We are fully aware that the gap between the neural control circuits and the semantic networks in the brain on the one hand, and the artificial controllers and the HSM boxes in our system on the other hand is enormous, but we believe that pursuing the road of dual interaction is promising to bring us a bit closer to artificial systems that can act in a more "brain-like" manner.

Acknowledgements

Among many people who contributed to the robot system, we would like to thank particularly F. Röthling who developed the grasping strategy for our robot hands, S. Wrede who coordinated and implemented many parts of the *Active Memory Infrastructure*, and R. Kõiva whose technical input was invaluable for attaining a smooth and reliable operation of many subsystems. This work was supported by Deutsche Forschungsgemeinschaft DFG.

References

1. Arbib, M.A.: From monkey-like action recognition to human languaga: An evolutionary framework for neurolinguistics. Behavioral and Brain Sciences **28**(2) (2005) 105-124
2. Fink, G., Fritsch, J., Leßmann, N., Ritter, H., Sagerer, G., Steil, J., Wachsmuth, I.: Architectures of situated communicators: From perception to cognition to learning. In Rickheit, G., Wachsmuth, I., eds.: Situated communication. Trends in linguistics. Mouton de Gruyter, Berlin (2006) p.357–376
3. Bauckhage, C., Wachsmuth, S., Hanheide, M., Wrede, S., Sagerer, G., Heidemann, G., Ritter, H.: The visual active memory perspective on integrated recognition systems. Image and Vision Computing, In Press (2006)
4. Li, S., Kleinehagenbrock, M., Fritsch, J., Wrede, B., Sagerer, G.: 'BIRON, let me show you something': Evaluating the interaction with a robot companion. In Thissen, W., Wieringa, P., Pantic, M., Ludema, M., eds.: Proc. IEEE Int. Conf. on Systems, Man, and Cybernetics, Special Session on Human-Robot Interaction. (2004) 2827–2834

5. Asoh, H., Motomura, Y., Asano, F., Hara, I., Hayamizu, S., Itou, K., Kurita, T., Matsui, T., Vlassis, N., Bunschoten, R., Krose, B.: Jijo-2: An office robot that communicates and learns. IEEE Intelligent Systems **16**(5) (2001) 46–55
6. Steil, J., Röthling, F., Haschke, R., Ritter, H.: Situated robot learning for multi-modal instruction and imitation of grasping. Robotics and Autonomous Systems, Special Issue on "Robot Learning by Demonstration" (47) (2004) 129–141
7. Ritter, H., Steil, J., Nölker, C., Röthling, F., McGuire, P.: Neural architectures for robotic intelligence. Reviews in the Neurosciences **14**(1–2) (2003) 121–143
8. Samek, M.: Practical Statecharts in C/C++: Quantum Programming for Embedded Systems. CMP Books (2002)
9. Peterson, J.L.: Petri Net Theory and The Modeling of Systems. Prentice Hall (1981)
10. Harel, D.: Statecharts: A visual formalism for complex systems. Science of Computer Programming (8) (1987) 231–274
11. Cutkosky, M., Howe, R.D.: Human grasp choice and robotic grasp analysis. In Venkataraman, S.T., Iberall, T., eds.: Dextrous Robot Hands. Springer (1990)
12. Zhu, X., Ding, H., Wang, J.: Grasp analysis and synthesis based on a new quantitative measure. IEEE Trans. Robotics and Automation **19**(6) (2003) 942–953
13. Xu, J., Liu, G., X.Wang, Li, Z.: A study on quality functions for grasp synthesis and fixture planning. In: Proc. IROS. Volume 4., Las Vegas (2003) 3429–3434
14. Ch. Borst, M.F., Hirzinger, G.: Grasping the dice by dicing the grasp. In: Proc. IROS. Volume 3., Las Vegas (2003) 3692–3697
15. Helmke, U., Hüper, K., Moore, J.B.: Quadratically convergent algorithms for optimal dextrous hand grasping. IEEE Trans. Robotics and Automation **18**(2) (2002) 138–146
16. Han, L., Trinkle, J.C., Li, Z.: Grasp analysis as linear matrix inequality problems. IEEE Trans. Robotics and Automation **16**(6) (2000) 663–674
17. Wrede, S., Hanheide, M., Bauckhage, C., Sagerer, G.: An Active Memory as a Model for Information Fusion. In: Proc. 7th Int. Conf. on Information Fusion. Number 1 (2004) 198–205
18. Wrede, S., Fritsch, J., Bauckhage, C., Sagerer, G.: An XML Based Framework for Cognitive Vision Architectures. In: Proc. Int. Conf. on Pattern Recognition. Number 1 (2004) 757–760
19. Wrede, S., Hanheide, M., Wachsmuth, S., Sagerer, G.: Integration and coordination in a cognitive vision system. In: Proc. Int. Conf. on Computer Vision Systems, St. Johns University, Manhattan, New York City, USA, IEEE (2006)
20. McGuire, P., Fritsch, J., Ritter, H., Steil, J., Röthling, F., Fink, G.A., Wachsmut, S., Sagerer, G.: Multi-modal human-machine communication for instructing robot grasping tasks. In: Proc. IROS. (2002) 1082–1089
21. Steil, J., Röthling, F., Haschke, R., Ritter, H.: Learning issues in a multi-modal robot-instruction scenario. In: Workshop on Imitation Learning, Proc. IROS. (2003)

Part II

Logic and Neural Networks

Introduction: Logic and Neural Networks

The second part of this book addresses the integration of logical knowledge representation, learning, and reasoning into the neural networks paradigm. The theme can be traced back to the seminal paper by McCulloch and Pitts [1], where the well-known intimate relationship between propositional logic and artificial neural networks with step activation functions was initially presented.[1]

Extending on these results, however, has proved to be a rather challenging task. While the propositional case has been reasonably well-researched (see e.g. [3, 4] for overviews), the integration of more expressive logics with neural networks appeared to be a bottleneck which was difficult to address. Only recently, new research results indicate a number of lines of investigation which have the potential to lay a foundation for significant advance in this area.

Consequently, this second part of the book focusses on these new developments. The chapters constitute a representative overview of recent lines of work on the integration of expressive logics and neural networks. The selection is certainly subjective, and we are aware that we could not cover all important aspects of this intriguing area of research. The reader interested in a comprehensive survey shall be referred to [4], and in particular to the extensive list of references contained therein.

Chapters 8 through 10 deal with the integration of first-order logic with neural networks, a task which still constitutes major challenges to the community [5]. Chapters 11 and 12 deal with different dimensions of expressivity, namely with modal, temporal, and multi-valued logic in the propositional case.

Chapter 8 features a survey and recent advances of the well-known SHRUTI system, which was one of the first systems developed for connectionist reasoning with first-order knowledge.

---

[1] It was Kleene [2], who later extended this study to show that such neural networks are equivalent to finite-state automata.

Chapter 9 presents the Core Method for first-order neural-symbolic integration, together with the description and evaluation of a learning system based on it.

Chapter 10 contains an alternative approach to first-order neural-symbolic integration based on topos theory, and also the description and evaluation of a corresponding learning system.

Chapter 11 is a survey on recent advances on the connectionist handling of modal and temporal propositional logics.

Finally, chapter 12 presents foundational results on the representation of multi-valued propositional logic programs by neural networks.

References

1. W.S. McCulloch and W. Pitts. A logical calculus of the ideas immanent in nervous activity. *Bulletin of Mathematical Biophysics*, 5 : 115–133, 1943.
2. S.C. Kleene. Representation of events in nerve nets and finite automata. In C.E. Shannon and J. McCarthy, editors, *Automata Studies*, volume 34 of *Annals of Mathematics Studies*, pages 3–41. Princeton University Press, Princeton, NJ, 1956.
3. A.S. d'Avila Garcez, K.B. Broda, and D.M. Gabbay. *Neural-Symbolic Learning Systems — Foundations and Applications*. Perspectives in Neural Computing. Springer, Berlin, 2002.
4. S. Bader and P. Hitzler. Dimensions of neural-symbolic integration – a structured survey. In: S. Artemov, H. Barringer, A.S. d'Avila Garcez, L.C. Lamb, and J. Woods, editors, *We Will Show Them: Essays in Honour of Dov Gabbay, Volume 1*, pages 167–194. College Publications, London, 2005.
5. S. Bader, P. Hitzler, and S. Hölldobler. The Integration of Connectionism and First-Order Knowledge Representation and Reasoning as a Challenge for Artificial Intelligence. *Information*, 9(1), 2006.

8

SHRUTI: A Neurally Motivated Architecture for Rapid, Scalable Inference

Lokendra Shastri

International Computer Science Institute
1947 Center Street, Suite 600
Berkeley, CA 94704 USA
shastri@icsi.berkeley.edu

Summary. The ability to reason effectively with a large body of knowledge is a cornerstone of human intelligence. Consequently, the development of efficient, large-scale reasoning systems has been a central research goal in computer science and artificial intelligence. Although there has been notable progress toward this goal, an efficient, large-scale reasoning system has remained elusive. Given that the human brain is the only extant system capable of supporting a broad range of efficient, large-scale reasoning, it seems reasonable to expect that an understanding of how the brain represents knowledge and performs inferences might lead to critical insights into the structure and design of large-scale inference systems. This article provides an overview of SHRUTI, a long-term research project on understanding the neural basis of knowledge representation, reasoning, and memory, with an emphasis on the representation and processing of relational (first-order) knowledge. Next, it examines some of the insights about the characteristics of rapid human reasoning stemming from this work. Finally, it describes the performance of a pilot system for limited first-order reasoning that incorporates these insights. The pilot system demonstrates that it is possible to encode very large knowledge bases containing several million rules and facts on an off-the-shelf workstation and carry out an interesting class of inference in under a second.

8.1 Introduction

The ability to reason effectively with a large body of knowledge is a cornerstone of human intelligence and is critical for decision-making, problem solving, planning, and language understanding. Consequently, the development of efficient, large-scale reasoning systems has been a central research goal in computer science and artificial intelligence. Although there has been notable progress toward this goal (see, for example, [1, 2, 3, 4]), an efficient, large-scale, reasoning system has largely remained elusive.

Results in computational complexity inform us that any generalized notion of inference is intractable, and hence, it is impossible to develop effective, large-scale reasoning systems of unrestricted expressiveness and inferential power. But at the same time, we also know that the human mind is capable of encoding a large and diverse body of common sense knowledge and rapidly performing a wide range of inferences [5, 6, 7, 8]. The human reasoning capacity is amply illustrated by our ability to understand language in real-time. Consider the simple narrative:

John fell in the hallway. Tom had cleaned it. He got hurt.

Upon reading this narrative, most of us would rapidly and effortlessly infer that Tom had cleaned the hallway, therefore, the hallway floor was wet; John slipped and fell, because the floor was wet; John got hurt because of the fall. Such "bridging" inferences help in establishing referential and causal coherence and are essential for language understanding. Since we understand language at the rate of several hundred words per minute, it follows that we draw such inferences within hundreds of milliseconds. Furthermore, we draw these inferences spontaneously, and without conscious effort – as though they are a *reflex* response of our cognitive apparatus. In view of this, it seems appropriate to refer to such reasoning as *reflexive reasoning* [9].

Our ability to understand language in real-time entails that there must exist computationally tractable classes of inference that, though limited in certain ways, are remarkably inclusive and cover a domain as rich as language understanding and common sense reasoning. At this time, however, we do not have an adequate characterization of such classes of inference.

Given that the human brain is the only extant system capable of supporting a broad range of efficient, large-scale reasoning, it seems appropriate to assume that understanding how the brain represents knowledge and performs inferences might lead to critical insights into the structure and design of large-scale inference systems.

This article reviews the current state of a long-term research project on understanding the neural basis of knowledge representation, reasoning, and memory, with an emphasis on the representation and processing of relational (first-order) knowledge. In particular, it describes SHRUTI, a neurally plausible model that demonstrates how a suitably structured network of simple nodes and links can encode several hundred thousand episodic and semantic facts, causal rules, entities, types, and utilities and yet perform a wide range of explanatory and predictive inferences within a few hundred milliseconds. Next, it examines some of the key insights about the characteristics of reflexive reasoning stemming from this work. Finally, it describes the performance of a pilot system for limited first-order reasoning with very large knowledge bases (KBs) that incorporates some of these insights. This pilot system demonstrates that it may be possible to encode very large KBs containing several million rules and facts on an off-the-shelf workstation and carry out an interesting class of inference in under a second.

8.2 Some Requirements for Supporting Reflexive Reasoning

Any system that attempts to explain our reflexive reasoning ability must possess a number of properties: First, such a system must be representationally adequate. It must be capable of encoding specific facts and events (e.g., John slipped in the hallway) as well as statistical summaries (e.g., It often rains in Seattle) and general regularities that capture the causal structure of the environment (e.g., If one falls down, one gets hurt). Often these regularities are context-dependent and evidential in nature. Second, the system should be inferentially adequate, that is, it should be capable of drawing inferences by combining evidence and arriving at *coherent* interpretations of perceptual or linguistic input. In order to do so, the system should be capable of establishing referential coherence. In particular, it should be able to posit the existence of appropriate entities, and it should be able to unify entities and events by recognizing that multiple designations refer to the same entity or event. Third, the system should be capable of learning and fine-tuning its causal model based on experience, instruction, and exploration. Finally, the system should be scalable and computationally effective; the causal model underlying human language understanding is extremely large, and hence, a system for establishing causal and referential coherence should be capable of encoding a large causal model and performing the requisite inferences within fractions of a second.

8.2.1 Relational Knowledge and the Binding Problem

Since connectionist models compute by spreading activation, they are well-suited for performing evidential and parallel computations. But representing events and situations and reasoning about them require the representation of *relational structures* and a solution to the *dynamic binding* problem [10, 11, 9].

Consider the representation of the event (E1) "John gave Mary a book in the hallway." This event is an instance of a specific sort of interaction involving John, Mary and a book that occurs in a particular location (the hallway). John and Mary are performing specific roles in this interaction: John is the one who is doing the giving, and Mary is the one who is doing the receiving. Moreover, a book is the object being given by John to Mary. Consequently, this event cannot be represented by simply activating the conceptual roles giver, recipient, and given-object, and the entities John, Mary, and book. Such a representation would be identical to that of of an event wherein Mary gave John a book in the hallway.

An unambiguous representation of E1 requires the representation of bindings between the roles of E1 (e.g., giver) and the entities that fill these roles in the event (e.g., John). Note that *two* levels of bindings are involved:

1. Entities occurring in the event are bound to the respective roles they fill in the event.

2. All of the role–entity bindings pertaining to the event are grouped together in order to distinguish them from role–entity bindings pertaining to other events.

Moreover, inferring that Mary has the book as a result of John giving the book to her involves rapidly establishing additional bindings between the role owner and the entity Mary and between the role owned-object and the entity book. Thus reasoning about events and situations also requires a solution to the problem of *propagating* bindings.

It is straightforward to represent role–entity bindings using additional nodes and links (e.g., using binder nodes that encode conjunctions of roles and entities). But while it is feasible to use additional nodes and links to encode long-term knowledge, it is implausible to assume that binder nodes can be recruited for representing large numbers of *dynamic* bindings arising rapidly during language understanding and visual processing. In standard computing within the von Neumann architecture, bindings are expressed using variables and pointers, but these techniques have no direct analogs in connectionist networks.

Connectionist modelers have devised several solutions to the binding problem. One of the earliest solutions is due to Feldman [10] who showed how any element of a group of N entities could be dynamically associated with any element of another group of N entities using an interconnection network. Several researchers have proposed formal as well as computational models for representing bindings using various types of convolution, tensor product, matrix multiplication, and XOR operations (e.g., [11, 12]. Touretzky and Hinton [13] proposed a solution to the binding problem in the context of a distributed connectionist production system (DCPS). The above models, however, do not address how multiple bindings can be expressed and propagated simultaneously in order to support inference. For example, DCPS can only deal with rules involving a *single* role and its reasoning process can only apply a *single* rule at a time. Thus such models are not adequate for modeling reflexive reasoning [9].

Another solution to the dynamic binding problem by Barnden and Srinivas [14] makes use of the relative position of active nodes and the similarity of their firing patterns to encode bindings. A solution proposed by Lange and Dyer [15] and Sun [16] assigns a distinct activation pattern (a signature) to each entity and propagates these signatures to establish role–entity bindings. Ajjanagadde and Shastri [17, 9] proposed a biologically plausible solution for expressing and propagating role–entity bindings; they suggested that a role–entity binding such as (John=giver) be expressed by the synchronous firing of the role (giver) and entity (John) nodes (also see [18, 19, 20]).

While the use of synchrony for binding perceptual features during visual processing had been suggested earlier by von der Malsburg [21], the SHRUTI model proposed by Shastri and Ajjanagadde [9] was the first model to offer a detailed account of how synchronous activity can be harnessed to represent complex relational knowledge and carry out rapid inference with respect to such knowledge. The advantages of using temporal synchrony to solve the dynamic binding problem over other solutions to this problem are discussed in [9, 22].

8.3 SHRUTI

SHRUTI is a fully implemented, neurally plausible model that demonstrates how a network of simple nodes and links can encode a large body of semantic and episodic facts, systematic rules, and knowledge about entities and types, and yet perform a wide range of explanatory and predictive inferences within a few hundred milliseconds [9, 23, 24, 25, 26, 27, 28].

SHRUTI can represent and process probabilistic, relational knowledge and deal with incomplete and inconsistent beliefs [23, 29]. At the same time, SHRUTI's representational machinery can encode parameterized action schemas, operators, and reactive plans [30]. These process-based representations can encode compositional and hierarchically organized actions and support control behaviors such as conditional execution, interruption, resumption and termination, partial ordering, concurrency, and iteration.

SHRUTI propagates both beliefs and utilities over a single underlying "causal structure" to make predictions, seek explanations, and identify actions that increase expected future utility. The computations performed by SHRUTI can also be seen as an approximation of probabilistic reasoning in belief-nets [31] and of dynamic programming, wherein a multi-step approximation of the Bellman equation [32] is computed using the causal domain model, prior probabilities, and expected future utilities associated with (partial) world states [33].

8.3.1 An Overview of SHRUTI's Architecture

The encoding of relational information in SHRUTI is mediated by structured clusters of nodes, referred to as *functional focal-clusters* (or FFCs). Each relational schema (or frame) has an associated functional focal-cluster. Such an FFC for the relation *fall* is depicted in Fig. 8.1 as the dotted ellipse labeled FALL. Each label within this ellipse denotes a connectionist node that serves a specific function outlined below. As explained in [23], each connectionist node within an FFC is a computational abstraction and corresponds to an ensemble of neurons. In general, a relational FFC consists of the following nodes:

- A *role* node for each role in the relational schema. For simplicity, Fig. 8.1 assumes that *fall* has only two roles: *patient* and *location*. The *synchronous* firing of a role node (e.g., patient) with an entity node (e.g., John) encodes a dynamic role–filler binding (e.g., [fall:patient=John]). Such dynamic role–filler bindings pertaining to the roles of a given relational schema constitute the currently active instance of that relation.

- Positive (+) and negative (-) *collector* nodes whose activation levels signify the levels of belief and disbelief, respectively, in the currently active relational instance. This can range from absolute belief (+ fully active; - inactive) to absolute disbelief (- fully active; + inactive); from ignorance (neither + nor - active) to contradiction (both + and - active); and various graded beliefs in between. The + and - collector nodes are mutually inhibitory (inhibitory links terminate in a filled blob).
- An *enabler* node (?) whose activation signifies a search for support or explanation for the currently active relational instance. This node's level of activity signals the strength with which support for this relational instance is being sought.
- A pair of *utility* nodes, +$p and +$n, associated with the positive collector, and a pair of utility nodes, -$p and -$n, associated with the negative collector. The activation levels of +$p and -$p indicate the desirability (i.e., *p*ositive utility) of the occurrence and non-occurrence, respectively, of the currently active relational instance. Similarly, the activation levels of +$n and -$n indicate the undesirability (i.e., *n*egative utility) of the occurrence and non-occurrence, respectively, of the currently active relational instance.

There exist links from a FFC's collector nodes to its enabler node. This link converts a dynamic assertion of a relational instance into a query about that assertion. Consequently, the system continually seeks an explanation for all active relational instances.

The FFC associated with a relational schema acts as an anchor for encoding and attaching various kinds of knowledge about the relation. This includes motor and perceptual schemas associated with the relational schema, causal connections between this relational schema ad other relational schemas, lexical and naming information, and episodic and semantic facts involving this generic relation. Information pertaining to a relation converges on its FFC, and this information can be accessed by fanning out from the FFC. It has been argued in [9] that it may be essential to associate such a FFC with each relational schema in order to process relational information without cross-talk and at speeds required by cognitive processing.

The FFC for an entity, say John, consists of an enabler node, ?:John, and a collector node, +:John. Persistent information about various perceptual and semantic features of John, his relationship with other concepts, and the roles he fills in various events are encoded via links between the FFC of John and appropriate circuits and FFCs representing sensory, perceptual, and semantic "knowledge" distributed across various neural structures and regions. The FFC of a type, T, contains a pair of enabler nodes ?e:T and ?v:T, and a pair of collector nodes +e:T and +v:T. The activation levels of +:John, +v:T, and +e:T signify the degrees of belief that entity John, type T, and an instance of T, respectively, play a role in a currently active situation. The activation

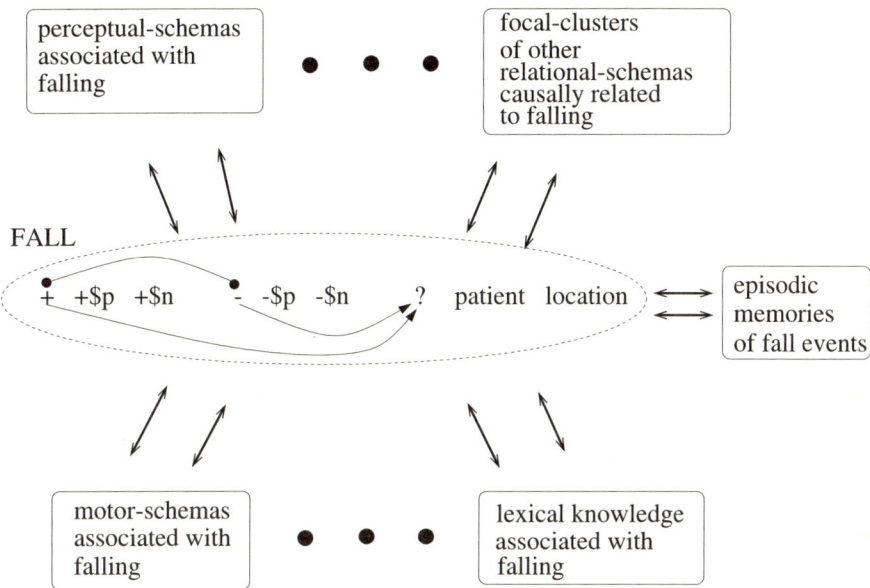

Fig. 8.1. An idealized depiction of the functional focal-cluster (FFC) of the relational schema *fall*. The FFC is enclosed within the dotted ellipse. Each label within the ellipse denotes a connectionist node. Each node is an abstraction of a small but physically dispersed ensemble of cells. The relation *fall* is assumed to have two roles. These are encoded by the nodes *patient* and *location*. The FFC also includes an *enabler* node (*?*), *collector* nodes (+ and -) and utility nodes (+$p, +$n, -$p and -$n). The grouping together of nodes within a functional focal-cluster is only meant to highlight their functional cohesiveness; their depicted proximity does not imply physical proximity. See text for additional details

of an enabler node signifies a search for an explanation about the appropriate entity, type, or instance.

The active representation of the relational instance "a man fell in the hallway" (i.e., [fall:patient=a man], [location=hallway]) is encoded by the firing of the +:fall node together with the synchronous firing of the (role) node *fall:patient* and the (filler) node +e:Man and the synchronous firing of the (role) node *fall:location* and the (filler) node +:Hallway (see Fig. 8.2).

Long-term facts are encoded as temporal pattern matching circuits, or 'fact' structures. All facts pertaining to a relational schema are attached to its FFC. If the enabler of a FFC is active, and if the dynamic role–filler bindings currently active in the FFC match the (static) bindings encoded by the long-term fact, the fact structure becomes active and, in turn, activates the positive collector of the relation's FFC and re-activates the role–filler bindings of the matching fact. SHRUTI encodes four different types of facts. These are:

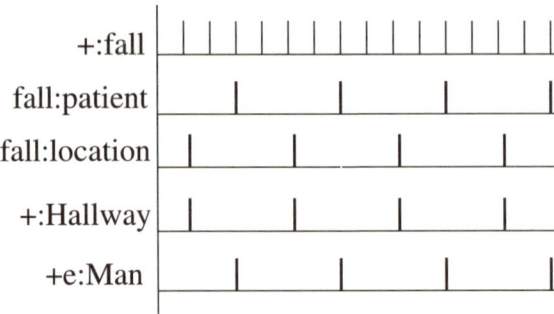

Fig. 8.2. The rhythmic pattern of activation representing the dynamic bindings ([*fall:patient* = *a Man*], [*fall:location* = *Hallway*]). Bindings are expressed by the synchronous firing of bound role and entity nodes

1. *Episodic facts (E-facts)* that record relational instances corresponding to specific events and situations (e.g., I saw John in the library on Tuesday);
2. *Taxon facts (T-facts)* that record statistical knowledge about a relation (e.g., Soccer moms own minivans) and can be viewed as generalizations defined over multiple E-facts. Both E-facts and T-facts respond to partial cues, but while E-facts distinguish between highly similar events, T-facts respond to similarities.
3. *Reward facts (R-facts)* that record rewards or punishments associated with specific events and situations. R-facts are analogous to E-facts. Activation of a positive (negative) reward fact indicates the real or imagined attainment of some reward (punishment).
4. *Value facts (V-facts)* that associate a reward or a punishment with a generic event or situation. V-facts are analogous to T-facts and hold a statistical summary of past activity. Thus they can predict future reward or punishment.

The distinction between different types of facts is motivated by biological considerations. For example, converging evidence from neuropsychology and neuroimaging suggests that while E-facts are encoded in cortico-hippocampal circuits, T-facts are encoded in purely cortical circuits [34, 35]. A connectionist encoding of E-facts and -facts is described in [23] and that of R-facts and V-facts in [28]. An anatomically realistic and physiologically grounded model of one-shot learning that is fully compatible with SHRUTI, and that can rapidly transform a transient pattern-of-activity-based representation of an event into a persistent structure-based memory trace of the event is described in [34, 35]. This model of one-shot learning based on synaptic long-term potentiation [36, 37] provides a biologically grounded explanation for the rapid memorization of E-facts and R-facts.

A rule is encoded in SHRUTI as follows (refer to Fig. 8.3):

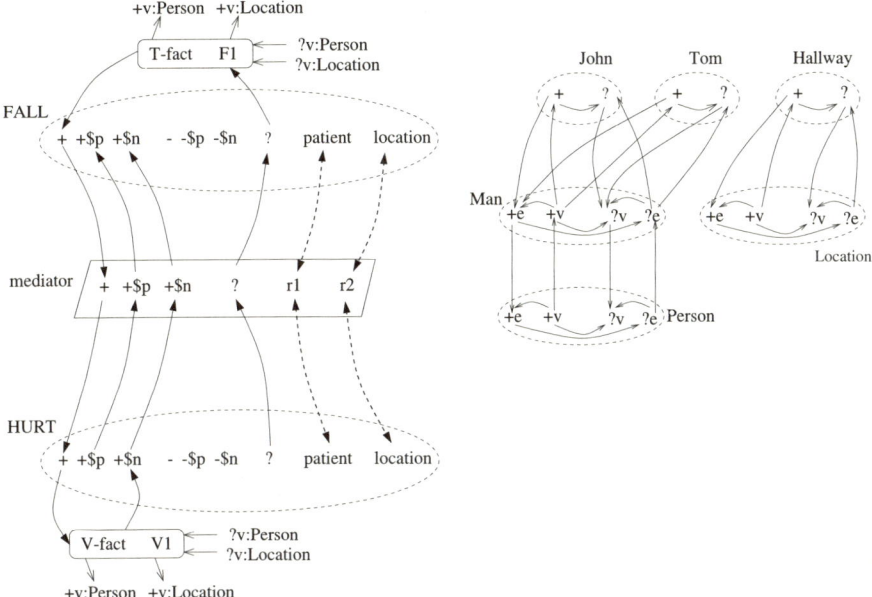

Fig. 8.3. A network fragment illustrating SHRUTI encoding of the causal rule-like knowledge "If one falls, one is likely to fall." See text for details

- By linking appropriate (+ or -) collectors of the antecedent FFCs to appropriate collectors of the consequent FFCs. These links facilitate predictive (forward) inferences.
- By linking enablers of the consequent FFCs to enablers of the antecedent FFCs. These links facilitate explanatory (backward) inferences.
- By linking the corresponding roles of the antecedent and consequent FFCs (in both directions). If roles a and b are linked, the firing of a induces synchronous firing in b, thereby propagating role–entity bindings between inferentially related FFCs.
- Links from the appropriate utility nodes of the consequent FFCs to the appropriate utility nodes of the antecedent FFCs. These links facilitate the propagation of utility.

Links between antecedent and consequent FFCs are realized through an intervening FFC termed the mediator FFC (shown as a parallelogram in Fig. 8.3).[1] Type restrictions and the instantiation of unbound variables are handled via connections between the rule mediator and the type hierarchy. These connections are not shown in Fig. 8.3. The weights of links encoding a rule are determined by the probabilistic or evidential strengths of the associated rule. Additional connectionist machinery exists to enable evidence

---

[1] The inclusion of a mediator was motivated, in part, by discussions the author had with Jerry Hobbs.

combination and *explaining away*. Details of the above representations and mechanisms are described in [23, 26, 27, 28].

8.3.2 Inference in SHRUTI

With reference to the network shown in Fig. 8.3, the dynamic assertion of the fact "John fell in the hallway" corresponds to the firing of the node +:fall together with the synchronous firing of the nodes *fall:patient* and +:John and the synchronous firing of the nodes *fall:location* and +:Hallway. Given the network connectivity, this pattern of activity rapidly evolves so that the role *hurt:patient* starts firing in synchrony with the role *fall:patient* (and hence, with +:John), and the role *hurt:location* starts firing in synchrony with the role *fall:location* (and hence, with +:Hallway). The resulting firing pattern represents not only the event "John fell in the hallway," but also the *inferred* event "John got hurt in the hallway." Thus SHRUTI is able to infer that John got hurt in the hallway given that John fell in the hallway. Furthermore, given the propagation of activity in the entity and type structure (see Figure 8.3), the network also draws inferences such as "a man fell in the hallway" and "a person fell in some location."

Similarly, the pattern of activity encoding "John got hurt in the hallway" automatically evolves to a pattern of activity that also encodes "Did John fall in the hallway?" At this time, the T-fact encoding the prior probability of people falling down provides a path or activity to flow from ?:fall to +:fall, leading to the inference that John may have fallen down. This, in turn, leads to the activation of +:hurt and sets up a reverberatory loop of activity involving the FFCs for fall and hurt signaling that falling down is a coherent explanation for John getting hurt. The strength of this reverberatory activity would depend on the link weights encoding the rule (if you fall you get hurt) and the T-fact F1 (people are likely to fall).

In addition to the propagation of beliefs described above, the activation of +:hurt activates the V-fact V1 which encodes the negative utility associated with getting hurt. Note that the output of V1 is linked to +:$n, and this signifies that getting hurt has a negative utility. Next V1 activates +$n:hurt, and this in turn activates +$n:fall signaling that negative utility is associated with falling.

As illustrated by this simple example, the assertion of any relational instance automatically leads to predictive and explanatory inferences and the evaluation of utility via the propagation of rhythmic activity between connected FFCs. The activation levels of the collector and enabler nodes of a relation are the result of the activation incident on them from various FFCs and the evidence combination functions operational at these nodes.

SHRUTI combines predictive inferences with explanatory (or abductive) inferences, exhibits priming, instantiates new entities during inference (if John ran, there must be a *path* along which he ran), unifies multiple entities by merging their phases of firing (if John got hurt and if there is a man who got hurt, then the man is likely to be John), and allows multiple explanations to compete with one another to identify the most promising explanation (explaining away).

With all the above machinery in place, SHRUTI can rapidly draw inferences such as "Tom had cleaned the hallway," "John fell because he slipped on the wet hallway floor," and "John got hurt because of the fall" in response to the input "John fell in the hallway. Tom had cleaned it. He got hurt." Moreover, SHRUTI draws these inferences rapidly; the time taken to draw an inference is simply $l \times \alpha$, where l is the length of the chain of inference, and α is the time required for connected nodes to synchronize. Assuming that (i) *gamma* band activity underlines the propagation of bindings, α is around 25–50 milliseconds. As discussed in [9] the speed of reflexive reasoning in SHRUTI satisfies the demands of real-time language understanding.

The estimate of utility at some antecedent predicate is based on both the utility value of its consequent and the probability that it will be reached from the consequent. This structure has the effect that the activation of a particular goal – via activation of a utility node – automatically leads to the assertion of its potential causes as subgoals, via spreading activation along a causal chain. Belief in a relational instance that is related to an active goal via a causal chain leads to an internal reward or punishment (via the activation of an R-fact) or to the recognition that such a reward is likely (via activation of a V-fact).

It has been shown that with an appropriate assignment of link weights, inference in SHRUTI is close to the probabilistic norm [27]. Moreover, it has been shown that approximations of these probabilistic weights can be learned using a neurally plausible unsupervised learning mechanism, Causal Hebbian learning [27] wherein link weights are updated depending on whether the pre-synaptic cell fires before or after the post-synaptic cell [38].

8.4 Some Insights Acquired from Investigating the Neural Basis of Reasoning

The work on SHRUTI has lead to several insights about the characteristics of reflexive reasoning that make it possible to carry out such reasoning in an efficient manner. As we shall see in Section 8.5, these insights can be leveraged to design a scalable, high-speed inference engine running on off-the-shelf workstations.

Much of Real-World Knowledge is Persistent

The body of knowledge required for real-world reasoning in extremely large. While some of this knowledge is dynamic and time varying, much of it is *stable* and *persistent* and can be viewed as a long-term knowledge-base (LTKB). This stable and persistent part of a real-world KB includes, among other things, common sense knowledge about space and time, natural kinds, naive physics, folk psychology, and a significant portion of knowledge about history and geography. This notion of a stable and persistent LTKB is consistent with the notion of long-term memory in psychology [39].

Explicit Encoding of Inferential Dependencies

In standard computational models, there is a sharp distinction between data, on the one hand, and the processes that operate on this data, on the other. In neuronal systems, however, the neural circuits that encode data also realize many of the processes that retrieve and draw inferences using this data. Although this makes the task of encoding new data more complex (since a new data item must be assimilated into a complex network), it ensures that once a data item is acquired, it becomes part of an associative web of knowledge that explicitly encodes immediate inferential dependencies between items in the KB. In particular, the KB is stored, not as a collection of assertions, but as an *inferential dependency graph* (IDG) linking "antecedent" structures with "consequent" structures (here a structure could be a proposition or a predicate). Since the IDG explicitly encodes inferential dependencies, it directs computations to inferentially related components of the KB. Given the persistence of the LTKB, the cost associated with encoding the LTKB as an IDG is worthwhile because it leads to the efficient processing of a large number of queries over the life of a LTKB.

The insight that inferential dependencies should be coded explicitly in an "inferential dependency graph" is a property of semantic nets [40] and belief nets (Pearl, 2000), and, more generally, of graphical models [41]. But the full computational significance of explicitly coding inferential dependencies becomes apparent only when one tries to understand how the brain might perform inference using a *common substrate* for encoding and processing.

First-order Reasoning can be Performed by Spreading Rhythmic Activity over an IDG

The explicit coding of inferential dependencies via interconnections between "antecedent" and "consequent" structures makes it possible to carry out inference via spreading activation. Indeed, spreading activation [42] and its variants such as marker passing [43, 44, 45] have long been used to support restricted and specialized forms of inference that can be reformulated as problems of computing set intersections and transitive closure (e.g., property inheritance and attribute-based classification). Such systems, however, were restricted to propositional reasoning or to "almost propositional" reasoning involving unary predicates and a single variable binding (c.f., property inheritance). In contrast, Shastri and Ajjanagadde [9] showed that limited forms of first-order inference involving n-ary predicates, variables and quantifiers can also be carried out effectively by spreading rhythmic activity within suitably structured

IDGs. The use of temporal synchrony makes it possible to express and propagate dynamic bindings between entities and arguments using a small number of temporal phases (one for each active entity), and obviates the need for communicating full symbolic names, addresses, or pointers.

Temporal Synchrony and its Relation to Marker Passing

The representation of dynamic bindings via temporal synchrony, and the propagating of bindings via the propagation of rhythmic activity can be related to marker passing [9]. Note that nodes firing in synchrony may be viewed as being marked with the *same* marker, and the propagation of synchronous activity along a chain of connected nodes can be viewed as the propagation of markers. Thus marker passing can also support limited first-order reasoning.

Inference Requires a Small Number of Phases or Markers

During an episode of first-order reasoning, the number, m, of distinct entities serving as role–fillers within active predicate instances remains very small relative to the total number, N, of entities in the KB. Note that the total number of predicate instances active during an episode of reasoning can be very large even though m remains quite small.

The above characteristic of reasoning can be leveraged to significantly reduce the amount of bits communicated in order to propagate bindings during first-order reasoning. Specifically, it enables the reasoning system to express dynamic bindings between entities and arguments using only m temporal phases (alternately, by using markers containing only $\log m$ bits) as against $\log N$ bits that would be required to encode symbolic names of N entities.

The use of temporal synchrony for encoding dynamic bindings leads to the prediction that a plausible value of m for rapid reasoning is likely to be $\leq$ 10 [9]. This prediction is motivated by the following biological consideration: Each entity participating in dynamic bindings occupies a distinct phase, and hence, the number of distinct entities that can occur as role–fillers in dynamic facts cannot exceed $\lfloor \pi_{max}/\omega \rfloor$. Here π_{max} is the maximum delay between consecutive firings of synchronous cell-clusters (about 25 milliseconds assuming that *gamma* band activity underlies the encoding of dynamic bindings), and ω equals the allowable jitter in synchrony (ca. $\pm$ 2 milliseconds). While more complex forms of reasoning are likely to require a value of m greater than 10, m is likely to be significantly smaller than N in almost all, if not all, plausible scenarios and domains.

Limited Instantiations of a Relation/Predicate

During an episode of reasoning, certain predicates may get instantiated *multiple* times (e.g., *left-of*(a,b) and *left-of*(b,c)). The number of times any given predicate is instantiated would vary depending on the rules in the KB and the query. Shastri and Ajjanagadde [9] have argued that in the case of rapid, biologically plausible reasoning, this *multiple instantiation limit* (k) is quite small

(≈ 3). Henderson [46] has shown that such a limit on k and the biologically motivated limit on m explain several properties of human parsing including our limited ability to deal with center embedding and local ambiguities.

As in the case of m, the biologically motivated limit $k \leq 3$ is quite tight, and must be set higher to support more complex inferences. But extensive simulations with several large synthetic KBs suggest that a limit of about $k = 15$ suffices for performing a broad range of derivations with inferential depths up to 16 (see Section 8.5).

The low bound on k can be leveraged in designing an efficient reasoning system by pre-allocating k of banks for each predicate and suitably interconnecting them to reflect the IDG. This obviates the need for expensive runtime memory allocation and dynamic structure building for representing predicate instantiations.

Form of Rules Amenable to Efficient Reasoning – Threaded versus Non-threaded Derivations

Let us refer to a variable appearing in multiple argument positions in the antecedent or in the consequent of a rule as a *pivotal* variable. A rule containing a pivotal variable introduces a co-reference constraint on argument bindings. Since bindings are propagated during reasoning, this constraint may have to be tracked and enforced at distal points in the IDG. Consequently, rules containing pivotal variables can significantly increase the complexity of reasoning. This motivates the following distinction between *threaded* and *non-threaded* derivations:

A derivation of Q obtained via backward chaining is called *threaded* iff either (i) the derivation does not involve any pivotal variables OR (ii) all pivotal variables occurring in the derivation get bound to constants introduced in Q.

While threaded derivations can be computed efficiently by propagating simple markers (or rhythmic activity) over an IDG [47], the computation of non-threaded derivations requires more complex computations.

8.5 A Pilot System for First-order Reasoning

The previous section enumerated several insights acquired from work on understanding how the brain might represent knowledge and perform reasoning. These insights have been used to design and implement a pilot, high-performance reasoning system for computing threaded derivations. The system is implemented in C and runs under Solaris and Linux on off the shelf workstations. It is a modification of SHRUTI-CM5, a reasoning system developed by D.R. Mani for the highly parallel CM-5 platform [48, 49].

Measuring Performance

The performance of any reasoning system is highly sensitive to the structural properties of a KB. Moreover, it varies widely from one query to another depending on the properties of the derivation required to answer the query. Thus a system's performance is best specified in terms of mean times required to process a large number of queries with respect to several KBs of varying size and structure.

Knowledge Base Characteristics

Since we did not have access to very large KBs, we generated several synthetic first-order KBs using a tool for generating synthetic KBs [48] from a specification of KB size (number of predicates, rules, facts, entities, etc.) and gross structure (number of domains, density of inter and intra-partition connectivity, etc.). In general, it is assumed that most of the KB predicates can be "partitioned" into domains such that while predicates within a domain are richly interconnected by rules, predicates across domains are interconnected only sparsely (cf. [50]). Table 8.1 displays key parameters of KBs used in pilot experiments.

Table 8.1. The size and structure of four synthetic KBs used in the experiments performed with the pilot reasoning system. "% Mult-Ant rules" refers to the percentage of rules with multiple antecedents. "% Mult-Inst rules" refers to the percentage of rules in which a predicate occurs in both the antecedent and the consequent. Such rules lead to multiple instantiations of a predicate. "% inter-domain rules" refers to the percentage of rules that connect predicates to other predicates outside their domain

Parameter	KB-S1	KB-S2	KB-S3	KB-S4
Size of KB	210,875	500,000	1,300,000	2,300,000
# Rules	50,875	150,000	400,000	600,000
# Facts	100,000	225,000	600,000	1,000,000
# Predicates	25,000	75,000	200,000	300,000
# Concepts/entities	35,000	50,000	100,000	200,000
% Mult-Ant rules	67	67	67	67
% Mult-Inst rules	25	25	25	25
# Domains	135	250	410	500
% inter-domain rules	2	2	2	2
Type-structure depth	8	8	10	10

Pilot System's Performance

The pilot reasoning system was tested on a Sun Fire V60x 2.8GHz Intel Xeon processor with 3GB of memory running Linux. The system ran on a single processor and did not make use of the dual processor configuration of the Sun Fire V60x.

Four hundred to five hundred queries were generated for each of the four KBs and executed to obtain data about response time and memory usage for different depths of inference and about the relationship between k and inferential coverage. Figure 8.4 displays the average response time of the pilot system to queries containing a single literal ($|Q| = 1$) as a function of the depth of inference (d). The pilot system propagated 100 distinct markers ($m=100$) and supported 25 multiple instantiations per predicate ($k=25$). As discussed below, these parameter values were found to be sufficient for processing queries up to an inferential depth of 16 (the maximum depth tested). As seen in the graph, the system can draw threaded first-order inferences extremely rapidly with KBs containing more than two million items.

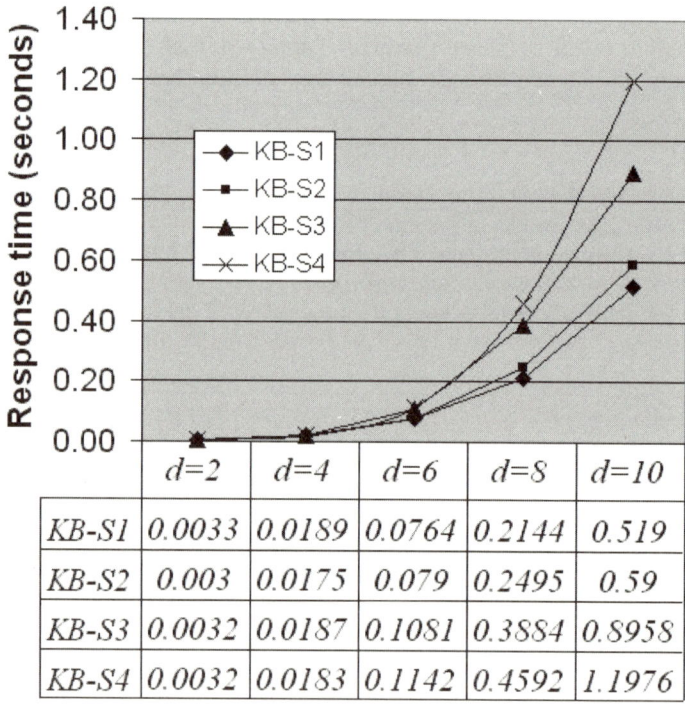

Fig. 8.4. The average response time of the pilot reasoning system for different depths of inference (d) for four KBs characterized in Table 8.1. To reduce clutter, standard deviations are not plotted. The standard deviations (σ) of response times for KB-S3 are 0.002, 0.014, 0.084, 0.244, and 0.516 seconds, respectively, for depths 2, 4, 6, 8, and 10. The σs of other KBs are analogous

Memory Requirement

The memory usage in all of the experiments remained well under 1 GB, and memory is not expected to be a limiting factor in the proposed reasoning system.

Inferentially Adequate Value of k

The experimental results indicate that $k \approx 15$ suffices for achieving inferential coverage for a wide range of queries and KBs. This is evident in Fig. 8.5, which displays the effect that increasing the value of k has on the average number of rule applications in response to a set of queries. Increasing k from 1 to 15 enables more rule applications, but increasing k beyond 15 has no impact on inferential coverage in any of the experiments conducted by us using a variety of KBs and queries run to a depth of 16. In view of this, a limit of $k = 25$ seems quite adequate.

While a limit of $k = 25$ may not work in, say, mathematics, where a small number of predicates are richly axiomatized, and hence, some predicates may be instantiated numerous times in a single derivation, we believe that this limit holds in common sense reasoning and in natural language understanding.

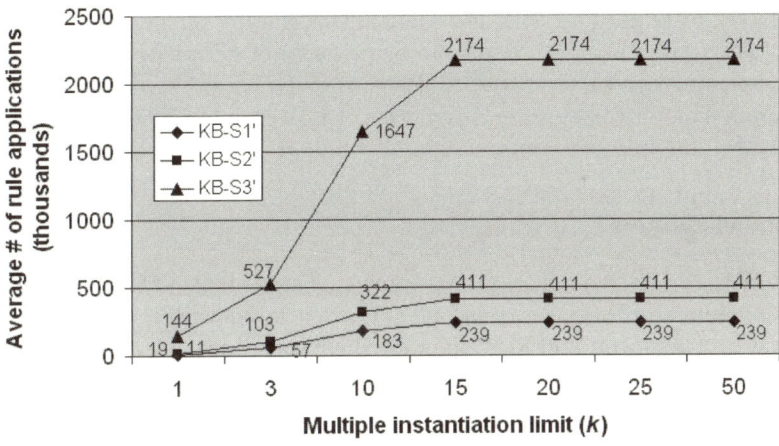

Fig. 8.5. Increasing the multiple instantiation limit, k, permits more rules to fire, and hence, leads to more inferences. This is evident as k increases from 1 to 15. Increasing k beyond 15, however, does not lead to any more rule firings. Data shown is for three variants of KBs KB-S1, KB-S2, and KB-S3 that are more likely to give rise to multiple instantiations due to a greater percentage of intra-domain rules (99%) and a greater percentage of rules likely to lead to multiple instantiations. Detailed data was collected for inferences of depth 10 ($d=10$). These results were validated for $d = 16$ by setting k to 15 and 25 and observing no change in the number of rule firings

An Algorithm for Computing Non-threaded Derivations

A marker-passing algorithm for the efficient computation of certain types of non-threaded derivations is being developed by the author. The algorithm involves the concurrent propagation of markers in the rule-base and the type hierarchy for the enforcement of equality constraints.

8.6 Conclusion

Over the past fifty years, AI has made remarkable progress and achieved impressive success in tackling many tasks that are quite challenging for humans (e.g., chess playing and data mining). But, paradoxically, it has recorded only limited success in replicating many abilities such as language understanding and common sense reasoning that are arguably trivial for humans. In particular, AI remains far from achieving its goal of developing systems whose intelligence compares favorably to that of humans along multiple dimensions.

Some of us have long argued that since the brain is the only extant physical system known to embody a broad range of cognitive abilities, it is very likely that understanding how the brain supports cognitive functions such as reasoning, memory, and language processing will provide crucial insights into the computational principles underlying cognition and into the architecture of intelligent systems. The work described in this chapter offers a tangible example of how an understanding of the neural basis of knowledge representation and reasoning can be leveraged to design a scalable inference system capable of rapidly computing an interesting class of inference of potential significance to natural language understanding and common sense reasoning.

Acknowledgments

Thanks to V. Ajjanagadde, C. Wendelken, D.R. Mani, D.G. Grannes, and M. Garagnani for their contributions to the development of the SHRUTI model, and to B. Thompson and M. Cohen for their contribution to the encoding of utilities in SHRUTI. Research on SHRUTI has been supported by the National Science Foundation, ONR, ARI, and ICSI. The implementation of the marker-passing version of SHRUTI for computing threaded derivations was supported by the National Science Foundation.

References

1. Stickel, M., Waldinger, R., Chaudhri, V.: A guide to snark. Technical report, Artificial Intelligence Center, SRI International, Menlo Park CA (2000)
2. Kalman, J.: Automated reasoning with Otter. Rinton Press, Princeton NJ (2001)

3. McCune, W.: Otter 3.3 reference manual. Technical Report Technical Memorandum No. 263, Mathematical and Computer Science Division, Argonne National Laboratory, Argonne IL (2003)
4. Chalupsky, H., MacGregor, R., Russ, T.: PowerLoom Manual. Information Sciences Institute, Marina del Rey CA. (2006)
5. Carpenter, P., Just, M.: Reading comprehension as eyes see it. In Just, M., Carpenter, P., eds.: Cognitive Processes in Comprehension. Erlbaum, Marwah NJ (1977)
6. McKoon, G., Ratcliff, R.: Inference during reading. Psychological Review **99** (1992) 440–466
7. Keenan, J., Baillet, S., Brown, P.: The effects of causal cohesion on comprehension and memory. Journal of Verbal Learning and Verbal Behaavior **23** (1984) 115–126
8. Kintsch, W.: The role of knowledge discourse comprehension: A construction-integration model. Psychological Review **95** (1988) 163–182
9. Shastri, L., Ajjanagadde, V.: From simple associations to systematic reasoning. Behavioral and Brain Sciences **16** (1993) 417–494
10. Feldman, J.: Dynamic connections in neural networks. Biological Cybernetics **46** (1982) 27–39
11. Smolensky, P.: Tensor product variable binding and the representation of symbolic structures in connectionist systems. Artificial Intelligence **46** (1990) 159–216
12. Plate, T.: Holographic reduced representations. IEEE Transactions on Neural Networks **6** (1995) 623–641
13. Touretzky, D., Hinton, G. Cognitive Science **12** (1988) 423–466
14. Barnden, J.A., Srinivas, K.: Encoding techniques for complex information structures in connectionist systems. Connection Science **3** (1991) 269–315
15. Lange, T., Dyer, M.: High-level inferencing in a connectionist network. Connection Science **1** (1989) 181–217
16. Sun, R.: On variable binding in connectionist networks. Connection Science **4** (1992) 93–124
17. Ajjanagadde, V., Shastri, L.: Efficient inference with multi-place predicates and variables in a connectionist system. In: Proceedings of the Eleventh Conference of the Cognitive Science Society. (1989) 396–403
18. Park, N., Robertson, D., Stenning, K.: An extension of the temporal synchrony approach to dynamic variable binding. Knowledge-Based Systems **8** (1995) 345–358
19. Sougne, J.: Connectionism and the problem of multiple instantiation. Trends in Cognitive Sciences **2** (1998) 183–189
20. Hummel, J., Holyoak, K.: Distributed representations of structure: a theory of analogical access and mapping. Psychological Review **104** (1997) 427–466
21. von der Malsburg, C.: The correlation theory of brain function. Technical Report 81-2, Max-Planck Institute for Biophysical Chemistry, Gottingen Germany (1981)
22. Shastri, L.: Temporal synchrony, dynamic bindings, and shruti – a representational but non-classical model of reflexive reasoning. Behavioral and Brain Sciences **19** (1996) 331–337
23. Shastri, L.: Advances in shruti – a neurally motivated model of relational knowledge representation and rapid inference using temporal synchrony. Applied Intelligence **11** (1999) 79–108

24. Shastri, L.: Types and quantifiers in shruti — a connectionist model of rapid reasoning and relational processing. In Werter, S., Sun, R., eds.: Hybrid Information Processing in Adaptive Autonomous Vehicles. Springer-Verlag, Heidelberg Germany (2000) 28–45
25. Mani, D., Shastri, L.: Reflexive reasoning with multiple-instantiation in a connectionist reasoning system with a type hierarchy. Connection Science **5** (1993) 205–242
26. Shastri, L., Wendelken, C.: Seeking coherent explanations – a fusion of structured connectionism, temporal synchrony, and evidential reasoning. In: Proceedings of the Twenty-Second Conference of the Cognitive Science Society, Philadelphia PA (2000)
27. Wendelken, C., Shastri, L.: Probabilistic inference and learning in a connectionist causal network. In: Proceedings of the Second International Symposium on Neural Computation, Berlin Germany (2000)
28. Wendelken, C., Shastri, L.: Combining belief and utility in a structured connectionist agent architecture. In: Proceedings of the Twenty-Fourth Annual Conference of the Cognitive Science Society, Fairfax VA (2002)
29. Shastri, L., Grannes, D.: A connectionist treatment of negation and inconsistency. In: Proceedings of the Eighteenth Conference of the Cognitive Science Society. (1996)
30. Shastri, L., Grannes, D., Narayanan, S., Feldman, J.: A connectionist encoding of parameterized schemas and reactive plans. Technical Report TR-02-008, International Computer Science Institute, Berkeley, California, 94704 (2002)
31. Pearl, J.: Probabilistic Reasoning in Intelligent Systems. Morgan Kaufmann, San Francisco CA (1988)
32. Bellman, R.: Dynamic Programming. Princeton University Press, Princeton NJ (1957)
33. Thompson, B., Cohen, M.: Naturalistic decision making and models of computational intelligence. Neural Computing Surveys **2** (1999) 26–28
34. Shastri, L.: Episodic memory trace formation in the hippocampal system: a model of cortico-hippocampal interactions. Technical Report TR-01-004, International Computer Science Institute, Berkeley, California, 94704 (2001)
35. Shastri, L.: Episodic memory and cortico-hippocampal interactions. Trends in Cognitive Sciences **6** (2002) 162–168
36. Bliss, T., Collingridge, G.: A synaptic model of memory: long-term potentiation in the hippocampus. Nature **361** (1993) 31–39
37. Shastri, L.: A computationally efficient abstraction of long-term potentiation. Neurocomputing **44–46** (2002) 33–41
38. Bi, G., Poo, M.: Synaptic modification by correlated activity: Hebb's postulate revisited. Annual review of neuroscience **24** (2001) 139–166
39. Wickelgren, W.A.: Learning and Memory. Prentice Hall, Englewood Cliffs NJ (1977)
40. Sowa, J., ed.: Principles of Semantic Networks. Morgan Kaufmann, San Francisco (1991)
41. Jordan, M., ed.: Learning in Graphical Models. MIT Press, Cambridge MA (1998)
42. Quillian, M.: Semantic memory. In Minsky, M., ed.: Semantic Information Processing. MIT Press, Cambridge MA (1968) 227–270
43. Fahlman, S.: NETL: A System for Representing and Using Real-World Knowledge. MIT Press, Cambridge MA (1979)

44. Charniak, E.: Passing markers: A theory of contextual influence in language comprehension. Cognitive Science **7** (1983) 171–190
45. Hendler, J.: Integrating marker-passing and problem solving: a spreading activation approach to improved choice in problem solving. Erlbaum, Marwah NJ (1987)
46. Henderson, J.: Connectionist syntactic parsing using temporal variable binding. Journal of Psycholinguistic Research **23** (1994) 353–379
47. Shastri, L.: A computational model of tractable reasoning – taking inspiration from cognition. In: Proceedings of the Thirteenth International Joint Conference on Artificial Intelligence, Chambery France (1993) 202–207
48. Mani, D.: The design and implementation of massively parallel knowledge representation and reasoning systems: a connectionist approach. PhD thesis, University of Pennsylvania (1995)
49. Shastri, L., Mani, D.: Massively parallel knowledge representation and reasoning: Taking a cue from the brain. In Geller, J., Kitano, H., Suttner, C., eds.: Parallel Processing for Artificial Intelligence 3. Elseveir Science, Amsterdam Netherland (1997) 3–40
50. McIlraith, S., Amir, E.: Theorem proving with structured theories. In: Proc. of the Seventeenth International Joint Conference on Artificial Intelligence, Seattle WA (2001)

9

The Core Method: Connectionist Model Generation for First-Order Logic Programs

Sebastian Bader,[1][*] Pascal Hitzler[2], Steffen Hölldobler[1] and Andreas Witzel[3][†]

[1] International Center for Computational Logic
Technische Universität Dresden
01062 Dresden, Germany
Sebastian.Bader@inf.tu-dresden.de, sh@iccl.tu-dresden.de
[2] Institute AIFB
Universität Karlsruhe
76131 Karlsruhe, Germany
hitzler@aifb.uni-karlsruhe.de
[3] Institute for Logic, Language and Computation
Universiteit van Amsterdam
1018 TV Amsterdam, The Netherlands
awitzel@illc.uva.nl

Summary. Research into the processing of symbolic knowledge by means of connectionist networks aims at systems which combine the declarative nature of logic-based artificial intelligence with the robustness and trainability of artificial neural networks. This endeavour has been addressed quite successfully in the past for propositional knowledge representation and reasoning tasks. However, as soon as these tasks are extended beyond propositional logic, it is not obvious at all what neural-symbolic systems should look like such that they are truly connectionist and allow for a declarative reading at the same time.

The Core Method – which we present here – aims at such an integration. It is a method for connectionist model generation using recurrent networks with feed-forward core. These networks can be trained by standard algorithms to learn symbolic knowledge, and they can be used for reasoning about this knowledge.

9.1 Introduction

From the very beginning, artificial neural networks have been related to propositional logic. McCulloch-Pitts networks, as presented in the seminal paper

---

[*] Sebastian Bader was supported by the GK334 of the German Research Foundation.
[†] Andreas Witzel was supported by a Marie Curie Early Stage Research fellowship in the project GloRiClass (MEST-CT-2005-020841).

from 1943 [1], represent propositional formulae. Finding a global minimum of the energy function modeling a symmetric network corresponds to finding a model of a propositional logic formula and vice versa [2]. These are just two examples for the strong relation between neural networks and propositional logic, which is well-known and well-studied. However, similar research concerning more expressive logics did not start until the late 1980s, which prompted John McCarthy to talk about the *propositional fixation* of connectionist systems in [3].

Since then, there have been numerous attempts to model first-order fragments in connectionist systems.[4] In [5] energy minimization was used to model inference processes involving unary relations. In [6] and [7] multi-place predicates and rules over such predicates are modeled. In [8] a connectionist inference system for a limited class of logic programs was developed. But a deeper analysis of these and other systems reveals that the systems are in fact propositional, and their capabilities were limited at best. Recursive auto-associative memories based on ideas first presented in [9], holographic reduced representations [10] or the networks used in [11] have considerable problems with deeply nested structures. We are unaware of any connectionist system that fully incorporates the power of symbolic knowledge and computation as argued for in e.g. [12], and indeed there remain many open challenges on the way to realising this vision [13].

In this chapter we are mainly interested in knowledge based artificial neural networks, i.e., networks which are initialized by available background knowledge before training methods are applied. In [14] it has been shown that such networks perform better than purely empirical and hand-built classifiers. [14] uses background knowledge in the form of propositional rules and encodes these rules in multi-layer feed-forward networks. Independently, we have developed a connectionist system for computing the least model of propositional logic programs if such a model exists [15]. This system has been further developed to the so-called *Core Method*: background knowledge represented as logic programs is encoded in a feed-forward network, recurrent connections allow for a computation or approximation of the least model of the logic program (if it exists), training methods can be applied to the feed-forward kernel in order to improve the performance of the network, and, finally, an improved program can be extracted from the trained kernel closing the neural-symbolic cycle as depicted in Figure 9.1.

In this chapter, we will present the Core Method in Section 9.3. In particular, we will discuss its propositional version including its relation to [14] and its extensions. The main focus of this paper will be on extending the Core Method to deal with first-order logic programs in Section 9.4. In particular, we will give a feasibility result, present a first practical implementation, and discuss preliminary experimental data in Section 9.5. These main sections

---

[4] See also the survey [4].

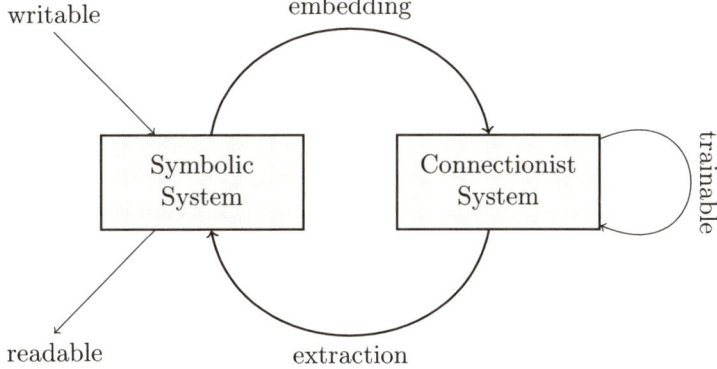

Fig. 9.1. The Neural-Symbolic Cycle.

are framed by introducing basic notions and notations in Section 9.2 and an outlook in Section 9.6.

9.2 Preliminaries

We assume the reader to be familiar with basic notions from artificial neural networks and logic programs and refer to e.g. [16] and [17], respectively. Nevertheless, we repeat some basic notions.

9.2.1 Logic Programs

A *logic program* over some first order language $\mathcal{L}$ is a set of *clauses* of the form $A \leftarrow L_1 \wedge \cdots \wedge L_n$, where A is an *atom* in $\mathcal{L}$, and the L_i are *literals* in $\mathcal{L}$, that is, atoms or negated atoms. A is called the *head* of the clause, the L_i are called *body literals*, and their conjunction $L_1 \wedge \cdots \wedge L_n$ is called the *body* of the clause. If $n = 0$, A is called a *fact*. A clause is *ground* if it does not contain any variables. *Local variables* are those variables occurring in some body but not in the corresponding head. A logic program is *covered* if none of the clauses contain local variables. A logic program is *propositional* if all predicate letters occurring in the program are nullary.

Example 1. The following propositional logic program will serve as our running example in Section 9.3.

$$\mathcal{P}_1 = \{\, p, \qquad\qquad \text{\% } p \text{ is always true.}$$
$$r \leftarrow p \wedge \neg q, \qquad \text{\% } r \text{ is true if } p \text{ is true and } q \text{ is false.}$$
$$r \leftarrow \neg p \wedge q \,\} \qquad \text{\% } r \text{ is true if } p \text{ is false and } q \text{ is true.}$$

Example 2. The following (first order) logic program will serve as our running example in Section 9.4.

$$\mathcal{P}_2 = \{\ e(0). \qquad \%\ 0\ is\ even$$
$$e(s(X)) \leftarrow o(X). \qquad \%\ the\ successor\ s(X)\ of\ an\ odd\ X\ is\ even$$
$$o(X) \leftarrow \neg e(X).\ \} \qquad \%\ X\ is\ odd\ if\ it\ is\ not\ even$$

The *Herbrand universe* $\mathcal{U}_\mathcal{L}$ is the set of all ground terms of $\mathcal{L}$, the *Herbrand base* $\mathcal{B}_\mathcal{L}$ is the set of all ground atoms, which we assume to be infinite – indeed the case of a finite $\mathcal{B}_\mathcal{L}$ can be reduced to a propositional setting. A *ground instance* of a literal or a clause is obtained by replacing all variables by terms from $\mathcal{U}_\mathcal{L}$. For a logic program P, $\mathcal{G}(P)$ denotes the set of all ground instances of clauses from P.

A *level mapping* is a function assigning a natural number $|A| \geq 1$ to each ground atom A. For negative ground literals we define $|\neg A| := |A|$. A logic program P is called *acyclic* if there exists a level mapping $|\cdot|$ such that for all clauses $A \leftarrow L_1 \wedge \cdots \wedge L_n \in \mathcal{G}(P)$ we have $|A| > |L_i|$ for $1 \leq i \leq n$.

Example 3. Consider the program from Example 2 and let s^n denote the n-fold application of s. One possible level mapping for which we find that $\mathcal{P}_2$ is acyclic is given as:

$$|\cdot| : \mathcal{B}_\mathcal{L} \to \mathbb{N}^+$$
$$e(s^n(0)) \mapsto 2n+1$$
$$o(s^n(0)) \mapsto 2n+2.$$

A *(Herbrand) interpretation* I is a subset of $\mathcal{B}_\mathcal{L}$. Those atoms A with $A \in I$ are said to be *true* under I, those with $A \notin I$ are said to be *false* under I. $\mathcal{I}_\mathcal{L}$ denotes the set of all interpretations. An interpretation I is a *(Herbrand) model* of a logic program P (in symbols: $I \models P$) if I is a model for each clause in $\mathcal{G}(P)$ in the usual sense.

Example 4. For the program $\mathcal{P}_2$ from Example 2 we have

$$M_2 := \{e(s^n(0)) \mid n\ \text{even}\} \cup \{o(s^m(0)) \mid m\ \text{odd}\} \models P.$$

Given a logic program P, the *single-step operator* $T_P : \mathcal{I}_\mathcal{L} \to \mathcal{I}_\mathcal{L}$ maps an interpretation I to the set of exactly those atoms A for which there is a clause $A \leftarrow \text{body} \in \mathcal{G}(P)$ such that the body is true under I. The operator T_P captures the semantics of P as the Herbrand models of the latter are exactly the pre-fixed points of the former, i.e. those interpretations I with $T_P(I) \subseteq I$. For logic programming purposes it is usually preferable to consider fixed points of T_P, instead of pre-fixed points, as the intended meaning of programs. These fixed points are called *supported models* of the program [18]. In Example 2, the (obviously intended) model M_2 is supported, while $\mathcal{B}_\mathcal{L}$ is a model but not supported.

9.2.2 Artificial Neural Networks

Artificial neural networks consist of simple computational units (neurons), which receive real numbers as inputs via weighted connections and perform *simple* operations: the weighted inputs are added and simple functions (like threshold, sigmoidal) are applied to the sum. We will consider networks, where the units are organized in layers. Neurons which do not receive input from other neurons are called *input neurons*, and those without outgoing connections to other neurons are called *output neurons*. Such so-called *feed-forward networks* compute functions from $\mathbb{R}^n$ to $\mathbb{R}^m$, where n and m are the number of input and output units, respectively. The gray shaded area in Figure 9.2 shows a simple feed-forward network. In this paper we will construct recurrent networks by connecting the output units of a feed-forward network N to the input units of N. Figure 9.2 shows a blueprint of such a recurrent network.

9.3 The Core Method for Propositional Logic

In a nutshell, the idea behind the Core Method is to use feed-forward connectionist networks – called core – to compute or approximate the meaning function of logic programs. If the output layer of a core is connected to its input layer then these recurrent connections allow for an iteration of the meaning function leading to a stable state, corresponding to the least model of the logic program provided that such a least model exists (see Figure 9.2). Moreover, the core can be trained using standard methods from connectionist systems. In other words, we are considering connectionist model generation using recurrent networks with feed-forward core.

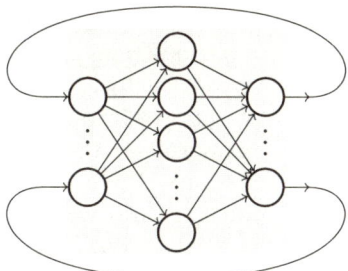

Fig. 9.2. The blueprint of a recurrent network used by the Core Method.

The ideas behind the Core Method for propositional logic programs were first presented in [15] (see also [19]). Consider the logic program from Example 1. A translation algorithm turns such a program into a core of logical threshold units. Because the program contains the predicate letters p, q and

r only, it suffices to consider interpretations of these three letters. Such interpretations can be represented by triples of logical threshold units. The input and the output layer of the core consist exactly of such triples. For each rule of the program a logical threshold unit is added to the hidden layer such that the unit becomes active iff the preconditions of the rule are met by the current activation pattern of the input layer; moreover this unit activates the output layer unit corresponding to the postcondition of the rule. Figure 9.3 shows the network obtained by the translation algorithm if applied to $\mathcal{P}_1$.

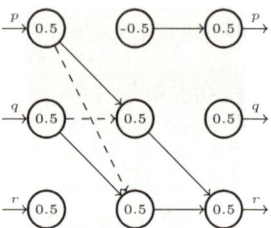

Fig. 9.3. The core corresponding to $\mathcal{P}_1 = \{p,\ r \leftarrow p \wedge \neg q,\ r \leftarrow \neg p \wedge q\}$. Solid connections have weight 1.0, dashed connections have weight -1.0. The numbers within the units denote the thresholds.

In [15] we proved – among other results – that for each propositional logic program $\mathcal{P}$ there exists a core computing its meaning function $T_\mathcal{P}$ and that for each acyclic propositional logic program $\mathcal{P}$ there exists a core with recurrent connections such that the computation with an arbitrary initial input converges and yields the unique fixed point of $T_\mathcal{P}$.

The use of logical threshold units in [15] made it easy to prove these results. However, it prevented the application of standard training methods like backpropagation to the kernel. This problem was solved in [20] by showing that the same results can be achieved if bipolar sigmoidal units are used instead (see also [21]). [20] also overcomes a restriction of the KBANN method originally presented in [14]: rules may now have arbitrarily many preconditions and programs may have arbitrarily many rules with the same postcondition.

The propositional Core Method has been extended in many directions. In [22] three-valued logic programs are discussed; This approach has been extended in [23] (see also the chapter by Komendantskaya, Lane and Seda in this volume) to finitely determined sets of truth values. Modal logic programs have been considered in [24] (see also the chapter by Garcez in this volume). Answer set programming and meta-level priorities are discussed in [21]. The Core Method has been applied to intuitionistic logic programs in [25].

To summarize, the propositional Core Method allows for model generation with respect to a variety of logics in a connectionist setting. Given logic programs are translated into recurrent connectionist networks with feed-forward cores, such that the cores compute the meaning functions associated with the programs. The cores can be trained using standard learning methods leading to improved logic programs. These improved programs must be extracted from the trained cores in order to complete the neural-symbolic cycle. The extraction process is outside the scope of this chapter and interested readers are referred to e.g. [26] or [21].

9.4 The Core Method and First-Order Logic

First- or higher-order logics are the primary choice if structured objects and structure-sensitive processes are to be modeled in Artificial Intelligence. In particular, first-order logic plays a prominent role because any computable function can be expressed by first-order logic programs. The extension of the Core Method to first-order logic poses a considerable problem because first-order interpretations usually do not map a finite but a countably infinite set of ground atoms to the set the truth values. Hence, they cannot be represented by a finite vector of units, each of which represents the value assigned to a particular ground atom.

We will first show in Section 9.4.1 that an extension of the core method to first-order logic programs is feasible. However, the result will be purely theoretical and thus the question remains how core-networks can be constructed for first-order programs. In Section 9.4.2, a first practical solution is discussed, which approximates the meaning functions of logic programs by means of piecewise constant functions. This approach is extended to a multi-dimensional setting in Section 9.4.3 allowing for arbitrary precision, even if implemented on a real computer. A novel training method, tailored for our specific setting, is discussed in Section 9.4.4. Some preliminary experimental data are presented in Section 9.5.

9.4.1 Feasibility

It is well known that multilayer feed-forward networks are universal approximators [27, 28] for certain functions of the type $\mathbb{R}^n \to \mathbb{R}^m$. Hence, if we find a way to represent interpretations of first-order logic programs by (finite vectors of) real numbers, then feed-forward networks can be used to approximate the meaning functions of such programs.

As proposed in [29], we use level mappings to bridge the gap between the space of interpretations and the real numbers.

Definition 1. *Let $I \in \mathcal{I}_\mathcal{L}$ be a Herbrand interpretation over $\mathcal{B}_\mathcal{L}$, $|\cdot|$ be an injective level mapping from $\mathcal{B}_\mathcal{L}$ to $\mathbb{N}^+$ and $b > 2$. Then we define the embedding function ι as follows:*

$$\iota : \mathcal{I}_\mathcal{L} \to \mathbb{R}$$
$$I \mapsto \sum_{A \in I} b^{-|A|}.$$

We will use $\mathfrak{C}$ to denote the set of all embedded interpretations:

$$\mathfrak{C} := \{\iota(I) \mid I \in \mathcal{I}_\mathcal{L}\} \subset \mathbb{R}$$

For $b > 2$, we find that ι is an injective mapping from $\mathcal{I}_\mathcal{L}$ to $\mathbb{R}$ and a bijection between $\mathcal{I}_\mathcal{L}$ and $\mathfrak{C}$.[5] Hence, we have a sound and complete encoding of interpretations.

Definition 2. *Let $\mathcal{P}$ be a logic program and $T_\mathcal{P}$ its associated meaning operator. We define a sound and complete encoding $f_P : \mathfrak{C} \to \mathfrak{C}$ of $T_\mathcal{P}$ as follows:*

$$f_P(r) = \iota(T_\mathcal{P}(\iota^{-1}(r))).$$

In [29] we proved – among other results – that for each logic program $\mathcal{P}$ which is acyclic wrt. a bijective level mapping the function f_P is contractive,[6] hence continuous. Moreover, $\mathfrak{C}$ is a compact subset of the real numbers. This has various implications: (i) We can apply Funahashi's result, viz. that every continuous function on (a compact subset of) the reals can be uniformly approximated by feed-forward networks with sigmoidal units in the hidden layer [28]. This shows that the meaning function of a logic program (of the kind discussed before) can be approximated by a core. (ii) Considering an appropriate metric, which will be discussed in a moment, we can apply Banach's contraction mapping theorem (see e.g. [30]) to conclude that the meaning function has a unique fixed point, which is obtained from an arbitrary initial interpretation by iterating the application of the meaning function. Using (i) and (ii) we were able to prove in [29] that the least model of logic programs which are acyclic wrt. a bijective level mapping can be approximated arbitrarily well by recurrent networks with feed-forward core.

But what exactly is the approximation of an interpretation or a model in this context? Let $\mathcal{P}$ be a logic program and l a level mapping. We can define a metric d on interpretations I and J as follows:

$$d(I, J) = \begin{cases} 0 & \text{if } I = J, \\ 2^{-n} & \text{if } n \text{ is the smallest level on which } I \text{ and } J \text{ disagree.} \end{cases}$$

As shown in [31] the set of all interpretations together with d is a complete metric space. Moreover, an interpretation I *approximates* an interpretation J *to degree* $n \in \mathbb{N}$ iff $d(I, J) \leq 2^{-n}$. In other words, if a recurrent network approximates the least model I of an acyclic logic program to a degree $n \in \mathbb{N}$ and outputs r then for all ground atoms A whose level is equal or less than n we find that $I(A) = \iota^{-1}(r)(A)$.

Theorem 1. *Let $\mathcal{P}$ be an acyclic logic program with respect to some bijective level mapping ι. Then there exists a 3-layer core-network with sigmoidal activation function in the hidden layer approximating $T_\mathcal{P}$ up to a given accuracy $\varepsilon > 0$.*

---

[5] For $b = 2$, ι is not injective, as $0.0\bar{1}_2 = 0.1_2$.
[6] This holds for $b > 3$ only. Therefore, we will use $b = 4$ throughout this article.

Proof (sketch). This theorem follows directly from the following facts:

1. $T_\mathcal{P}$ for an acyclic logic program is contractive [31].
2. f_P is a sound and complete embedding of $T_\mathcal{P}$ into the real numbers which is contractive and hence continuous for acyclic logic programs [29].
3. $\mathfrak{C} \subset \mathbb{R}$ is compact [29].
4. Every continuous function on a compact subset of $\mathbb{R}$ can be uniformly approximated by feed-forward networks with sigmoidal units in the hidden layer (Funahashi's theorem [28]).

Theorem 1 shows the existence of an approximating core network. Similar results for more general programs can also be obtained and are reported in [19]. Moreover, in [29] it is also shown that a recurrent network with a core network as stated in Theorem 1 approximates the least fixed point of $T_\mathcal{P}$ for an acyclic logic program $\mathcal{P}$. But as mentioned above, these results are purely theoretical. Networks are known to exist, but we do not yet know how to construct them given a logic program. We will address this next.

9.4.2 Embedding

In this section, we will show how to construct a core network approximating the meaning operator of a given logic program. As above, we will consider logic programs $\mathcal{P}$ which are acyclic wrt. a bijective level mapping. We will construct sigmoidal networks and RBF networks with a raised cosine activation function. All ideas presented here can be found in detail in [32]. To illustrate the ideas, we will use the program $\mathcal{P}_2$ given in Example 2 as a running example. The construction consists of five steps:

1. Construct f_P as a real embedding of $T_\mathcal{P}$.
2. Approximate f_P using a piecewise constant function $\bar{f}_P$.
3. Implement $\bar{f}_P$ using (a) step and (b) triangular functions.
4. Implement $\bar{f}_P$ using (a) sigmoidal and (b) raised cosine functions.
5. Construct the core network approximating f_P.

In the sequel we will describe the ideas underlying the construction. A rigorous development including all proofs can be found in [32, 33]. One should observe that f_P is a function on $\mathfrak{C}$ and not on $\mathbb{R}$. Although the functions constructed below will be defined on intervals of $\mathbb{R}$, we are concerned with accuracy on $\mathfrak{C}$ only.

Construct f_P as a real embedding of $T_\mathcal{P}$:

We use $f_P(r) = \iota(T_\mathcal{P}(\iota^{-1}(r)))$ as a real-valued version for $T_\mathcal{P}$. As mentioned above, f_P is a sound and complete encoding of the immediate consequence operator. But first, we will have a closer look at its domain $\mathfrak{C}$. For readers familiar with fractal geometry, we note that $\mathfrak{C}$ is a variant of the classical

Cantor set [34]. The interval $[0, \frac{1}{b-1}]$ is split into b equal parts, where b is the natural number used in the definition of the embedding function ι. All except the left- and rightmost subintervals are removed. The remaining two parts are split again and the subintervals except the first and the last are removed, etc. This process is repeated ad infinitum and we find $\mathfrak{C}$ to be its limit, i.e. it is the intersection of all $\mathfrak{C}_n$, where $\mathfrak{C}_n$ denotes the result after n splits and removals. The first four iterations (i.e. $\mathfrak{C}_0$, $\mathfrak{C}_1$, $\mathfrak{C}_2$ and $\mathfrak{C}_3$) are depicted in Figure 9.4.

Fig. 9.4. The first four iterations ($\mathfrak{C}_0$, $\mathfrak{C}_1$, $\mathfrak{C}_2$ and $\mathfrak{C}_3$) of the construction of $\mathfrak{C}$ for $b = 4$.

Example 5. Considering program $\mathcal{P}_2$, and the level mapping from Example 3 ($|e(s^n(0))| := 2n + 1$, $|o(s^n(0))| := 2n + 2$), we obtain $f_{\mathcal{P}_2}$ as depicted in Figure 9.5 on the left.

Approximate $f_\mathcal{P}$ using a piecewise constant function $\bar{f}_\mathcal{P}$:

Under the assumption that $\mathcal{P}$ is acyclic, we find that all variables occurring in the precondition of a rule are also contained in its postcondition. Hence, for each level n and two interpretations I and J, we find that whenever $d(I, J) \leq 2^{-n}$ holds, $d(T_\mathcal{P}(I), T_\mathcal{P}(J)) \leq 2^{-n}$ follows. Therefore, we can approximate $T_\mathcal{P}$ to degree n by some function $\bar{T}_\mathcal{P}$ which considers ground atoms with a level less or equal to n only. Due to the acyclicity of $\mathcal{P}$, we can construct a finite ground program $\bar{\mathcal{P}} \subseteq \mathcal{G}(\mathcal{P})$ containing those clauses of $\mathcal{G}(\mathcal{P})$ with literals of level less or equal n only and find $T_{\bar{\mathcal{P}}} = \bar{T}_\mathcal{P}$. We will use $\bar{f}_\mathcal{P}$ to denote the embedding of $\bar{T}_\mathcal{P}$ and we find that $\bar{f}_\mathcal{P} := f_{\bar{\mathcal{P}}}$ is a piecewise constant function, being constant on each connected interval of $\mathfrak{C}_{n-1}$. Furthermore, we find that $|f_\mathcal{P}(x) - \bar{f}_\mathcal{P}(x)| \leq 2^{-n}$ for all $x \in \mathfrak{C}$, i.e., $\bar{f}_\mathcal{P}$ is a constant piecewise approximation of $f_\mathcal{P}$.

Example 6. For our running example $\mathcal{P}_2$ and $n = 3$, we obtain

$$\bar{\mathcal{P}}_2 = \{\ e(0).$$
$$e(s(0)) \leftarrow o(0).$$
$$o(0) \leftarrow \neg e(0).\ \}$$

and $\bar{f}_{\mathcal{P}_2}$ as depicted in Figure 9.5 on the right.

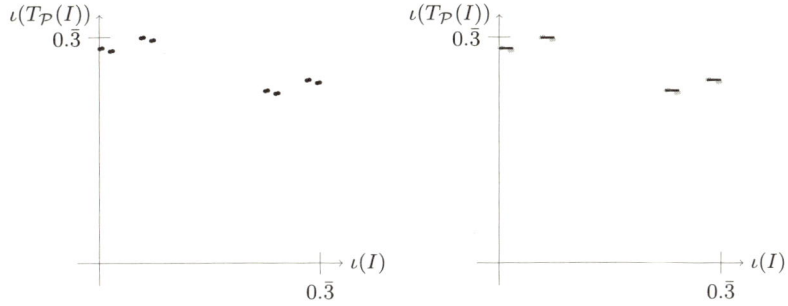

Fig. 9.5. On the left is the plot of $f_{\mathcal{P}_2}$. On the right a piecewise constant approximation $\bar{f}_{\mathcal{P}_2}$ (for level $n = 3$) of $f_{\mathcal{P}_2}$ (depicted in light gray) is shown.

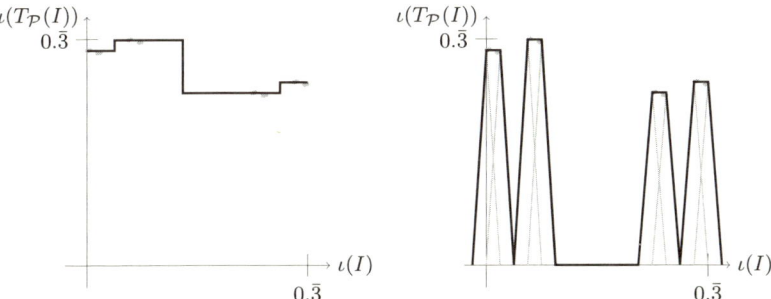

Fig. 9.6. Two linear approximations of $\bar{f}_{\mathcal{P}_2}$. On the left, three step functions were used; On the right, eight triangular functions (depicted in gray) add up to the approximation, which is shown using thick lines.

Implement $\bar{f}_\mathcal{P}$ using (a) step and (b) triangular functions:

As a next step, we will show how to implement $\bar{f}_\mathcal{P}$ using step and triangular functions. Those functions are the linear counterparts of the functions actually used in the networks constructed below. If $\bar{f}_\mathcal{P}$ consists of k intervals, then we can implement it using $k - 1$ step functions which are placed such that the steps are between two neighboring intervals. This is depicted in Figure 9.6 on the left.

Each constant piece of length $\lambda := \frac{1}{b-1} \cdot \frac{1}{b^n}$ could also be implemented using two triangular functions with width λ which are centered at the endpoints. Those two triangles add up to the constant piece. For base b, we find that the gaps between two intervals have a length of at least $(b-2)\lambda$. Therefore, the triangular functions of two different intervals will never interfere. The triangular implementation is depicted in Figure 9.6 on the right.

Implement $\bar{f}_\mathcal{P}$ using (a) sigmoidal and (b) raised cosine functions:

To obtain a sigmoidal approximation, we replace each step function with a sigmoidal function. Unfortunately, those add some further approximation error, which can be dealt with by increasing the accuracy in the constructions above. By dividing the desired accuracy by two, we can use one half as accuracy for the constructions so far and the other half as a margin to approximate the constant pieces by sigmoidal functions. This is possible because we are concerned with the approximation on $\mathfrak{C}$ only.

The triangular functions described above can simply be replaced by raised cosine activation functions, as those add up exactly as the triangles do and do not interfere with other intervals either.

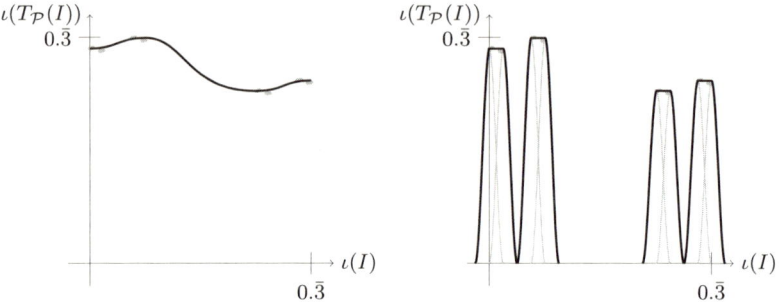

Fig. 9.7. Two non-linear approximations of $\bar{f}_{\mathcal{P}_2}$. On the left, sigmoidal functions were used and on the right, raised cosines.

Construct the core network approximating $f_\mathcal{P}$:

A standard sigmoidal core network approximating the $T_\mathcal{P}$-operator of a given program $\mathcal{P}$ consists of:

- An input layer containing one input unit whose activation will represent an interpretation I.
- A hidden layer containing a unit with sigmoidal activation function for each sigmoidal function constructed above.
- An output layer containing one unit whose activation will represent the approximation of $T_\mathcal{P}(I)$.

The weights from input to hidden layer together with the bias of the hidden units define the positions of the sigmoidals. The weights from hidden to output layer represent the heights of the single functions. An RBF network can be constructed analogously, but will contain more hidden layer units, one for each raised cosine functions. Detailed constructions can be found in [32].

A constructive proof for Theorem 1 is now possible. It follows directly from the fact that the network constructed as above approximates a given $T_\mathcal{P}$-operator up to any given accuracy. As the proof of the correctness of the construction is rather straightforward but tedious, we omit it here and refer the interested reader to [32, 33].

In this first approach we used a one-dimensional embedding to obtain a unique real number $\iota(I)$ for a given interpretation I. Unfortunately, the precision of a real computer is limited, which implies that using e.g. a 32-bit computer we could embed the first 16 atoms only. We address this problem in the next section.

9.4.3 Multi-Dimensional Embedding

We have just seen how to construct a core network for a given program and some desired level of accuracy. Due to the one-dimensional embedding, the precision of a real implementation is very limited. This limitation can be overcome by distributing an interpretation over more than one real number. In our running example $\mathcal{P}_2$, we could embed all *even*-atoms into one real number and all *odd*-atoms into another one, thereby obtaining a two-dimensional vector for each interpretation, hence doubling the accuracy. By embedding interpretations into higher-dimensional vectors, we can approximate meaning functions of logic programs arbitrarily well. For various reasons, spelled out in [35, 36], we will use an RBF network approach, inspired by vector-based networks as described in [37]. Analogously to the previous section, we will proceed as follows:

1. Construct f_P as a real embedding of $T_\mathcal{P}$.
2. Approximate f_P using a piecewise constant functions $\bar{f}_P$.
3. Construct the core network approximating f_P.

I.e., after discussing a new embedding of interpretations into vectors of real numbers, we will show how to approximate the embedded $T_\mathcal{P}$-operator using a piecewise constant function. This piecewise function will then be implemented using a connectionist system.

Additionally, we will present a novel training method, tailored for our specific setting. The system presented here is a fine blend of techniques from the *Supervised Growing Neural Gas (SGNG)* [37] and our embedding.

Construct f_P as a real embedding of $T_\mathcal{P}$:

We will first extend level mappings to a multi-dimensional setting, and then use them to represent interpretations as real vectors. This leads to a multi-dimensional embedding of $T_\mathcal{P}$.

Definition 3. *An m-dimensional level mapping is a bijective function* $\|\cdot\| : \mathcal{B}_\mathcal{L} \to (\mathbb{N}^+, \{1, \ldots, m\})$. *For $A \in \mathcal{B}_\mathcal{L}$ and $\|A\| = (l, d)$, we call l and d the* level *and* dimension *of A, respectively. Again, we define $\|\neg A\| := \|A\|$.*

Example 7. Reconsider the program from Example 2. A possible 2-dimensional level mapping is given as:

$$\|\cdot\| : \mathcal{B}_\mathcal{L} \to (\mathbb{N}^+, \{1, 2\})$$
$$e(s^n(0)) \mapsto (n+1, 1)$$
$$o(s^n(0)) \mapsto (n+1, 2).$$

I.e., all *even*-atoms are mapped to the first dimension, whereas the *odd*-atoms are mapped to the second.

Definition 4. *Let $b \geq 3$ and let $A \in \mathcal{B}_\mathcal{L}$ be an atom with $\|A\| = (l, d)$. The m-dimensional embedding $\iota : \mathcal{B}_\mathcal{L} \to \mathbb{R}^m$ is defined as:*

$$\iota(A) := (\iota_1(A), \ldots, \iota_m(A)) \quad \text{with} \quad \iota_j(A) := \begin{cases} b^{-l} & \text{if } j = d \\ 0 & \text{otherwise} \end{cases}$$

The extension $\iota : \mathcal{I}_\mathcal{L} \to \mathbb{R}^m$ is obtained as:

$$\iota(I) := \sum_{A \in I} \iota(A).$$

We will use $\mathfrak{C}^m$ to denote the set of all embedded interpretations:

$$\mathfrak{C}^m := \{\iota(I) \mid I \in \mathcal{I}_\mathcal{L}\} \subset \mathbb{R}^m.$$

As mentioned above, $\mathfrak{C}^1$ is the classical Cantor set and $\mathfrak{C}^2$ the 2-dimensional variant of it [34]. Obviously, ι is injective for a bijective level mapping and it is bijective on $\mathfrak{C}^m$. Without going into detail, Figure 9.8 shows the first 4 steps in the construction of $\mathfrak{C}^2$. The big square is first replaced by 2^m shrunken copies of itself, the result is again replaced by 2^m smaller copies and so on. The limit of this iterative replacement is $\mathfrak{C}^2$. Again, we will use $\mathfrak{C}^m_i$ to denote the result of the i-th replacement, i.e. Figure 9.8 depicts $\mathfrak{C}^2_0, \mathfrak{C}^2_1, \mathfrak{C}^2_2$ and $\mathfrak{C}^2_3$. For readers with background in fractal geometry we note, that these are the first 4 applications of an appropriately set up iterated function system [34].

Example 8. Using the 1-dimensional level mapping from Example 3, we obtain $\mathfrak{C}^1$ as depicted in Figure 9.8 at the top. Using the 2-dimensional from above, we obtain $\mathfrak{C}^2$ and $\iota(M_2) = (0.10\overline{10}_4, 0.01\overline{01}_4) \approx (0.2666667, 0.0666667)$ for the embedding of the intended model M_2.

Using the m-dimensional embedding, the $T_\mathcal{P}$-operator can be embedded into the real vectors to obtain a real-valued function $f_\mathcal{P}$.

Definition 5. *Let ι be an m-dimensional embedding as introduced in Definition 4. The m-dimensional embedding $f_\mathcal{P}$ of a given $T_\mathcal{P}$-operator is defined as:*

$$f_\mathcal{P} : \mathfrak{C}^m \to \mathfrak{C}^m$$
$$\mathbf{x} \mapsto \iota\left(T_\mathcal{P}\left(\iota^{-1}(\mathbf{x})\right)\right).$$

This embedding of $T_\mathcal{P}$ is preferable to the one presented above, because it allows for better approximation precision on real computers.

9 The Core Method for First-Order Logic Programs   219

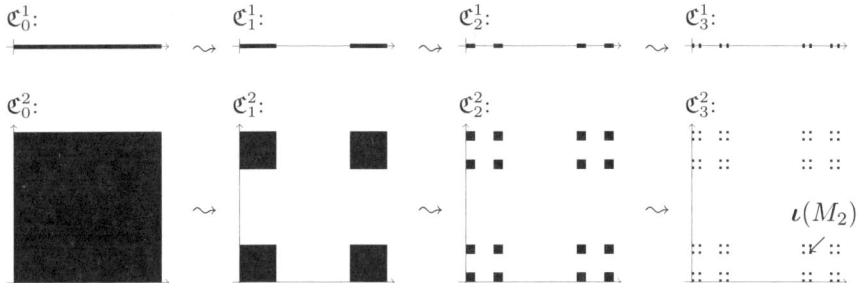

Fig. 9.8. The first four iterations of the construction of $\mathfrak{C}^1$ are depicted on the top. The construction of $\mathfrak{C}^2$ is depicted below containing $\iota(M_2)$ from Example 8. Both are constructed using $b = 4$.

Approximate $f_\mathcal{P}$ using a piecewise constant function $\bar{f}_\mathcal{P}$:

As above and under the assumption that $\mathcal{P}$ is acyclic, we can approximate $T_\mathcal{P}$ up to some level n by some $\bar{T}_\mathcal{P}$. After embedding $\bar{T}_\mathcal{P}$ into $\mathbb{R}^m$, we find that it is constant on certain regions, namely on all connected intervals in $\mathfrak{C}^m_{n-1}$. Those intervals will be referred to as *hyper-squares* in the sequel. We will use H_l to denote a hyper-square of level l, i.e. one of the squares occurring in $\mathfrak{C}^m_l$. An approximation of $T_\mathcal{P}$ up to some level n will yield a function which is constant on all hyper-squares of level $n-1$.

Example 9. Considering program $\mathcal{P}_2$, and the level mapping from Example 7, we obtain $\bar{f}_\mathcal{P}$ for $n = 3$ as depicted in Figure 9.9.

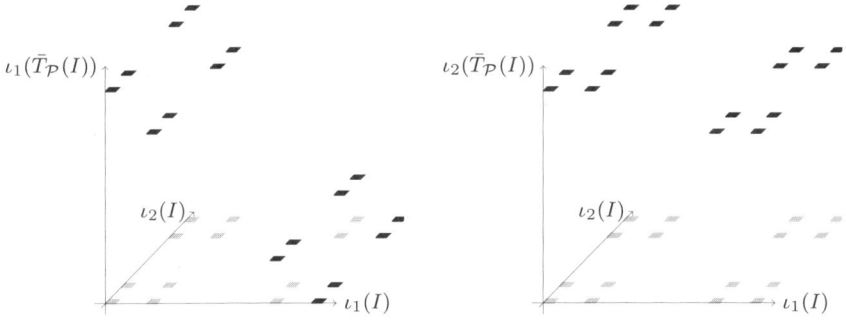

Fig. 9.9. $\bar{f}_\mathcal{P}$ for $\mathcal{P}_2$, the 2-dimensional level mapping from Example 7 and $n = 3$. The outputs for dimension 1 and 2 are shown on the left and on the right, respectively.

To simplify the notations later, we introduce the *largest exclusive hyper-square* and the *smallest inclusive hyper-square* as follows.

Definition 6. *The* largest exclusive hyper-square *of a vector* $\mathbf{u} \in \mathfrak{C}_0^m$ *and a set of vectors* $V = \{\mathbf{v}_1, \ldots, \mathbf{v}_k\} \subseteq \mathfrak{C}_0^m$, *denoted by* $H_{ex}(\mathbf{u}, V)$, *either does not exist or is the hyper-square* H *of least level for which* $\mathbf{u} \in H$ *and* $V \cap H = \emptyset$. *The* smallest inclusive hyper-square *of a non-empty set of vectors* $U = \{\mathbf{u}_1, \ldots, \mathbf{u}_k\} \subseteq \mathfrak{C}_0^m$, *denoted by* $H_{in}(U)$, *is the hyper-square* H *of greatest level for which* $U \subseteq H$.

Example 10. Let $v_1 = (0.07, 0.07)$, $v_2 = (0.27, 0.03)$, $v_3 = (0.13, 0.27)$ and $v_4 = (0.27, 0.13)$ as depicted in Figure 9.10. The largest exclusive hyper-square of v_1 with respect to $\{v_2, v_3, v_4\}$ is shown in light gray on the left. That of v_3 with respect to $\{v_1, v_2, v_4\}$ does not exists, because there is no hyper-square which contains only v_3. The smallest inclusive hyper-square of all four vectors is shown on the right, and is in this case $\mathfrak{C}_0$.

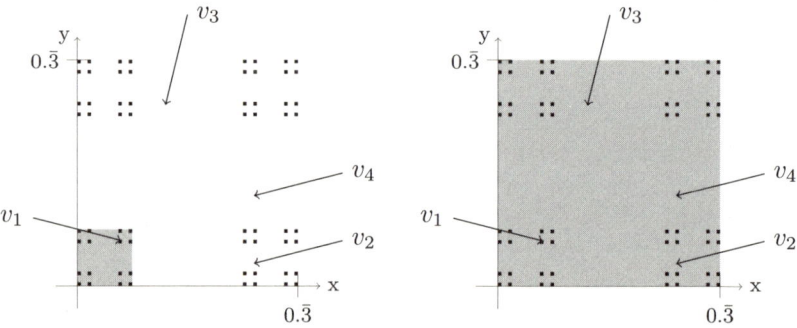

Fig. 9.10. The largest exclusive hyper-square of v_1 with respect to the set $\{v_2, v_3, v_4\}$ is shown on the left and the smallest inclusive hyper-square of the set $\{v_1, v_2, v_3, v_4\}$ is shown on the right. Both are depicted in light gray together with $\mathfrak{C}^2$ in black.

Construct the core network approximating f_P:

We will use a 3-layered network with a winner-take-all hidden layer. For each hyper-square H of level $n-1$, we add a unit to the hidden layer, such that the input weights encode the position of the center of H. The unit shall output 1 if it is selected as winner, and 0 otherwise. The weight associated with the output connections of this unit is the value of $\bar{f}_P$ on that hyper-square. Thus, we obtain a connectionist network approximating the semantic operator T_P up to the given accuracy ε.

To determine the winner for a given input, we designed an activation function such that its outcome is greatest for the closest "responsible" unit. Responsible units are defined as follows: Given some hyper-square H, units which are positioned in H but not in any of its sub-hyper-squares are called

Input: $\mathbf{x}, \mathbf{y} \in \mathfrak{C}_0^m$
Output: Activation $d_{\mathfrak{C}}(\mathbf{x}, \mathbf{y}) \in (\mathbb{N}, \{1, 2, 3\}, \mathbb{R})$

if $\mathbf{x} = \mathbf{y}$ **then return** $(\infty, 0, 0)$
$l := $ level of $H_{in}(\{\mathbf{x}, \mathbf{y}\})$
Compute k according to the following 3 cases:

$$k := \begin{cases} 3 & \text{if } H_{in}(\{\mathbf{x}\}) \text{ and } H_{in}(\{\mathbf{y}\}) \text{ are of level greater than } l \\ 2 & \text{if } H_{in}(\{\mathbf{x}\}) \text{ or } H_{in}(\{\mathbf{y}\}) \text{ is of level greater than } l \\ 1 & \text{otherwise} \end{cases}$$

$m := \frac{1}{|\mathbf{x}-\mathbf{y}|}$, i.e., m is the inverse of the Euclidean distance
return (l, k, m)

Fig. 9.11. Algorithm for the activation function for the Fine Blend

default units of H, and they are responsible for inputs from H except for inputs from sub-hyper-squares containing other units. If H does not have any default units, the units positioned in its sub-hyper-squares are responsible for all inputs from H as well. After all units' activations have been computed, the unit with the greatest value is selected as the winner. The details of this function $d_{\mathfrak{C}}$ are given in Algorithm 9.11. Please note that the algorithm outputs a 3-tuple, which is compared component wise, i.e. the first component is most important. If for two activations this first component is equal, the second component is used and the third, if the first two are equal.

Example 11. Let v_1, v_2, v_3 and v_4 from Example 10 be the incoming connections for the units u_1, u_2, u_3 and u_4 respectively. The different input regions for which each of the units are responsible are depicted in Figure 9.12. For the vector $i = (0.05, 0.02)$ (also shown in Figure 9.12), we obtain the following activations:

$$d_{\mathfrak{C}}(v_1, i) = (1, 2, 20.18)$$
$$d_{\mathfrak{C}}(v_2, i) = (0, 1, 4.61)$$
$$d_{\mathfrak{C}}(v_3, i) = (0, 2, 3.84)$$
$$d_{\mathfrak{C}}(v_4, i) = (0, 2, 4.09)$$

I.e., we find $d_{\mathfrak{C}}(v_1, i) > d_{\mathfrak{C}}(v_4, i) > d_{\mathfrak{C}}(v_3, i) > d_{\mathfrak{C}}(v_2, i)$. Even though, v_2 is euclidically closer to i than v_3 and v_4 it is further away according to our "distance" function. This is due to the fact, that it is considered to be the default unit for the south-east hyper-square, whereas v_3 and v_4 are responsible for parts of the big square.

9.4.4 Training

In this section, we will describe the adaptation of the system during training, i.e. how the weights and the structure of a network are changed, given

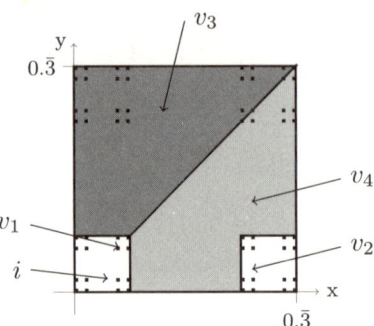

Fig. 9.12. The areas of responsibility for v_1, v_2, v_3 and v_4. For each of the four regions one of the units is responsible.

training samples with input and desired output. This process can be used to refine a network resulting from an incorrect program, or to train a network from scratch.[7] The training samples in our case come from the original (non-approximated) program, but might also be observed in the real world or given by experts. First we discuss the adaptation of the weights and then the adaptation of the structure by adding and removing hidden-layer units. Some of the methods used here are adaptations of ideas described in [37]. For a more detailed discussion of the training algorithms and modifications we refer to [35, 36].

Adapting the weights

Let $\mathbf{x}$ be the input, $\mathbf{y}$ be the desired output and u be the winner-unit from the hidden layer. Let $\mathbf{w}_{in}$ denote the weights of the incoming connections of u and $\mathbf{w}_{out}$ be the weights of the outgoing connections. To adapt the system, we move u towards the center $\mathbf{c}$ of $H_{in}(\{\mathbf{x}, u\})$, i.e.:

$$\mathbf{w}_{in} \leftarrow \mu \cdot \mathbf{c} + (1 - \mu) \cdot \mathbf{w}_{in}.$$

Furthermore, we change the output weights for u towards the desired output:

$$\mathbf{w}_{out} \leftarrow \eta \cdot \mathbf{y} + (1 - \eta) \cdot \mathbf{w}_{out}.$$

η and μ are predefined learning rates. Note that (in contrast to the methods described in [37]) the winner unit is not moved towards the input, but towards the center of the smallest hyper-square including the unit and the input. The intention is that units should be positioned in the center of the hyper-square for which they are responsible. Figure 9.13 depicts the adaptation of the incoming connections.

---

[7] E.g., using an initial network with a single hidden layer unit.

9 The Core Method for First-Order Logic Programs   223

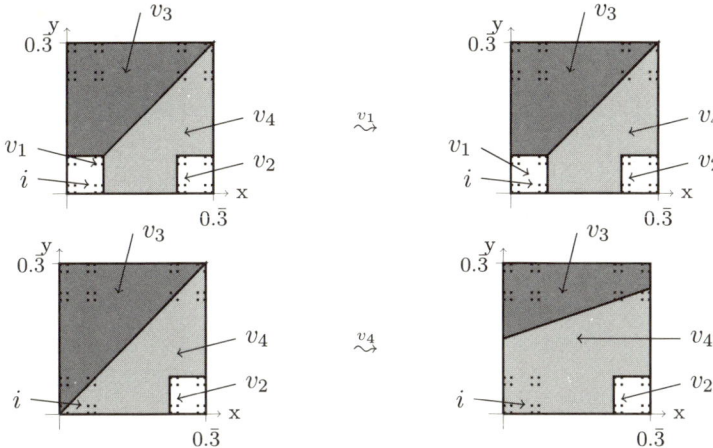

Fig. 9.13. The adaptation of the input weights for a given input i. The first row shows the result of adapting v_1. The second row shows the result if v_1 would not be there and therefore v_4 would be selected as winner. To emphasize the effect, we used a learning rate $\mu = 1.0$, i.e., the winning unit is moved directly into the center of the hyper-square.

Adding new units

The adjustment described above enables a certain kind of expansion of the network by allowing units to move to positions where they are responsible for larger areas of the input space. A refinement now should take care of densifying the network in areas where a great error is caused. Every unit will accumulate the error for those training samples it is winner for. If this accumulated error exceeds a given threshold, the unit will be selected for refinement. I.e., we try to figure out the area it is responsible for and a suitable position to add a new unit.

Let u be a unit selected for refinement. If it occupies a hyper-square on its own, then the largest such hyper-square is considered to be u's responsibility area. Otherwise, we take the smallest hyper-square containing u. Now u is moved to the center of this area. Information gathered by u during the training process is used to determine a sub-hyper-square into whose center a new unit is placed, and to set up the output weights for the new unit. All units collect statistics to guide the refinement process. E.g., the error per sub-hyper-square or the average direction between the center of the hyper-square and the training samples contributing to the error could be used (weighted by the error). This process is depicted in Figure 9.14.

Removing inutile units

Each unit maintains a utility value, initially set to 1, which decreases over time and increases only if the unit contributes to the network's output. The

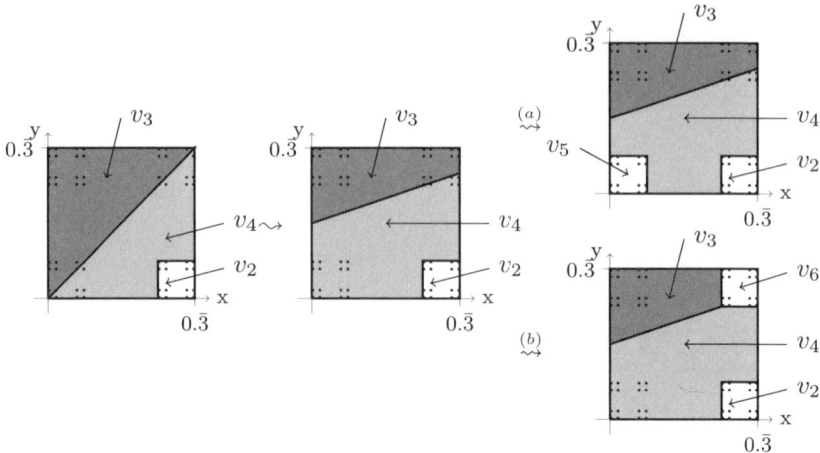

Fig. 9.14. Adding a new unit to support v_4. First, v_4 is moved to the center of the hyper-square it is responsible for. There are four possible sub-hyper-squares to add a new unit. Because v_4 is neither responsible for the north-western, nor for the south-eastern sub-hyper-square, there are two cases left. If most error was caused in the south western sub-hyper-square (a), a new unit (v_5) is added there. If most error was caused in the north-eastern area (b), a new unit (v_6) would be added there.

contribution of a unit is the expected increase of error if the unit would be removed [37]. If a unit's utility drops below a threshold, the unit will be removed as depicted in Figure 9.15.

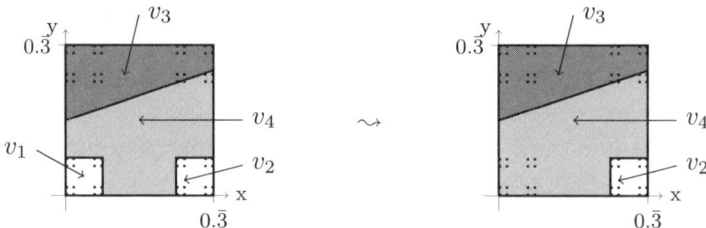

Fig. 9.15. Removing an inutile unit. Let us assume that the outgoing weights of v_1 and v_4 are equal. In this case we would find that the over-all error would not increase if we remove v_1. Therefore its utility would decrease over time until it drops below the threshold and the unit is removed.

The methods described above, i.e. the adaptation of the weights, the addition of new units and the removal of inutile ones, allows the network to learn from examples. While the idea of growing and shrinking the network using utility values was taken from vector-based networks [37], the adaptation of the weights and the positioning of new units are specifically tailored for the type of function we like to represent, namely functions on $\mathfrak{C}_0^m$. The preliminary experiments described in the following section, will show that our method actually works.

9.5 Evaluation

In this section we will discuss some preliminary experiments. Those are not meant to be exhaustive, but rather to provide a proof of concept. An in-depth anaylsis of all required parameters will be done in the future. In the diagrams, we use a logarithmic scale for the error axis, and the error values are relative to ε, i.e. a value of 1 designates an absolute error of ε.

To initialize the network we used the following wrong program:

$$e(s(X)) \leftarrow \neg o(X).$$
$$o(X) \leftarrow e(X).$$

Training samples were created randomly using the semantic operator of the correct program given in Example 2, viz.:

$$e(0).$$
$$e(s(X)) \leftarrow o(X).$$
$$o(X) \leftarrow \neg e(X).$$

Variants of Fine Blend

To illustrate the effects of varying the parameters, we use two setups: One with softer utility criteria (FineBlend 1) and one with stricter ones (FineBlend 2). Figure 9.16 shows that, starting from the incorrect initialization, the former decreases the initial error, paying with an increasing number of units, while the latter significantly decreases the number of units, paying with an increasing error. Hence, the performance of the network critically depends on the choice of the parameters. The optimal parameters obviously depend on the concrete setting, e.g. the kind and amount of noise present in the training data, and methods for finding them will be investigated in the future. For our further experiments we will use the FineBlend 1 parameters, which resulted from a mixture of intuition and (non-exhaustive) comparative simulations.

Fine Blend versus SGNG

Figure 9.17 compares FineBlend 1 with SGNG [37]. Both start off similarly, but soon SGNG fails to improve further. The increasing number of units is partly due to the fact that no error threshold is used to inhibit refinement, but this should not be the cause for the constantly high error level. The choice of

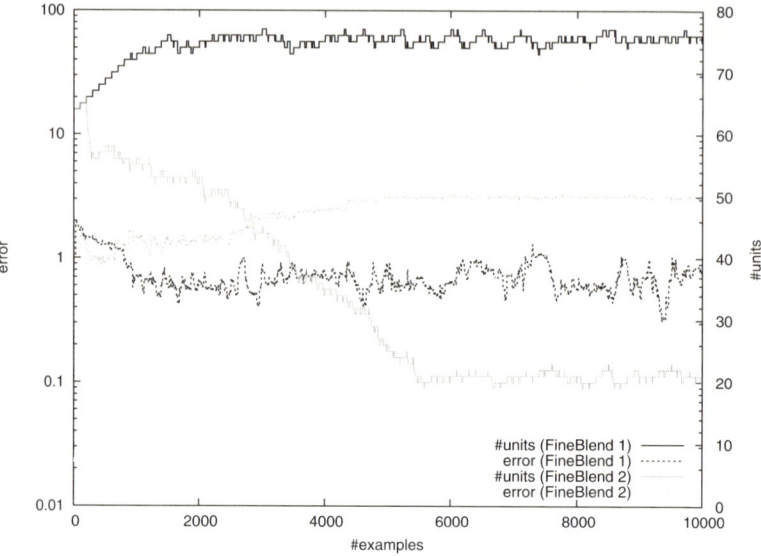

Fig. 9.16. FineBlend 1 versus FineBlend 2.

SGNG parameters is rather subjective, and even though some testing was done to find them, they might be far from optimal. Finding the optimal parameters for SGNG is beyond the scope of this paper; however, it should be clear that it is not perfectly suited for our specific application. This comparison to an established generic architecture shows that our specialized architecture actually works, i.e. it is able to learn, and that it achieves the goal of specialization, i.e. it outperforms the generic architecture in our specific setting.

Robustness

The described system is able to handle noisy data and to cope with damage. Indeed, the effects of damage to the system are quite obvious: If a hidden unit u fails, the receptive area is taken over by other units, thus only the specific results learned for u's receptive area are lost. While a corruption of the input weights may cause no changes at all in the network function, generally it can alter the unit's receptive area. If the output weights are corrupted, only certain inputs are effected. If the damage to the system occurs during training, it will be repaired very quickly as indicated by the following experiment. Noise is generally handled gracefully, because wrong or unnecessary adjustments or refinements can be undone in the further training process.

Unit Failure

Figure 9.18 shows the effects of unit failure. A FineBlend 1 network is initialized and refined through training with 5000 samples, then one third of its

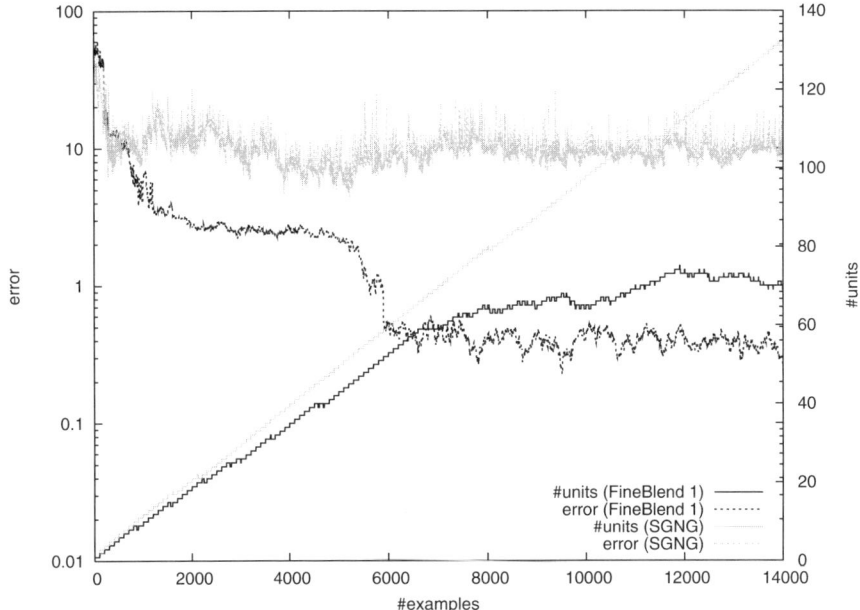

Fig. 9.17. FineBlend 1 versus SGNG.

hidden units are removed randomly, and then training is continued as if nothing had happened. The network proves to handle the damage gracefully and to recover quickly. The relative error exceeds 1 only slightly and drops back very soon; the number of units continues to increase to the previous level, recreating the redundancy necessary for robustness.

Iterating Random Inputs

One of the original aims of the Core Method is to obtain connectionist systems for logic programs which, when iteratively feeding their output back as input, settle to a stable state corresponding to an approximation of a desired model of the program, or more precisely to a fixed point of the program's single-step operator. In this sense, the Core Method allows to reason with the acquired knowledge. For our system, this model generation also serves as a sanity check: if the model can be reproduced successfully in the connectionist setting then this shows that the network was indeed able to acquire symbolic knowledge during training.

In our running example program, a unique fixed point is known to exist. To check whether our system reflects this, we proceed as follows:

1. Train a network from scratch until the relative error caused by the network is below 1, i.e. network outputs are in the ε-neighborhood of the desired outputs.

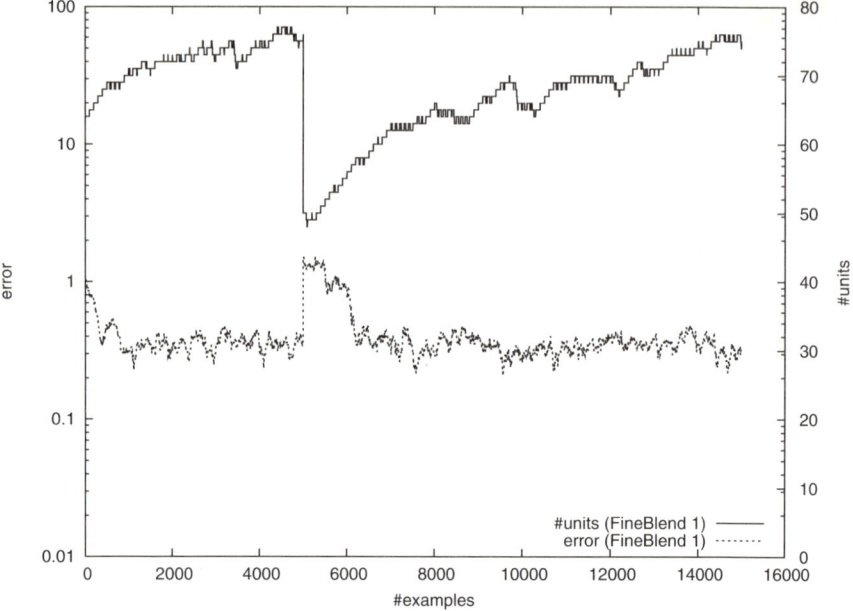

Fig. 9.18. The effects of unit failure.

2. Transform the obtained network into a recurrent one by connecting the outputs to the corresponding inputs.
3. Choose a random input vector $\in \mathfrak{C}_0^m$ (which is not necessarily a valid embedded interpretation) and use it as initial input to the network.
4. Iterate the network until it reaches a stable state, i.e. until the outputs stay inside an ε-neighborhood.

For our example program, the unique fixed point of $T_{\mathcal{P}_2}$ is M_2 as given in Example 4. Figure 9.19 shows the input space and the ε-neighborhood of M, along with all intermediate results of the iteration for 5 random initial inputs. The example computations converge, because the underlying program is acyclic [35, 36, 29]. After at most 6 steps, the network is stable in all cases, in fact it is completely stable in the sense that all outputs stay exactly the same and not only within an ε-neighborhood. This corresponds roughly to the number of applications of our program's $T_{\mathcal{P}_2}$ operator required to fix the significant atoms, which confirms that the training method really implements our intention of learning $T_{\mathcal{P}_2}$. The fact that even a network obtained through training from scratch converges in this sense further underlines the efficiency of our training method.

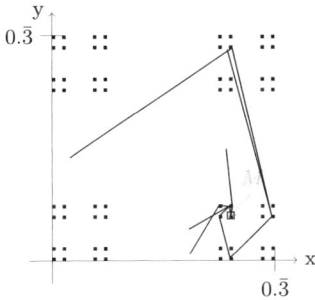

Fig. 9.19. Iterating random inputs. The two dimensions of the input vectors are plotted against each other. The ε-neighborhood of the fixed point M is shown as a small box.

9.6 Conclusion

After reviewing the connection between the T_P-operator associated with propositional logic programs and simple three-layer feed-forward neural networks, we reported on extensions to first order logic programs. By restating results from [29], we showed that a representation of semantic operators is possible. Afterwards, we described a first approach [32] to actually construct approximating networks. This approach was extended to a multi-dimensional setting allowing for arbitrary precision, even if implemented on a real computer [35, 36]. Finally, we reported on some preliminary experiments which show that our approach actually works.

Our system realises part of the neural-symbolic cycle in that it is able to learn first-order logic programs, and to outperform other approaches in this specific learning task. The trained network is also able to generate the model of the target program which shows that it has acquired the desired declarative knowledge. At the same time, the network is robust in that it can recover from substantial unit failure. Summarising, our system combines the power of connectionist learning and robustness with the processing of declarative knowledge, and thus retains the best of both worlds. It is fair to say that this system provides a major step in the overcoming what McCarthy called *propositional fixation*.

We are currently re-implementing the first-order Core Method in order to further evaluate and test it using real world examples. An in-depth analysis of the system shall provide heuristics to determine optimal values for the parameters of the system. Furthermore, we intend to compare our approach with other methods. Concerning a complete neural-symbolic cycle we note that whereas the extraction of propositional rules from trained networks is well understood, the extraction of first-order rules is still an open question, which will be addressed in the future.

Acknowledgments:

Many thanks to Sven-Erik Bornscheuer, Artur d'Avila Garcez, Yvonne McIntyre (formerly Kalinke), Anthony K. Seda, Hans-Peter Störr and Jörg Wunderlich who all contributed to the Core Method.

References

1. McCulloch, W.S., Pitts, W.: A logical calculus of the ideas immanent in nervous activity. Bulletin of Mathematical Biophysics **5** (1943) 115–133
2. Pinkas, G.: Symmetric neural networks and logic satisfiability. Neural Computation **3** (1991) 282–291
3. McCarthy, J.: Epistemological challanges for connectionism. Behavioural and Brain Sciences **11** (1988) 44 Commentary to [12].
4. Bader, S., Hitzler, P.: Dimensions of neural-symbolic integration - a structured survey. In: S. Artemov, H. Barringer, A. S. d'Avila Garcez, L. C. Lamb and J. Woods (eds).: We Will Show Them: Essays in Honour of Dov Gabbay, Volume 1. International Federation for Computational Logic, College Publications (2005) 167–194
5. Ballard, D.H.: Parallel logic inference and energy minimization. In: Proceedings of the AAAI National Conference on Artificial Intelligence. (1986) 203 – 208
6. Lange, T.E., Dyer, M.G.: High-level inferencing in a connectionist network. Connection Science **1** (1989) 181–217
7. Shastri, L., Ajjanagadde, V.: From associations to systematic reasoning: A connectionist representation of rules, variables and dynamic bindings using temporal synchrony. Behavioural and Brain Sciences **16** (1993) 417–494
8. Hölldobler, S., Kurfess, F.: CHCL – A connectionist inference system. In Fronhöfer, B., Wrightson, G., eds.: Parallelization in Inference Systems. Springer, LNAI **590** (1992) 318 – 342
9. Pollack, J.B.: Recursive distributed representations. AIJ **46** (1990) 77–105
10. Plate, T.A.: Holographic reduced networks. In Giles, C.L., Hanson, S.J., Cowan, J.D., eds.: Advances in Neural Information Processing Systems 5. Morgan Kaufmann (1992)
11. Elman, J.L.: Structured representations and connectionist models. In: Proceedings of the Annual Conference of the Cognitive Science Society. (1989) 17–25
12. Smolensky, P.: On variable binding and the representation of symbolic structures in connectionist systems. Technical Report CU-CS-355-87, Department of Computer Science & Institute of Cognitive Science, University of Colorado, Boulder, CO 80309-0430 (1987)
13. Bader, S., Hitzler, P., Hölldobler, S.: The Integration of Connectionism and First-Order Knowledge Representation and Reasoning as a Challenge for Artificial Intelligence. Information **9** (2006).
14. Towell, G.G., Shavlik, J.W.: Extracting refined rules from knowledge-based neural networks. Machine Learning **13** (1993) 71–101
15. Hölldobler, S., Kalinke, Y.: Towards a massively parallel computational model for logic programming. In: Proceedings ECAI94 Workshop on Combining Symbolic and Connectionist Processing, ECCAI (1994) 68–77

16. Bishop, C.M.: Neural Networks for Pattern Recognition. Oxford University Press (1995)
17. Lloyd, J.W.: Foundations of Logic Programming. Springer, Berlin (1988)
18. Apt, K.R., Blair, H.A., Walker, A.: Towards a theory of declarative knowledge. In Minker, J., ed.: Foundations of Deductive Databases and Logic Programming. Morgan Kaufmann, Los Altos, CA (1988) 89–148
19. Hitzler, P., Hölldobler, S., Seda, A.K.: Logic programs and connectionist networks. Journal of Applied Logic **3** (2004) 245–272
20. d'Avila Garcez, A.S., Zaverucha, G., de Carvalho, L.A.V.: Logical inference and inductive learning in artificial neural networks. In Hermann, C., Reine, F., Strohmaier, A., eds.: Knowledge Representation in Neural networks. Logos Verlag, Berlin (1997) 33–46
21. d'Avila Garcez, A.S., Broda, K.B., Gabbay, D.M.: Neural-Symbolic Learning Systems — Foundations and Applications. Perspectives in Neural Computing. Springer, Berlin (2002)
22. Kalinke, Y.: Ein massiv paralleles Berechnungsmodell für normale logische Programme. Master's thesis, TU Dresden, Fakultät Informatik (1994) (in German).
23. Seda, A., Lane, M.: Some aspects of the integration of connectionist and logic-based systems. In: Proceedings of the Third International Conference on Information, International Information Institute, Tokyo, Japan (2004) 297–300
24. d'Avila Garcez, A.S., Lamb, L.C., Gabbay, D.M.: A connectionist inductive learning system for modal logic programming. In: Proceedings of the IEEE International Conference on Neural Information Processing ICONIP'02, Singapore. (2002)
25. d'Avila Garcez, A.S., Lamb, L.C., Gabbay, D.M.: Neural-symbolic intuitionistic reasoning. In A. Abraham, M.K., Franke, K., eds.: Frontiers in Artificial Intelligence and Applications, Melbourne, Australia, IOS Press (2003) Proceedings of the Third International Conference on Hybrid Intelligent Systems (HIS'03).
26. Andrews, R., Diederich, J., Tickle, A.: A survey and critique of techniques for extracting rules from trained artificial neural networks. Knowledge–Based Systems **8** (1995)
27. Hornik, K., Stinchcombe, M., White, H.: Multilayer feedforward networks are universal approximators. Neural Networks **2** (1989) 359–366
28. Funahashi, K.I.: On the approximate realization of continuous mappings by neural networks. Neural Networks **2** (1989) 183–192
29. Hölldobler, S., Kalinke, Y., Störr, H.P.: Approximating the semantics of logic programs by recurrent neural networks. Applied Intelligence **11** (1999) 45–58
30. Willard, S.: General Topology. Addison–Wesley (1970)
31. Fitting, M.: Metric methods, three examples and a theorem. Journal of Logic Programming **21** (1994) 113–127
32. Bader, S., Hitzler, P., Witzel, A.: Integrating first-order logic programs and connectionist systems – a constructive approach. In d'Avila Garcez, A.S., Elman, J., Hitzler, P., eds.: Proceedings of the IJCAI-05 Workshop on Neural-Symbolic Learning and Reasoning, NeSy'05, Edinburgh, UK. (2005)
33. Witzel, A.: Integrating first-order logic programs and connectionist systems – a constructive approach. Project thesis, Department of Computer Science, Technische Universität Dresden, Dresden, Germany (2005)
34. Barnsley, M.: Fractals Everywhere. Academic Press, San Diego, CA, USA (1993)

35. Witzel, A.: Neural-symbolic integration – constructive approaches. Master's thesis, Department of Computer Science, Technische Universität Dresden, Dresden, Germany (2006)
36. Bader, S., Hitzler, P., Hölldobler, S., Witzel, A.: A fully connectionist model generator for covered first-order logic programs. In Veloso, M.M., ed.: Proceedings of the Twentieth International Joint Conference on Artificial Intelligence (IJCAI-07), Hyderabad, India, Menlo Park CA, AAAI Press (2007) 666–671
37. Fritzke, B.: Vektorbasierte Neuronale Netze. Habilitation, Technische Universität Dresden (1998)

10

Learning Models of Predicate Logical Theories with Neural Networks Based on Topos Theory

Helmar Gust, Kai-Uwe Kühnberger, and Peter Geibel

Institute of Cognitive Science, University of Osnabrück,
Albrechtstr. 28, 49076 Osnabrück
{hgust,kkuehnbe,pgeibel}@uos.de

Summary. This chapter presents an approach to learn first-order logical theories with neural networks. We discuss representation issues for this task in terms of a variable-free representation of predicate logic using topos theory and the possibility to use automatically generated equations (induced by the topos) as input for a neural network. Besides the translation of first-order logic into a variable-free representation, a programming language fragment for representing variable-free logic, the structure of the used neural network for learning, and the overall architecture of the system are discussed. Finally, an evaluation of the approach is presented by applying the framework to theorem proving problems.

10.1 Introduction

A classical problem in cognitive science and artificial intelligence concerns the gap between symbolic and subsymbolic representations. On the one hand, symbolic approaches have been successfully applied to a variety of domains for modeling higher cognitive abilities such as reasoning, theorem proving, planning, or problem solving. On the other hand, subsymbolic approaches have been proven to be extremely successful in other domains, in particular, with respect to lower cognitive abilities, such as learning from noisy data, controlling real world robots in a rapidly changing environment, or detecting visual patterns, just to mention some of these various applications.

On the symbolic level, recursion principles ensure that the formalisms are productive and allow very compact representations: due to the compositionality principle of almost all logic- and algebra-based models commonly used in AI, it is possible to compute the meaning of a complex symbolic expression using the meaning of the embedded subexpressions. For example, the interpretation of a first-order formula like

$$\exists x_1 \forall x_2 : \phi(x_1, x_2) \rightarrow \forall y_1 \exists y_2 : \phi(y_2, y_1) \qquad (10.1)$$

can be computed from the interpretation of the underlying constituents, as for example, $\phi(x_1, x_2)$ and $\phi(y_2, y_1)$. Whereas compositionality and recursion allow symbolic approaches to compute potentially infinitely many expressions with different meanings (using a finite basis), it is usually assumed that neural networks are non-compositional on a principal basis making it difficult to represent and compute complex, i.e. recursive data structures like lists, trees, tables, well-formed formulas etc. Most types of highly structured data are to a certain extend incompatible with the required homogeneous input and output for neural networks.

From a cognitive science perspective it would be desirable to bring these two approaches together. For example, humans perceive sensory data on a subsymbolic level, process this data on the neural level, and finally reason on interpretations of the sensory input data on a conceptual level. There is no general way to explain how this transfer of sensory subsymbolic input data to a conceptual level is possible at all. From a dual perspective a similar problem arises: whereas conceptual reasoning can be modeled by symbolic approaches, it is far from being clear how this can be translated into an input of a neural network and which types of processing corresponds on the neural level to these symbolic manipulations.

The described gap between symbolic and subsymbolic processing has been tackled for more than twenty years. Two aspects can be distinguished: the representation problem [1] and the inference problem [2]. The first problem states that, if at all, complex data structures can only implicitly be used and the representation of structured objects is a non-trivial challenge for connectionist networks [3] – and prerequisite for reasoning and learning. The second problem concerns possibilities to model inferences and reasoning processes based on logical systems with neural accounts.

A certain endeavor has been invested to solve the representation problem as well as the inference problem. It is well-known that classical logical connectives like conjunction, disjunction, or negation can be represented (and learned) by neural networks [4]. Furthermore it is known that every Boolean function can be learned by a neural network [5]. Therefore, it is rather straightforward to represent propositional logic with neural networks. Unfortunately the situation with respect to first-order logic (FOL) is more difficult due to the possibility to quantify over the universe (or over sub-domains of the universe). The corresponding problem, usually called the variable-binding problem, is caused by the usage of the quantifiers $\forall$ and $\exists$ which bind variables that may occur at different positions in one and the same formula. It is therefore no surprise that a number of attempts has been proposed to solve this problem for neural networks. Examples for such attempts are sign propagation [6], dynamic localist representations [1], tensor product representations [7], or holographic reduced representations [8]. Unfortunately these accounts have certain non-trivial side-effects. Whereas sign propagation as well as dynamic localist representations lack the ability of learning, the tensor product representation results in an exponentially increasing number of elements to represent variable bindings, only to mention some of the problems.

10 Learning Models with Neural Networks Based on Topos Theory

With respect to the inference problem of connectionist networks, the number of proposed solutions is rather small and relatively new. An attempt is [9] in which a logical deduction operator is approximated by a neural network and the fixed point of such an operator provides the semantics of a logical theory. Another approach is [10] where category theoretic methods are assigned to neural constructions. In [11], tractable fragments of predicate logic are learned by connectionist networks. Finally in [12, 13], a procedure is proposed how to translate predicate logic into variable-free logic that can be used as input for a neural network. To the knowledge of the authors, the latter account is the only one that does not require hard-wired networks designed for modeling a particular theory. Rather one network topology can be used for arbitrary first-order theories.

We will apply the account presented in [12, 13] to model first-order inferences of neural networks. The overall idea is to translate a first-order theory into a variable-free representation in a topos and to use this representation to generate equations induced by standard constructions in a topos. These equations in turn are now used to train a neural network. After the training phase the network has learned a model of the input theory.

The paper has the following structure: In Section 10.2, a variable-free logic based on topos theory is introduced and the translation of logical theories in topos theory is sketched. Section 10.3 presents the overall architecture of the present approach. In Section 10.4, certain principles are specified governing the translation of first-order logic into categorical constructions. These principles are intended to reduce the learning complexity. Section 10.5 introduces a fragment of a programming language used to represent the mentioned categorical constructions. Section 10.6 discusses roughly the topology of the neural network and how learning of first-order models with neural networks can be achieved. Section 10.7 introduces two examples from the domain of theorem proving: the Socrates example and the famous steamroller problem, the second being a well-known benchmark problem for theorem provers. Finally, an evaluation of the learned models can be found in Section 10.8 and Section 10.9 concludes the paper with some rough remarks concerning potential problems, future work, and possible applications.

10.2 Variable-Free Logic

10.2.1 Interpretation in an Abstract Topos

An interpretation I of a logic language $\mathcal{L}$ is a function mapping formulas of $\mathcal{L}$ to truth-values, function symbols and relation symbols of $\mathcal{L}$ to functions and relations (of appropriate arity) over a universe U, and terms to elements of U. Functions and relations together with the universe U form an algebra and I must be specified in terms of this algebra. Interpreting expressions of the algebra is done using a meta-language, which in most cases is nothing but natural language.

A more flexible way of interpreting logical languages $\mathcal{L}$ uses an abstract topos[1] $\mathcal{T}$ in such a way, that all logical formation rules of the language correspond to universal constructions in $\mathcal{T}$. Then, for defining the interpretation, there is no need to deal with special situations in intended model classes or special truth-value configurations. All this need not to be fixed until designing a concrete application.

The approach sketched in this paper combines a direct interpretation of the language in an abstract topos which gives a very abstract characterization of the semantics of the language, independent of the concrete logical basis of an application.

A topos is a comprehensive characterization of (order-)sorted higher-order many-valued predicate logic [14]. Factorizing an interpretation of a logical language $\mathcal{L}$ via the very general structure of a topos enables us to characterize the semantics of the constructions in $\mathcal{L}$ by constructions in an abstract topos $\mathcal{T}$ without the need of taking special properties, like the structure of the universe, the sortal system, and the truth-values into account. The constructions are general. By specifying a specific topos structure when applying such a system, all the constructions get their specific form automatically without any additional effort, even in cases where it might be very complex to spell out, for example, the interpretation of universal quantification in detail in a sorted higher-oder intensional logic like Montague's intentional logic [15]. All we have to do here is to specify the topos $\mathcal{T}$ as the slice (or comma) category $\mathcal{SET} \downarrow J$ of functions to an index set J of $\mathcal{SET}$ (the category of sets and functions). Due to the fundamental theorem of topoi [14], such a slice category of a topos is again a topos, and all required constructions are well defined.

10.2.2 Basic Topos Theory

Good introductions to topos theory and its connection to logic can be found in [14] and in [16] (where the second reference emphasizes a more algorithmic point of view). Certain special aspects of the relation between topoi and logic are discussed in [17] and [18]. A topos $\mathcal{T}$ is a category with nice properties. The paradigmatic case is the category of sets and functions $\mathcal{SET}$, which is a topos. The central properties of a topos are:

- All finite diagrams have limits.
- Exponents exist.
- There is a subobject classifier $true : t \longrightarrow \Omega$.

As a consequence, a topos has

- an initial object i,

---

[1] *Abstract* should mean here that we view $\mathcal{T}$ as an abstract structure of arrows and we use nothing else than the defining properties of a topos.

10 Learning Models with Neural Networks Based on Topos Theory   237

- a terminal object t,
- a truth-value object Ω,
- finite products and co-products.

The fundamental constructions are

- product arrows (denoted by $g \times f$) are given by the adjointness condition

$$\frac{g : c \to a, f : c \to b}{g \times f : c \to a \times b}$$

$$\pi_1 \circ g \times f = g$$
$$\pi_2 \circ g \times f = f$$

where π_i are the corresponding projections
- coproduct arrows (denoted by $\langle g, f \rangle$) are given by

$$\frac{g : a \to c, f : b \to c}{\langle g, f \rangle : a + b \to c}$$

$$\langle g, f \rangle \circ j_1 = g$$
$$\langle g, f \rangle \circ j_2 = f$$

where j_i are the corresponding injections
- exponents (denoted by $exp(f)$) by

$$\frac{f : c \times a \to b}{exp_{\pi_1}(f) : c \to b^a}$$

We omit the defining diagrams here, since they can be found in every textbook (cf. [14]).

The subobject classifier diagram generalizes the concept of subsets in $\mathcal{SET}$ and therefore characterizes the interpretation of a predicate by two corresponding arrows f and g, making the following diagram a pullback[2]:

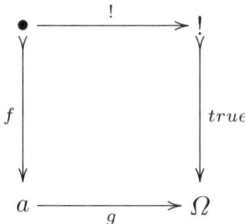

Given one of them, we introduce two functions ext and $char$, which refer to the corresponding other arrow:

$$f = ext(g), \quad g = char(f)$$

Other constructions we need are summarized in the following list:

---

[2] If we want to emphasize certain attributes of arrows, $\rightarrowtail$ denote monic arrows (in $\mathcal{SET}$ injections) and $\twoheadrightarrow$ denote epic arrows (in $\mathcal{SET}$ surjections).

- Diagonals $\Delta : a \to a \times a$ defined by $\pi_1 \circ \Delta = id_a$ and $\pi_2 \circ \Delta = id_a$.

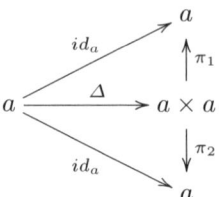

- Images: In a topos, for every arrow f, there is a unique epi-mono factorisation $f = img(f) \circ f^*$. Here $img(f)$ denotes the monic part and f^* the epic part of f.

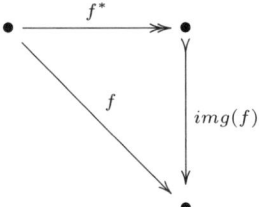

- Equalizers: For two parallel arrows f and g the equalizer $eql(f,g)$ makes both arrows equal:
$$f \circ eql(f,g) = g \circ eql(f,g)$$
Every other arrow h making f and g equal factors through $eql(f,g)$:

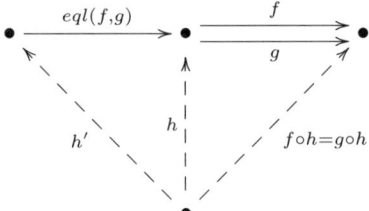

In a topos, $eql(f,g)$ is monic.

10.2.3 Translating Logic into Topos Theory

The first step to learn logical models with neural networks is to translate an input given in a first-order logical language into Topos theory.

Predicates, Functions and Constants

The interpretation of an n-ary predicate p maps an n-tuple of individuals to a truth value. This corresponds in a topos to an arrow $[p] : u^n \longrightarrow \Omega$, where u denotes an object representing the universe. For an n-ary function symbol f we get $[f] : u^n \longrightarrow u$ and for a constant c we get $[c] : ! \longrightarrow u$.

The Modeling of Basic Logical Operators

For the classical logical constants T, F and operators $\neg, \wedge, \vee, \rightarrow$ the interpretation in $\mathcal{T}$ can be given directly by the corresponding arrows or constructions. Figures 10.1, 10.2, and 10.3 depict the essential constructions for the classical logical operators. Figure 10.1 depicts the defining diagrams for the *false* and the *true* arrow.

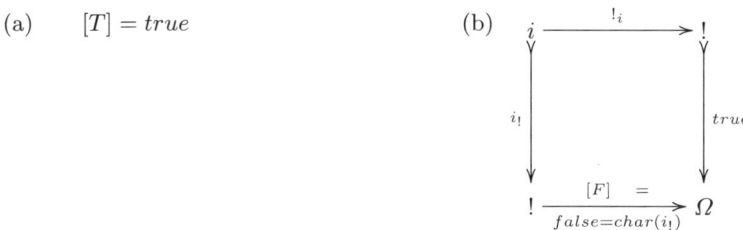

Fig. 10.1. Diagram (a) represent the interpretation of the logical constant T in a topos, and diagram (b) represents the logical constant F (represented as the arrow *false*).

We consider roughly the construction of some more complicated examples: Figure 10.2(a) depicts the construction of conjunction. Logically conjunctions map products of truth values into a truth value, i.e. conjunction is of type $\Omega \times \Omega$ into Ω. The construction of the implication arrow (Figure 10.2(b)) is slightly more complicated because an equalizer construction is needed.

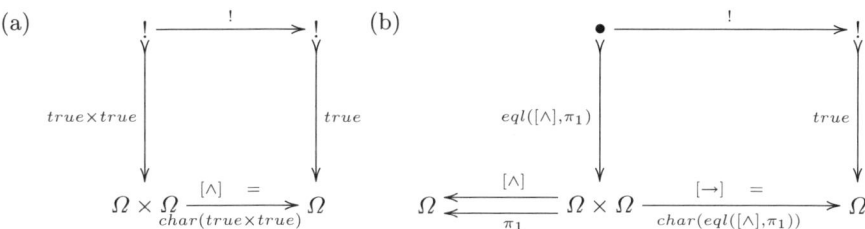

Fig. 10.2. Diagram (a) depicts the construction of the arrow for conjunction and (b) the construction for implication.

Figure 10.3(a) depicts the construction of negation and Figure 10.3(b) the construction of disjunction:

$$[\vee] = char(img(\langle id \times true \circ !, true \circ ! \times id\rangle))$$

For all domain specific operators (functions and predicates) the domain theory will only impose constraints via the axioms.

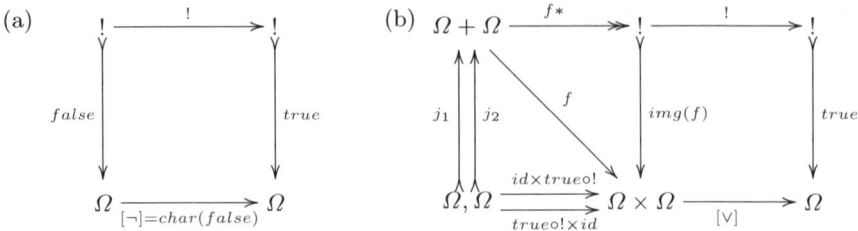

Fig. 10.3. Diagram (a) depicts the construction of the arrow for negation and diagram (b) the construction for disjunction, where f is a coproduct arrow defined as follows: $f = \langle id \times true \circ !, true \circ ! \times id \rangle$.

Quantifiers

The interpretation of quantifiers in a topos can in general be characterized by adjoint functors to the inverse of an arrow $[p]$ representing the predicate p (cf. [14]). In our case, where only projections are involved, we can directly construct the resulting arrows: For existential quantification, we get

$$\exists \pi_2 : \alpha = char(img(\pi_2 \circ ext(\alpha)))$$

which is illustrated in Figure 10.4. Notice: the bottom triangle does not commute and the left front square is not a pullback.

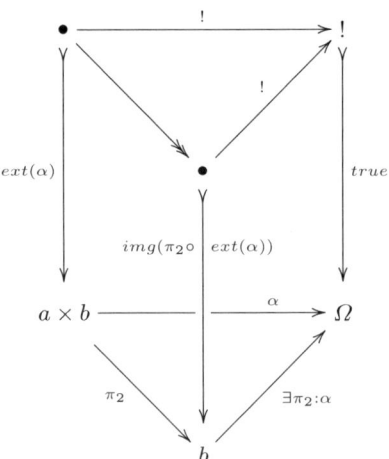

Fig. 10.4. The topos representation of existential quantification.

Consider the subformula $\exists y_2 : \phi(y_2, y_1)$ from Formula (10.1) from the Introduction. Since ϕ is a 2-ary predicate the interpretation of ϕ is an arrow

10 Learning Models with Neural Networks Based on Topos Theory   241

$[\phi] : u \times u \longrightarrow \Omega$. The interpretation of the quantified subformula refers to a 1-ary predicate $\lambda y_1.\exists y_2 : \phi(y_2, y_1)$. Variables do no longer occur in the interpretation of formulas. The "information" they code in the classical case is present in the domains of the predicates. Therefore, we get:

$$[\phi(y_2, y_1)] = [\phi] : u \times u \longrightarrow \Omega$$
$$[\exists y_2 : \phi(y_2, y_1)] = \exists \pi_2 : [\phi(y_2, y_1)] = \exists \pi_2 : [\phi]$$

This can be generalized to quantification along arbitrary arrows, since the diagram does not rely on π_2 being a projection. This is different in case of universal quantification. Only along projections we can give an explicit characterization without referring to a right adjoint functor to the inverse arrow of α. The interpretation of universal quantification along a projection is

$$(\forall \pi_1 : \alpha) = char(exp_!(true \circ ! \circ \pi_2)) \circ exp_{\pi_1}(\alpha)$$

Figure 10.5 provides the commuting diagram for universal quantification.

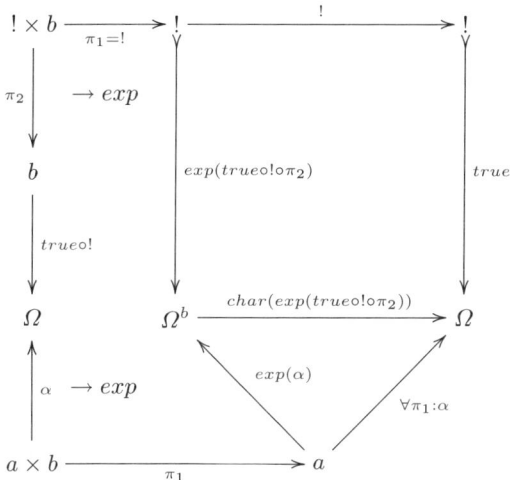

Fig. 10.5. The topos representation of universal quantification.

In order to exemplify universal quantification consider the subformula from Formula (10.1): $\forall x_2 : \phi(x_1, x_2)$. The interpretation of this subformula refers again to a 1-ary predicate $\lambda x_1.\forall x_2 : \phi(x_1, x_2)$. We get:

$$[\forall x_2 : \phi(x_1, x_2)] = \forall \pi_1 : [\phi(x_1, x_2)] = \forall \pi_1 : [\phi]$$

Although we get an explicit characterization for universal quantification along a projection, the construction is quite complex, since it uses exponents which can be viewed as higher order entities.

Interpretations and Models

Now we are able to translate an arbitrary FOL formula A into an arrow $[A]$ in a topos. The constructions used for the translation process introduce constraints in form of commuting triangles, existence of needed arrows, and uniqueness of arrows. Commuting triangles are simple constraints of the form $h = f \circ g$ and existence can be handled by skolemization introducing unique names. Uniqueness, of course, may be problematic. As in classical interpretations, we have some freedom in mapping atomic predicates function symbols and constants to arrows in $\mathcal{T}$.

In order to exemplify the approach consider again the example from the introduction:

$$\exists x_1 \forall x_2 : \phi(x_1, x_2) \rightarrow \forall y_1 \exists y_2 : \phi(y_2, y_1) \tag{10.2}$$

(10.2) translates in a topos to:

$$[\rightarrow] \circ ((\exists! : \forall \pi_1 : [\phi]) \times (\forall! : \exists \pi_2 : [\phi])) \tag{10.3}$$

The overall idea now is very simple: For every axiom A of a FOL theory we introduce an equation $[A] = true$. The translation process of the axioms itself introduces a lot of additional arrows and equations originating from the commuting triangles involved in the constructions. We interpret the equations as constraints for in interpretation to be a model of the FOL theory: $[.]$: $\mathcal{L} \longrightarrow \mathcal{T}$ is a model of a theory, if and only if it obeys all the constraints from translating the axioms of the theory.

Given a query formula Q, proving Q means checking if $[Q] = true$ holds in all models $[Q]$. This is the case, if and only if we can derive $[Q] = true$ solely from the constraints.

From a categorical point of view, objects and arrows have no inner structure, i.e. they are considered as atomic. Therefore it is possible to choose arbitrary representations of these entities as long as the internal structure of the representations does not play a role. Notice: although it seems to be the case that expressions like $\forall \pi_1 : [\phi]$ or $f \times g$ have an inner structure these expressions are just complex names for atomic entities.

10.3 Architecture

In this section, we describe how the translation process described in Section 10.2 can be implemented in a way such that neural networks can learn logical models of a theory. We have to collect all the constraints that are induced by

10 Learning Models with Neural Networks Based on Topos Theory   243

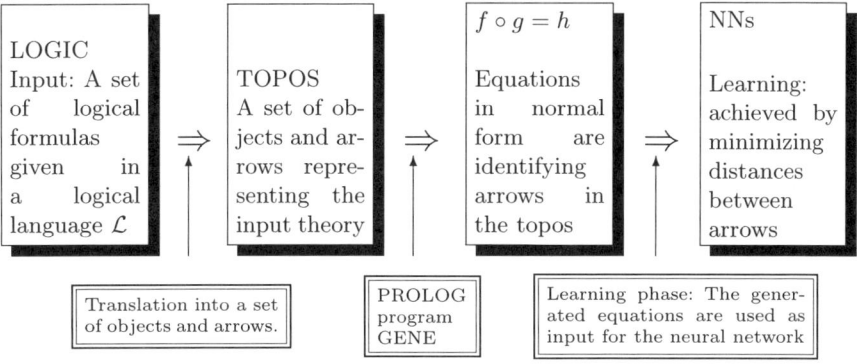

Fig. 10.6. The general architecture of the system.

the given theory and the constructions in form of equations and represent the corresponding entities in a form suitable for the neural network. Figure 10.6 depicts the general architecture of our approach in four steps:

- First, input data is given by a set of logical formulas (axioms and queries) relative to a given logical language $\mathcal{L}$. The language $\mathcal{L}$ is considered to be a classical first-order language.
- Second, this set of formulas is translated into objects and arrows of a topos based on the fact that FOL can be represented by a topos. The motivation for this step is to translate logical expressions into a homogeneous structure where only one operation is allowed (concatenation of arrows) and variables do no longer occur.
- Third, a PROLOG program generates equations in normal form $f \circ g = h$ identifying new arrows in the topos. This is possible because a topos allows limit constructions, exponentiations etc. In order to make this work, we developed a simple topos language $\mathcal{L}_T$ to code the definitions of objects and arrows in a way such that they can be processed by the program components (cf. Section 10.5). The idea is that a given arrow f can be used to generate new equations like $id \circ f = f$, $f \circ id = f$ and so on.
- Last but not least, these equations are used as input for the training of a neural network. The network topology of the underlying neural network will be described below (cf. Subsection 10.6.2).

The motivation of the proposed solution is based on the idea that we need to transform an interpretation function I of classical logic into a function $I' : \mathbb{R}^m \to \mathbb{R}^n$ in order to make it appropriate as input for a neural network.

The resulting trained network represents one concrete model of the underlying theory (provided the network can successively be trained). This means that the network should output *true* for the axioms and the consequences of the theory. On other formulas, in particular on formulas not related to the theory, it can produce arbitrary values.

10.4 Principles for Translating First-Order Logic to Categorical Constructions

As mentioned in the previous section, for every logical construction there is a corresponding categorical construction. Nevertheless there is some freedom how to implement these constructions and there are some possibilities for simplifications.

The general scheme for combining a subformula p with n arguments by a junctor J with a formula q with m arguments to a complex formula with k arguments is given by the following diagram:

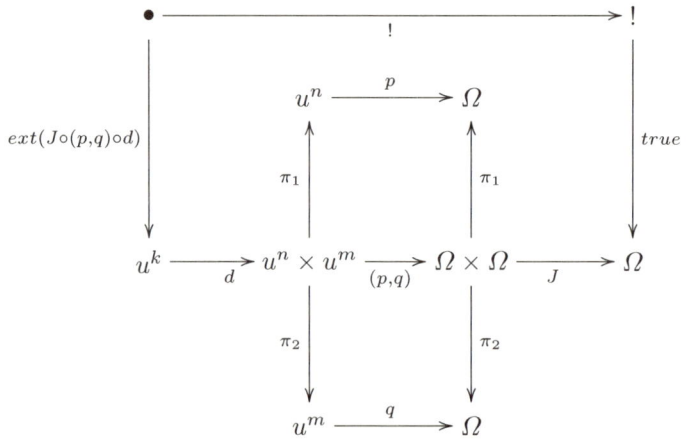

Here d denotes a generalized diagonal responsible for the variable bindings of the arguments. The arrow (p, q) is defined by the equation $(p, q) = p \circ \pi_1 \times q \circ \pi_2$.

For the frequently used special case that p and q have identical arguments this simplifies to

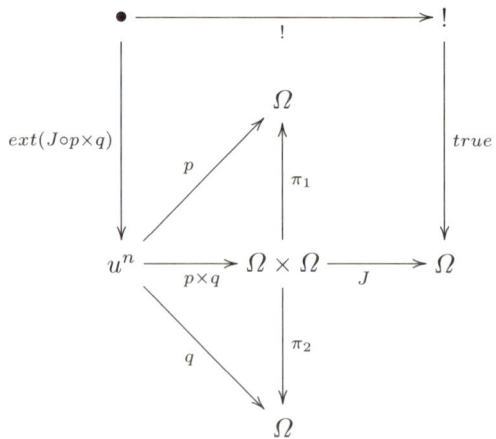

10 Learning Models with Neural Networks Based on Topos Theory

The equations for training the network can be obtained by looking at the commuting triangles of the diagrams associated with the used constructions. But this is not enough, because most of the constructions are universal constructions, i.e. that additional triangles must commute for all objects and suitable arrows.

As a simple example we consider the defining commuting diagram for negation $[\neg]$:

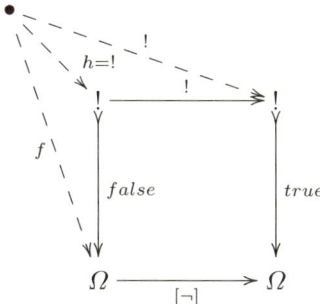

Due to the fact that the square needs to be a pullback we get: for all arrows f with $[\neg] \circ f = true \circ !$ the equation $f = false \circ !$ must hold. This seems to be natural for classical (bivalent) logic, but in a topos in general we cannot guarantee that Ω has exactly two elements. As a consequence there can only be one f with $[\neg] \circ f = true \circ !$. This property cannot be expressed in simple equations.

A related example is the commuting diagram defining $false$.

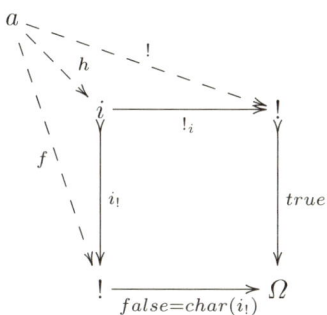

Since there is exactly one f (namely !) and $false \circ ! \neq true$ (at least in a non-degenerated topos) there cannot be an h at all. Similarly to the above example, non-existence of arrows cannot be expressed by simple equations. In Section 10.6, the problem is solved by explicitly defining $true$ and $false$ to be distinct.

An important and general case is the defining commuting diagram for implication $[\rightarrow]$:

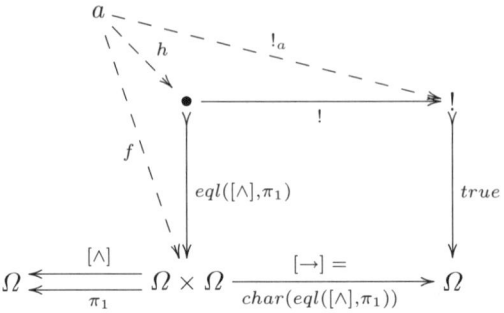

For every f with $[\to] \circ f = true \circ \,!$ there must a uniquely determined h with $eql([\wedge], \pi_1) \circ h = f$. We can account for this by introducing a function $univ_{pb}$ for each pullback pb (so we have to index the pullbacks) mapping the outer dashed arrows (in this case f and $!_a$) to the arrow h resembling the universal property of the pullback.

Since a topos is closed under products, coproducts and exponents, in non-degenerated cases there are infinitely many objects and arrows resulting in infinitely many commuting triangles inducing infinitely many new equations. This has two consequences: first, we are only able to approximate a logical model by using a finite subset of equations dealing with those objects and arrows that are essential for the modeling. Second, we should try to keep constructions easy by reducing the first-order formulas to more simple normal form-like expressions. For example, we can reduce quantification to universal closures (without existential quantification), because complex alternating prefixes of quantifiers imply quite complex categorical constructions.

In contrast to constructions discussed so far, universal closures are quite simple and do not even require a full implementation of pullbacks. Given a predicate $p : a \to \Omega$ which is true on a, then obviously the following diagram is a pullback:

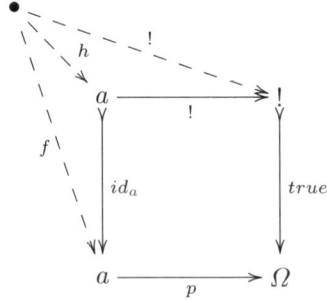

The left-most and upper-most triangles commute in a trivial sense (due to the arrow id_a) with $h = f$. Thus the universal property of this diagram holds without any additional equations. The fact that the universal closure of p is

10 Learning Models with Neural Networks Based on Topos Theory   247

true can therefore be expressed by the single equation $p = p \circ id_a = true \circ !_a$.
To simplify expressions even more we introduce the following two conventions:

1. Given an arrow f from the terminal object ! to an arbitrary object b, then for every other object a there exists the arrow $f_a = f \circ !_a$ according to the following diagram:

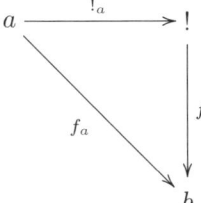

   We can index all arrows of this type by their domains and use the same name $f, !$ etc. in place of $f_a, !_a$ etc. to denote all these arrows. This is consistent with the convention to name all terminal arrows with the same symbol !. These conventions reduce the necessary representation space.
2. Due to the fact that objects and arrows will be viewed as different representation spaces, formally we identify an object with its identity arrow.[3]

With the described conventions the universal closure of a predicate p reduces to the equation $p = true$ without any additional constraint.

What remains is, that most of the logical connectives of Section 10.4 are defined by nontrivial pullbacks resulting in probably infinitely many equations. As a first approximation, instead of using the defining diagrams of Section 10.4 for generating equations for the logical operators, we use simple equations defining the effect of these operators on the two explicitly given truth values $true$ and $false$ (which implements the usual truth tables). As an example, consider negation and implication:

$[\neg] \circ true = false$
$[\neg] \circ false = true$
$[\rightarrow] \circ true \times true = true$
$[\rightarrow] \circ true \times false = false$
$[\rightarrow] \circ false \times true = true$
$[\rightarrow] \circ false \times false = true$

At least for implication (cf. Figure 10.2(b)) this is a very rough approximation and the results in Section 10.8 show that probably it is too rough for certain applications.

---

[3] It is well-known in category theory that one can abstract from objects and reduce constructions to arrows.

248   Helmar Gust, Kai-Uwe Kühnberger, and Peter Geibel

10.5 A Programming Language Fragment for Categorical Constructions

The coding of topos entities in $\mathcal{L}_T$ is straightforward: In the first part, we define objects and arrows and in the second part we specify the defining equations. Table 10.1 summarizes the important constructions. Furthermore $\mathcal{L}_T$ provides a macro mechanism to allow a compact coding for complex equations (cf. the definition of ==> in Table 10.2). All derived objects and arrows, e.g. identities and products, are recognized by the PROLOG program and the corresponding defining equations are automatically generated.

Table 10.1. The specification of language $\mathcal{L}_T$ encoding topos entities.

$\mathcal{L}_T$	Intended Interpretation
!	Terminal object !
@	Truth value objects Ω
u	The universe U
t	Truth value $true$
f	Truth value $false$
Y x Z	Product object of Y and Z
y x z	Product arrow of y and z
!	Terminal arrows
f: Y --> Z	Definition of an arrow
y o z	Composition of arrows y and z

Technically we use a subset of the term language of Prolog as the representation language for the translated logical entities. The operator concept and the variable concept of Prolog then provides a very powerful tool for defining macros for compact and readable translations: practically for all logical connectives there are corresponding macros which then expand to the spelled-out categorical constructions. In Table 10.2 the relevant macros are summarized: First it should be mentioned that **not, and, or, ->** are the categorical counterparts of the logical connectives $\neg, \wedge, \vee, \rightarrow$, respectively, as described in Section 10.4. The macro **not** expands a predicate P to its negation. For **and, or, ->** the situation is slightly different: they translate a pair of predicates to the product predicate, i.e. if for example, unary predicates P_1 and P_2 are given, then **P1 and P2** refers to a binary predicate which is true if P_1 is true on the first argument and P_2 is true on the second argument. Therefore, these operators are not symmetric. The macro **==>** expands to an equation defining the universal closure of an implication (Figure 10.2) where premise and consequence have identical arguments.

The introduced constructions need to be specified with respect to their semantics. Table 10.3 is an example of how certain logical properties of objects and arrows of the topos can be coded in $\mathcal{L}_T$. The terminal, the truth value

Table 10.2. Relevant macros

```
define not   X :: (not) o X.
define X and Y :: (and) o (X,Y).
define X or Y  :: (or)  o (X,Y).
define X -> Y  :: (->)  o (X,Y).
define X ==> Y :: (->)  o (X x Y) = t.
```

object, and the universe are specified straightforwardly as ordinary objects in the topos. The modifier **static** for the truth values will be interpreted by the network program to use fixed representations. The logical connectives are introduced as arrows mapping truth values or pairs of truth values to truth values. The defining equations realize the corresponding truth-value tables. This completes the propositional part of the translation.

Table 10.3. Example code of the objects and arrows

```
!.                       # the terminal object
@.                       # the truthvalue object
! x ! = !.
u.                       # the universe
static  t:: ! --> @,     # true
static  f:: ! --> @.     # false
not:: @       --> @,     # negation
->::  @ x @ --> @.       # implication
not t = f,
not f = t,
-> o t x t = t,
-> o t x f = f,
-> o f x t = t,
-> o f x f = t.
```

10.6 Learning Models by a Network

10.6.1 Representations of Categorical Entities: Vector Space Model and Equational Descriptions of Categorical Constructions

As mentioned already in Section 10.2 objects and arrows are atomic, i.e. they have no inner structure from a categorical point of view. Therefore we are free to choose a suitable representation. Since we want to train a neural network to learn the composition operation it is natural to use a representation which can directly be used for the input of neural networks, i.e. real-valued vectors are preferred representations.

Based on Kolmogorov's theorem, neural networks are often considered as universal function approximators [4]. We will not use neural networks to approximate a given function, but rather we will use a neural network to

250   Helmar Gust, Kai-Uwe Kühnberger, and Peter Geibel

approximate the composition process (and thereby the behavior) of functions and predicates. More precisely, not the structural properties of these entities will be represented (simply because all these entities are viewed as atomic in our account), rather the behavior in the composition process is modeled. This means that representations of arrows need to be learned.

We discussed in Section 10.2 and Section 10.4 that the result of translating first-order formulas into a topos results in a set of defining diagrams. The essential constraints of these diagrams are commuting triangles. The module *GENE* translating topos entities (objects and arrows) into the set of equations in normal form as depicted in the architecture of the system (compare Figure 10.6), has a crucial restriction: not all equations can be generated due to the fact that we can handle only finitely many equations. Practically this means that in the examples specified in Section 10.9 approx. 80 equations are used in the Socrates example and approx. 800 equations are used in the Steamroller example. The basic idea is to use at least those equations which comprise the objects and arrows explicitly occurring in the axioms.

10.6.2 Network Topology

Figure 10.7 depicts the structure of the neural network that is used in order to model the composition process of evaluating terms and formulas. Each arrow in the topos is represented as a point in the n-dimensional real-valued unit cube together with pointers to the respective domain and codomain. Each object of the topos can also be represented as a point in an n-dimensional real-valued unit cube. In fact we chose to identify objects with their identity arrows. In the examples used in this paper[4], we varied the values for n. The input of the network is represented by weights from the initial node with activation 1. This allows the backpropagation of the errors into the representation of the inputs of the network. The input of the network represents the two arrows to be composed by the following parts:

- The domain of the first arrow
- The representation of the first arrow
- The codomain of the first arrow which must be equal to the domain of the second arrow
- The representation of the second arrow
- The codomain of the second arrow

These requirements lead to a net with $5 \cdot n$ many input values (compare the first layer in Figure 10.7). The output of the network is the representation of the composed arrow. In the example, we use $h \cdot n$ many nodes for the hidden

---

[4] The choice of n depends on the number of objects and arrows which need to be represented. If n is too large, then overfitting of the network can occur. If n is too small, the network may not converge. Currently we do not know the precise correlation between the choice of n and the relative size of the logical theory.

10 Learning Models with Neural Networks Based on Topos Theory

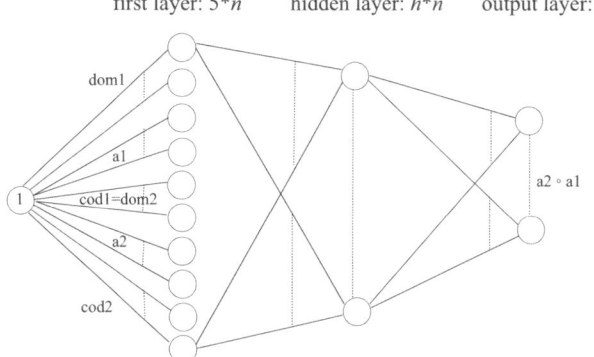

Fig. 10.7. The structure of the neural network that is used to learn the composition of first-order formulas.

layer where h is a parameter that can be varied. For simple examples like Example 1 described in Subsection 10.7.1 a value of $h = 2$ works well.

In order to enable the system to learn logical inferences, some basic arrows have static (fixed) representations. These representations correspond directly to truth values:[5]

- The truth value *true* : $(1.0, 0.0, 0.0, \ldots, 0.0)$
- The truth value *false* : $(0.0, 1.0, 0.0, \ldots, 0.0)$

Notice that the truth value *true* and the truth value *false* are maximally distinct in the first two dimensions. All other objects and arrows are initialized with the value $(0.5, 0.5, 0.5, \ldots, 0.5)$.[6] The defining equations of the theory and the equations generated by categorical constructions (like products) are used to train the neural network.

10.6.3 Backpropagation and Learning Representations of Categorical Entities

We use classical backpropagation with momentum to train this four layer network. The units use sigmoid transfer functions. In each step the weights and updates of the first layer (the representation layer) are loaded from a table storing the current representations of the topos entities. After a training

---

[5] The choice of the particular values for *true* and *false* are motivated by the usage of a simple projection to a two-dimensional plane for visualizing the results. The actual values used in the applications (cf. Sections 10.7 and 10.8) are 0.999954 and 0.000046.

[6] The actual value used in the applications (cf. Sections 10.7 and 10.8) is 0.622459.

252   Helmar Gust, Kai-Uwe Kühnberger, and Peter Geibel

step the updated weights and updates are stored back to the representation table.

Since the representations of the entities occur as node activities in the output layer and as weights in the input layer, the sigmoid transfer function of the nodes is used to translate between the two modes of representation.

10.6.4 Approximating Models

As mentioned in Section 10.2 each construct of a first-order logical language can be translated into a construction in a topos. If we provide concrete representations of the topos entities (objects and arrows) and an implementation of the concatenation operation such that the representations and the concatenation operation obey the topos constraints, such a system realizes a model: It assigns (categorical) elements of the universe (arrows $! \longrightarrow u$) to the (closed) terms of the language and truth-values to the closed formulas of the language (arrows $! \longrightarrow \Omega$) such that the interpretations of the axioms $\mathcal{A}$ give the truth-value *true*. The constraints, i.e. the set of equations coming from the axioms and the topos constructions, insure that (due Modus Ponens) the interpretation of the consequences of the axioms must be true, too. All other formulas might get arbitrary truth-values.

Since the set of equations might be infinite and we can only use finitely many equations for training the network, we will get only an approximation of a model. How good such an approximation works depends first, on the (finite) subset of equations chosen for training and second, on the error of the network, especially how errors propagate when the concatenation operation needs to be iterated.

Concerning the first issue, we minimize the problem by avoiding universal constructions (particularly pullbacks, equalizers, and exponents) and try to keep products small. Due to the fact that in the constructions for the logical operators, pullbacks and equalizers are involved clearly it is not possible to eliminate universal constructions completely (in a proper model). Nevertheless it should be possible to approximate a model in a sense that the approximation coincides with a proper model on a "relevant" set of formulas. Relevance is clearly hard to characterize. As a first idea in this context, relevance might be specified as follows:

- Given a deduction operator $\mathcal{D}$, all formulas that can be derived with n iterations of $\mathcal{D}$ (the elements of $\mathcal{D}^n(\mathcal{A})$) can be learned to be *true*.

Concerning the second issue (error propagation), up to now we don't have estimations about error propagation. Training of the networks minimizes the average error over the test equations. A possibility to judge the quality of the result is to look at the maximal error. If the maximal error does not converge, it cannot be expected that the net will give good results. A first idea would be to consider the variance of the output for varying the input in the range of the maximal error. Using these ideas it should be possible to estimate an value for n above which can assess the quality of the approximation.

10.7 Applications: Socrates' Mortality and the Steamroller Problem

10.7.1 Example 1: Socrates and Angels

Our first example is a quite simple theory which is intended to show how the present approach can be used to learn the transitivity of the logical implication operation. Consider the following theory consisting of two facts and three rules. Furthermore three objects are introduced: *Socrates, robot*, and *something*.

Rules:
> *All human beings are mortal.*
> *All mortal beings ascend to heaven.*
> *All beings in heaven are angels.*

Facts:
> *Socrates is human.*
> *Robot is not human.*

In Table 10.4, predicates of the given theory, individuals, and clauses are coded in the language $\mathcal{L}_T$. Queries are specified by test equations making statements about the properties of the introduced individuals.

10.7.2 Example 2: The Steamroller Problem

In theorem proving, the famous steamroller problem is considered as a benchmark problem [19]. Here is a natural language description of the steamroller:

- Wolves, foxes, birds, caterpillars, and snails are animals, and there are some of each of them.
- Also there are some grains, and grains are plants.
- Every animal either likes to eat all plants or all animals much smaller than itself that like to eat some plants.
- Caterpillars and snails are much smaller than birds, which are much smaller than foxes, which in turn are much smaller than wolves.
- Wolves do not like to eat foxes or grains, while birds like to eat caterpillars but not snails.
- Caterpillars and snails like to eat some plants.
- Prove: There is an animal that likes to eat a grain eating animal.

Table 10.4. Example code of Socrates' mortality.

```
# predicates of the theory
  human, mortal, heaven, angel:  u ---> @ .
  X ==> Y: -> o X x Y o d u = t o ! u .
  human   ==> mortal.
  mortal  ==> heaven.
  heaven  ==> angel.
#individuals
distinctive
  socrates, robot, something: ! ---> u.
  human o socrates = t.
  human o robot    = f.
# test the learned inferences
tests
  mortal o something = t,
  mortal o something = f,
  mortal o robot     = t,
  mortal o robot     = f,
  mortal o socrates  = t,
  mortal o socrates  = f,
  heaven o something = t,
  heaven o something = f,
  heaven o socrates  = t,
  heaven o socrates  = f,
  heaven o robot     = t,
  heaven o robot     = f,
  angel  o something = t,
  angel  o something = f,
  angel  o socrates  = t,
  angel  o socrates  = f,
  angel  o robot     = t,
  angel  o robot     = f.
```

A straightforward first-order representation of the given theory yields 27 clauses (containing also non-Horn clauses). Using a many-sorted logic we can reduce the total number of clauses to 12 clauses. It should be mentioned that these 12 clauses do also contain non-Horn clauses similarly to the naive representation.

We sketch the representation of the steamroller problem in our language $\mathcal{L}_T$. Table 10.5 represents some of the clauses of the steamroller problem. Predicates wolf, fox, bird etc. are introduced as predicates mapping elements of the universe u into the truth value object @, whereas individuals wo, fo, bi etc.[7] are introduced as morphisms from the terminal object ! to the universe u. Very similar to the situation concerning animals grain and plants are introduced.

Concerning the relation much_smaller some additional remarks are necessary. The inferences of the system are crucially based on two properties of this relation: much_smaller is considered as a transitive and asymmetric relation, i.e. the following two laws hold:

---

[7] Individuals result from skolemizing the existential quantifiers.

Table 10.5. Example code of some important clauses of the steamroller problem.

```
# predicates of the theory
% Wolves, foxes, birds, caterpillars, and snails are
% animals, and there are some of each of them.
  wolf,
  fox,
  bird,
  caterpillar,
  snail,
  animal
    :: u ---> @ .
  wolf          ==> animal.
  fox           ==> animal.
  bird          ==> animal.
  caterpillar   ==> animal.
  snail         ==> animal.
distinctive
  wo, fo, bi, ca, sn:: ! ---> u .
  wo is wolf.
  fo is fox.
  bi is bird.
  ca is caterpillar.
  sn is snail.

# Also there are some grains, and grains are plants.
  grain,
  plant
    :: u ---> @ .

  grain         ==> plant .
  gr is grain.

# Caterpillars and snails are much smaller than
# birds, which are much smaller than foxes, which in
# turn are much smaller than wolves.
  much_smaller:: u x u ---> @ .

  caterpillar and bird  ==> much_smaller.
  snail       and bird  ==> much_smaller.
  bird        and fox   ==> much_smaller.
  fox         and wolf  ==> much_smaller.
```

$$\forall x \forall y \forall z : ms(x,y) \land ms(y,z) \rightarrow ms(x,z)$$

$$\forall x \forall y : ms(x,y) \rightarrow \neg ms(y,x)$$

In our language $\mathcal{L}_T$, we can express transitivity of the much_smaller relation by the following clause.

```
(much_smaller and  much_smaller) o d1223
    ==> much_smaller o p13.
```

In this representation some non-standard diagonal functions are used. For example, d1223 is a morphism u x u x u ---> u x u x u x u which corresponds to product arrow $id_u \times d \times id_u$. d1223 guarantees that the distribution

Table 10.6. Some test equations of the steamroller problem.

```
tests
% negated and grouped
% to avoid existential quatification and to keep products small

animal and animal and grain ==>
      not (like_to_eat and like_to_eat) o d1223,

(not (and) o ((animal and animal and grain) x
      ((like_to_eat and like_to_eat) o d1223)
      )
    ) = t ,

(and) o ((animal and animal and grain) x
      ((like_to_eat and like_to_eat) o d1223))
      = t .

tests
animal o wo = t,
animal o fo = t,
animal o bi = t,
animal o ca = t,
animal o sn = t,
plant  o gr = t,
animal o gr = t.

tests
much_smaller o bi x wo = t,
much_smaller o ca x wo = t,
much_smaller o sn x wo = t,
much_smaller o ca x fo = t,
much_smaller o sn x fo = t.

tests
like_to_eat o wo x bi = t,
like_to_eat o bi x gr = t,
like_to_eat o fo x wo = t,
like_to_eat o bi x wo = t.
```

of variables is working in the correct way. Furthermore p13 corresponds to the product arrow $\pi_1 \times \pi_3$.

Due to the fact that products are not associative we need constructions to guarantee the correct projections of complex products. Because the most complex product used in the modeling is a product of 8 components, there are theoretically $\sum_{i=1}^{8} \binom{8}{i} \cdot i!$ many projections which must be specified for at least 7^8 many tuples (we have at least 7 elements of u). Clearly it is impossible to train a neural network with this number of equations.

In Table 10.6, some test equations that are used to query the system are summarized. Some remarks should be added concerning the complex queries at the beginning which represent the overall query of the steamroller problem. Here is again the natural language description of this query:

"Therefore there is an animal that likes to eat a grain eating animal".

10 Learning Models with Neural Networks Based on Topos Theory   257

Logically this can be represented by the following formula:

$$\exists x \exists y \exists z : animal(x) \wedge animal(y) \wedge grain(z) \wedge \\ like\_to\_eat(x,y) \wedge like\_to\_eat(y,z)$$

The negation of this formula yields the following formula:

$$\forall x \forall y \forall z : (animal(x) \wedge animal(y) \wedge grain(z)) \to \\ \neg(like\_to\_eat(x,y) \wedge like\_to\_eat(y,z))$$

The remaining queries in Table 10.6 are self-explanatory: these examples check whether *wolves, foxes, birds* etc. are animals and *grains* are plants. Furthermore some properties of the binary *much_smaller* relation and the binary *like_to_eat* relation are tested.

10.8 Evaluation

The input generated by the Prolog program is fed into the neural network. The results of an example run is then given by the errors of the test equations. Withe respect to the Socrates example (cf. Section 10.7.1), these test equations query whether the composition of *angel* and *robot* is *false*, whether the composition of *angel* and *robot* is *true*, whether the composition of *angel* and *socrates* is false etc. The numerical results of a test run of example 1 are summarized in Table 10.7.

The system convincingly learns that *Socrates* is mortal, is ascending to *heaven*, and is an *angel*. Furthermore it learns that the negations of these consequences are false. In other words, the system learns the transitivity property of the implication in universally quantified formulas. With respect to *robot* the system evaluates in the test run depicted in Table 10.7 with a rather high stability that *robot* is immortal, that it is not ascending to heaven, and it is not an angel. In the case of *something* the system evaluates that *something* is (more or less) neither in heaven, nor mortal, nor an angel.

We should have a closer look on how the neural network interprets queries. In the left diagram of Figure 10.8, the maximal error of the neural network of 10 runs with maximally 4250 iterations is depicted. The curves show four characteristic phases: in the first phase (up to a very few number of iterations), the maximal error increases. In a second phase, the maximal error remains stable on a high level. During the third phase the maximal error dramatically decreases due to the rearrangement of the input representations. In a fourth phase, the maximal error remains stable on a low level.

The right diagram of Figure 10.8 shows the stability behavior of the neural network. In a first phase, the instability of the weights increases dramatically. In a second phase, the stability of the network increases and the network remains (relatively) stable in the third phase. Interesting are certain fluctuations (of a certain run) of the stability behavior between, for example, around 3500 and 4000 iterations.

Table 10.7. Results of a test run of the Socrates example (example 1). The numerical values specify the error of the test equations (queries).

```
Equation:                          error:     representation of composition:
angel    o robot       = f         0.004262   0.090351  0.980911  0.002547  0.000084
angel    o robot       = t         0.894740   0.090351  0.980911  0.002547  0.000084
angel    o socrates    = f         0.982207   0.983796  0.001631  0.000693  0.000542
angel    o socrates    = t         0.000132   0.983796  0.001631  0.000693  0.000542
angel    o something   = f         0.127513   0.393812  0.683885  0.008322  0.002576
angel    o something   = t         0.417559   0.393812  0.683885  0.008322  0.002576
heaven   o robot       = f         0.003767   0.083263  0.975272  0.000339  0.000160
heaven   o robot       = t         0.895694   0.083263  0.975272  0.000339  0.000160
heaven   o socrates    = f         0.956884   0.959382  0.003238  0.000135  0.000762
heaven   o socrates    = t         0.000828   0.959382  0.003238  0.000135  0.000762
heaven   o something   = f         0.069778   0.297163  0.773610  0.002120  0.006547
heaven   o something   = t         0.546182   0.297163  0.773610  0.002120  0.006547
mortal   o socrates    = f         0.949548   0.950727  0.002307  0.000403  0.001527
mortal   o socrates    = t         0.001215   0.950727  0.002307  0.000403  0.001527
mortal   o robot       = f         0.003148   0.069418  0.961429  0.000529  0.000204
mortal   o robot       = t         0.895078   0.069418  0.961429  0.000529  0.000204
mortal   o something   = f         0.117132   0.348415  0.664452  0.007669  0.016891
mortal   o something   = t         0.433140   0.348415  0.664452  0.007669  0.016891
```

The left diagram of Figure 10.9 depicts the average error of the system (relative to the same runs as above). All runs except one nicely show the decrease of the average error. Furthermore a termination condition in terms of a small average error is depicted. The right diagram of Figure 10.9 shows the behavior of *Socrates is an angel*. The classification is as expected. *Socrates is classified as an angel* using the transitivity of the implication.

Figure 10.10 shows the behavior of *The robot is an angel* (left diagram) and *something is an angel*. Whereas the system shows a tendency to classify the *robot* as non-angel, the representations of *something* are quite arbitrarily distributed. Clearly the input only specifies that the *robot* is not human. It does not follow logically that the *robot* cannot be an angel. The result of the weight distribution of the neural network with respect to the *robot* can be interpreted as a support of something similar to a closed-world assumption. It is interesting that the interpretation of *something* (cf. Figure 10.10) differs from the interpretation of *robot*, because with respect to *something* there is no clear tendency how to classify this object.

The models approximated by the network behave as expected: Test equations which are logically derivable are interpreted as *true* by the system, whereas equations that correspond logically to a provably false statement are interpreted as *false*. Those equations for which no logical deduction of the truth value is possible, are more or less arbitrarily distributed between *false* and *true* in the set of models.

In [12], the Socrates example was already discussed. Whereas the behavior of the system in [12] is very similar to the behavior here, due to the simplifications we introduced in Section 10.4 it is possible to achieve convergence of the system in most cases with less than 800 iterations. In [12], more than 400,000 iterations were needed to achieve stability. Nevertheless the quality of the solutions is absolutely comparable with the results in [12].

Considering the steamroller problem (Section 10.7, Example 2), first, we should notice that even for symbolic theorem provers the steamroller it is a

10 Learning Models with Neural Networks Based on Topos Theory   259

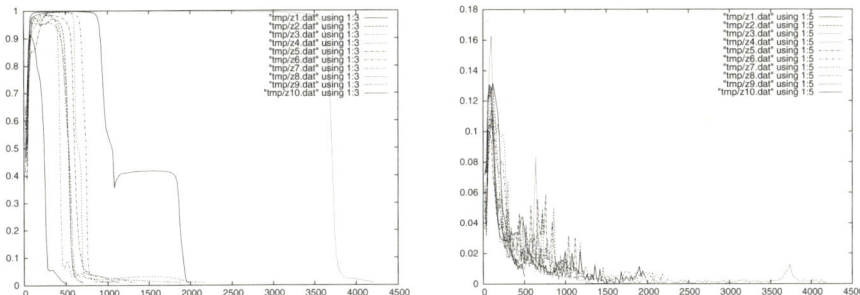

Fig. 10.8. The maximal error and the stability of the neural network with respect to the Socrates example (10 runs with maximally 4250 iterations). The left diagram depicts the maximal error and the right diagram shows the stability behavior of the network.

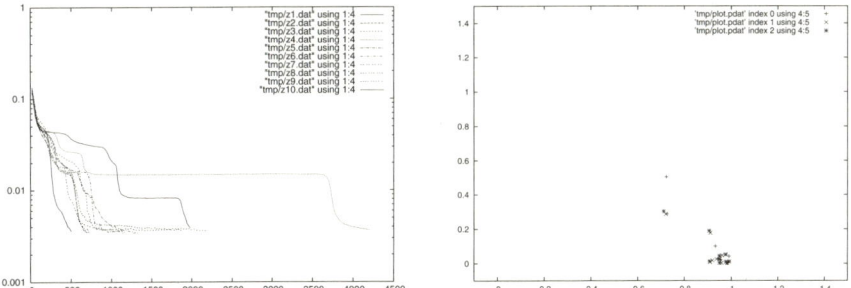

Fig. 10.9. The left diagram depicts the behavior of the average error of the system. A termination condition with respect to the average error determines the number of iterations of the whole system. The right diagram shows the distribution of the interpretations of *Socrates is an angel* (10 runs with maximally 4250 iterations). The diagram shows that Socrates is stably classified as an angel.

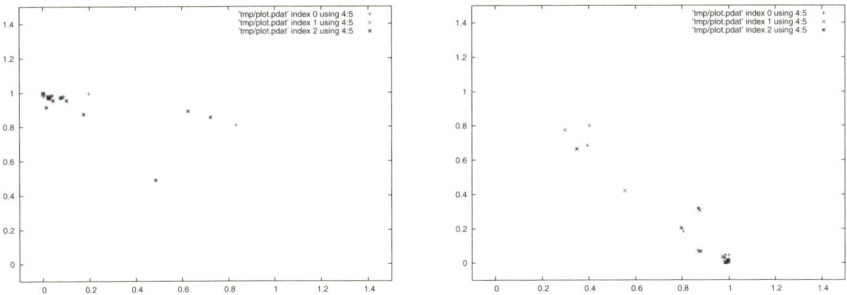

Fig. 10.10. The left diagram depicts the distribution of the interpretation of *The robot is an angel* (10 runs with maximally maximally 4250 iterations). The diagram shows that the robot is quite stably classified as a non-angel. The right diagram depicts the distribution of the representation of *Something is an angel* with respect to the Socrates example. The result is a quite equal distribution.

benchmark problem. Similarly it is a hard problem for the present approach as well. Therefore our modeling must be considered as a very first attempt to solve problems of a a comparable complexity and the current results should be considered as work in process.

A separation of the steamroller problem into different subproblems is helpful to identify potential difficulties. Some of these problems are summarized in the following list:

- Although the number of equations can be significantly reduced by the approximations introduced in Section 10.4, this number is still quite high (approx. 800 equations).
- Testing the learning of the underlying ontology, we figured out that the system has some problems in learning the associativity of concatenation in complex expressions. Table 10.8 shows the results of some queries of an example run of the system. Although the system classifies *wolf*, *fox* etc. tentatively as *animal* and *grain* more or less as *plant*, the significance of this classification is far from being clearly distinctive. We think that this is due to the fact that associativity of concatenation is not well-learned in more complex situations like the one of the steamroller problem.
- As discussed in Section 10.7 the number of possible projections is rather high in complex reasoning processes. It is clearly the case that the number of projections we can practically use is a tiny fraction of all possible projection functions. This can be exemplified by results concerning the transitive and asymmetric *much_smaller* relation (compare the discussion in Section 10.7). Table 10.9 shows the results of some queries of a run of the system. In the table, the first five queries are logically false (by asymmetry and transitivity of the *much_smaller* relation) and the last five queries are logically true (derivable by the transitivity of the *much_smaller* relation). Although the system shows a tendency to distinguish these two groups of queries, again the distinctiveness is far from being satisfactory. By adding more projection functions the results may be improved. The problem is the enormous number of these projection functions resulting in a much slower system.
- Dependent on the complexity of the underlying input theory the number of runs for which the system does not converge increases. A possible way to avoid this is to increase also the dimension of the representation space making the system clearly much slower.
- Unfortunately the current results for the steamroller problem are even in case of convergence of a run quite often not convincing. The reason might be the too rough approximations of the modeling. Notice that we currently use universal constructions in an extremely limited manner.

With the proposed neural approach we are currently not able to solve completely a benchmark problem for symbolic theorem provers like the steamroller. In the future, we hope to find possibilities to address the roughly sketched problems for complex inference processes.

Table 10.8. Results of a test run of the steamroller problem focusing on queries concerning the underlying ontology.

```
Equation:              error:     representation of composition:
animal  o gr   = t     0.523919   0.096594  0.406415  0.015048  0.257761
plant   o gr   = t     0.227860   0.651427  0.042832  0.489557  0.304671
animal  o sn   = t     0.072986   0.622445  0.022519  0.020093  0.050558
animal  o ca   = t     0.090758   0.588392  0.032222  0.036294  0.098955
animal  o bi   = t     0.076140   0.614318  0.035708  0.021240  0.042989
animal  o fo   = t     0.080184   0.602162  0.025493  0.013849  0.035990
animal  o wo   = t     0.101175   0.553314  0.039884  0.012865  0.033384
```

Table 10.9. Results of a test run of the steamroller problem focusing on queries concerning the *much_smaller* relation.

```
Equation:                       error:     representation of composition:
much_smaller o fo x sn   = t    0.017769   0.879243  0.098225  0.017152  0.105096
much_smaller o fo x ca   = t    0.082838   0.659993  0.204772  0.014686  0.089348
much_smaller o wo x sn   = t    0.011850   0.918014  0.056595  0.016021  0.116378
much_smaller o wo x ca   = t    0.062300   0.705555  0.169615  0.020756  0.093564
much_smaller o wo x bi   = t    0.043545   0.736651  0.113148  0.016931  0.068486
much_smaller o sn x fo   = t    0.000706   0.992366  0.002381  0.026206  0.025807
much_smaller o ca x fo   = t    0.001061   0.990853  0.002879  0.038597  0.023405
much_smaller o sn x wo   = t    0.001034   0.986824  0.004423  0.035339  0.025149
much_smaller o ca x wo   = t    0.001257   0.988095  0.003977  0.039950  0.027726
much_smaller o bi x wo   = t    0.001088   0.992163  0.002388  0.036556  0.027928
```

10.9 Conclusion

We presented an approach for modeling first-order inferences with neural networks that was inspired by two main ideas:

- First, the neural representation of structured data is distributed and uniform (in the sense that no hard-wired neural network needs to be developed for a particular application, but the same neural network can be used for different applications).
- Second, the network is able to learn models of simple first-order theories using standard learning techniques (backpropagation).

These two constraints were realized by translating axioms of first-order theories into a variable-free representation using topos theory. Induced by commuting triangles in the topos, equations $f \circ g = h$ in normal form were generated to train the network. The result of the learning process is a model of the input theory. The framework was applied to the Socrates example and the steamroller problem.

The translation of first-order formulas into training data of a neural network allows, in principle, to represent models of symbolic theories in artificial intelligence and cognitive science (that are based on full first-order logic) with neural networks.[8] In other words the account provides a recipe – and not just a general statement of the theoretical possibility – of how to learn models of theories based on FOL with neural networks. Notice further that the presented

---

[8] Notice that a large part of theories in artificial intelligence are formulated with tools taken from logic and are mostly based on FOL or subsystems of FOL.

approach tries to combine the advantages of connectionist networks and logical systems: Instead of representing symbols like constants or predicates using single neurons, the representation is rather distributed realizing the very idea of distributed computation in neural networks. Furthermore the neural network can be trained to learn a model without any hardcoded devices. The result is a distributed representation of a symbolic system.

Clearly some issues remain open at the moment. The following list roughly sketches open problem, related questions, and possible real-world applications that will be considered in the future.

- Clearly the account needs to be improved in order to be able to model complex reasoning processes. One aspect of future work will be to test the approach on non-trivial application scenarios and benchmark problems of theorem proving.
- From a dual perspective, it would be desirable to find a possibility to translate the distribution of weights in a trained neural network back to the symbolic level. The symbol grounding problem could then be analyzed in detail by translating the representation levels into each other.
- Taking into account time-critical applications, for example, in scenarios where agents need to act rapidly in a real-world environment, the usage of a trained neural network in a complex (perhaps hybrid) system would significantly facilitate the application, because there are no inference steps that need to be computed.
- Due to the fact that the neural network outputs an error for every query, an interesting aspect could be the possibility to introduce further truth values like n (neither true nor false) and b (both true and false). Such an extension is straightforward on the neural network side and corresponds nicely to well-known accounts of many-valued logic in symbolic systems [20].
- A theoretical characterization of the learned models would be desirable. These models would be dependent on various parameters, for example, the used training equations induced by the category theoretic constructions, the dimension of the representation space, the number of available projection functions, the number of possible truth values of the truth value object etc.

Although the presented modeling needs to be improved in various respects, we think that it is a promising approach for learning symbolic theories with subsymbolic means.

References

1. Barnden J. A. (1989). Neural net implementation of complex symbol processing in a mental model approach to syllogistic reasoning, in Proceedings of the International Joint Conference on Artificial Intelligence, 568-573.

2. Shastri L., Ajjanagadde V. (1990). From simple associations to systematic reasoning: A connectionist representation of rules, variables and dynamic bindings using temporal synchrony, *Behavioral and Brain Sciences* 16: 417-494.
3. Pollack J. (1990). Recursive Distributed Representations *Artificial Intelligence* 46(1):77-105.
4. Rojas, R. (1996). *Neural Networks – A Systematic Introduction*, Springer, Berlin, New York.
5. Steinbach B., Kohut R. (2002). Neural Networks – A Model of Boolean Functions. In Steinbach B (ed.): *Boolean Problems, Proceedings of the 5th International Workshop on Boolean Problems*, 223-240, Freiberg.
6. Lange T., Dyer M. G. (1989). *High-level inferencing in a connectionist network.* Technical report UCLA-AI-89-12.
7. Smolenski P. (1990). Tensor product variable binding and the representation of symbolic structures in connectionist systems, *Artificial Intelligence* 46(1–2): 159–216.
8. Plate T. (1994). *Distributed Representations and Nested Compositional Structure.* PhD thesis, University of Toronto.
9. Hitzler P., Hölldobler S., Seda A. (2004). Logic programs and connectionist networks. *Journal of Applied Logic*, 2(3):245-272.
10. Healy M., Caudell T. (2004). *Neural Networks, Knowledge and Cognition: A Mathematical Semantic Model Based upon Category Theory*, University of New Mexico, EECE-TR-04-020.
11. D'Avila Garcez A., Broda K., Gabbay D. (2002). *Neural-Symbolic Learning Systems: Foundations and Applications.* Springer-Verlag.
12. Gust H., Kühnberger K.-U. (2004). Cloning Composition and Logical Inference in Neural Networks Using Variable-Free Logic. In: AAAI Fall Symposium Series 2004, Symposium: Compositional Connectionism in Cognitive Science, Washington D.C., 25-30.
13. Gust H., Kühnberger K.-U. (2005.) Learning Symbolic Inferences with Neural Networks. In: Bara B., Barsalou L., Bucciarelli M. (eds): CogSci 2005, XXVII Annual Conference of the Cognitive Science Society. Lawrence Erlbaum, 875-880.
14. Goldblatt R. (1979). Topoi: The Categorial Analysis of Logic. Studies in Logic and the Foundations of Mathematics, 98 (1979), North-Holland, Amsterdam.
15. Montague R. (1970). Pragmatics and intensional logic. Synthèse 22:68-94. Reprinted in: Formal Philosophy. Selected Papers of Richard Montague. Edited and with an introduction by Richmond H. Thomason. New Haven/London: Yale University Press, 1974, 119-147.
16. Rydeheard D. E., Burstall R. M. (1988). Computational Category Theory, Prentice Hall.
17. Gust H. (2000). Quantificational Operators and their Interpretation as Higher Order Operators. In Böttner M., Thümmel W. (eds) Variable-free Semantics. Secolo, Osnabrück, 132-161.
18. Gust H. (2006). A sorted logical language with variable patterns. In Bab, S, Gulden, J, Noll, T, Wieczorek, T (eds): Models and Human Reasoning, Wissenschaft und Technik Verlag, Berlin, 35-62.
19. Walter C. (1985). A Mechanical Solution of Schubert's Steamroller by Many-Sorted Resolution. *Artificial Intelligence* 26: 217-224.

20. Urquhart A. (2001). Basic many-valued logic. In Gabbay D., Guenthner F. (eds.), *Handbook of Philosophical Logic*, 2nd ed., vol. 2, Kluwer Acad. Publ., Dordrecht, pp. 249-295.

11

Advances in Neural-Symbolic Learning Systems: Modal and Temporal Reasoning

Artur S. d'Avila Garcez

City University London, EC1V 0HB, UK, aag@soi.city.ac.uk

Summary. Three notable hallmarks of intelligent cognition are the ability to draw rational conclusions, the ability to make plausible assumptions, and the ability to generalise from experience. Although human cognition often involves the interaction of these three abilities, in Artificial Intelligence they are typically studied in isolation. In our research programme, we seek to integrate the three abilities within neural computation, offering a unified framework for learning and reasoning that exploits the parallelism and robustness of connectionism. A neural network can be the machine for computation, inductive learning, and effective reasoning, while logic provides rigour, modularity, and explanation capability to the network. We call such systems, combining a connectionist learning component with a logical reasoning component, neural-symbolic learning systems. In what follows, I review the work on neural-symbolic learning systems, starting with logic programming and then looking at how to represent modal logic and other forms of non-classical reasoning in neural networks. The model consists of a network ensemble, each network representing the knowledge of an agent (or possible world) at a particular time point. Ensembles may be seen as in different levels of abstraction so that networks may be fibred onto (combined with) other networks to form a structure combining different logical systems or, for example, object-level and meta-level knowledge. We claim that this quite powerful yet simple structure offers a basis for an expressive yet computationally tractable cognitive model of integrated reasoning and learning.

11.1 Introduction

Three notable hallmarks of intelligent cognition are the ability to draw rational conclusions, the ability to make plausible assumptions, and the ability to generalise from experience. In a logical setting, these abilities correspond to the processes of deduction, abduction, and induction, respectively. Although human cognition often involves the interaction of these three abilities, they are typically studied in isolation. For example, in Artificial Intelligence (AI), symbolic (logic-based) approaches have been mainly concerned with deductive reasoning, while connectionist (neural networks-based) approaches have focused on inductive learning. It is well known that this connectionist/symbolic dichotomy in AI reflects a distinction between brain and mind, but we argue this should not dissuade us from seeking a fruitful synthesis of these paradigms [1, 2].

In our research programme, we seek to integrate the processes of abduction, induction and deduction within the neural computation approach. When we think of neural networks, what springs to mind is their ability to learn from examples using efficient algorithms in a massively parallel fashion. When we think of symbolic logic, we recognise its rigour, semantic clarity, and the availability of automated proof methods which can provide explanations to the reasoning process, e.g. through a proof history. Our long-term goal is to produce biologically-motivated computational models with *the ability to learn from experience and the ability to reason from what has been learned* [3], offering a unified framework for learning and reasoning that exploits the parallelism and robustness of connectionist architectures. To this end, we choose to work with standard neural networks whose learning capabilities have been already demonstrated in significant practical applications, and to investigate how they can be enhanced with more advanced reasoning capabilities. The neural networks should provide the machinery for cognitive computation, inductive learning, and effective reasoning, while logic provides the rigour and explanation capability to the model, facilitating the interaction with the outside world. We call such systems, combining a connectionist learning component with a logical reasoning component, *neural-symbolic learning systems* [4].

In neural computation, induction is typically seen as the process of changing the weights of a network in ways that reflect the statistical properties of a dataset (set of examples), allowing for useful generalisations over unseen examples. In the same setting, deduction can be seen as the network computation of output values as a response to input values, given a particular set of weights. Standard feedforward and partially recurrent networks have been shown capable of deductive reasoning of various kinds depending on the network architecture, including nonmonotonic [4], modal [5], intuitionistic [6] and abductive reasoning [2].

In what follows, we review the work on the integration of logics and neural networks, starting with logic programming [4] and then looking at how to represent modal logic [5] and other non-classical logics [6, 7] in neural networks. We then look at how to combine different neural networks and their associated logics with the use of the fibring method [8]. The overall model consists of a set (an ensemble) of simple, single-hidden layer neural networks – each may represent the knowledge of an agent (or a possible world) at a particular time point – and connections between networks to represent the relationships between agents/possible worlds. Each ensemble may be seen as in a different level of abstraction so that networks in one level may be fibred onto networks in another level to form a structure combining meta-level and object-level information or combining different logics. As shown in [8], this structure is strictly more expressive than the network ensembles without fibring, and we claim this forms the basis for an expressive yet computationally tractable model of integrated learning and reasoning.

Throughout, I will try and place the recent advances in neural-symbolic learning systems in the context of John McCarthy's note *Epistemological challenges for connectionism* [9], written as a response to Paul Smolensky's paper *On the proper treatment of connectionism* [10] almost twenty years ago. McCarthy identifies four knowledge representation problems for neural networks: the problem of *elaboration tolerance* (the ability of a representation to be elaborated to take additional phenomena into account); the *propositional fixation* of neural networks (based on the assumption that neural networks cannot represent relational knowledge); the problem of how to make use of any available *background knowledge* as part of learning, and the problem of how to obtain domain *descriptions* from trained networks as opposed to mere discriminations. Neural-symbolic integration may address each of the above challenges. In a nutshell, the problem of elaboration tolerance can be resolved by having networks that are forming in a hierarchy, similarly to the idea of using self-organising maps [11, 12] for language processing, where the lower levels of abstraction are used for the formation of concepts that are then used at the higher levels of the hierarchy. Connectionist modal logic [5] deals with the so-called propositional fixation of neural networks by allowing them to encode relational knowledge in the form of accessibility relations. A number of other formalisms have also tackled this issue as early as 1990 [13, 14, 15, 16], the key question still remaining being that of how to have representations simple enough to promote an effective relational learning. Learning with background knowledge can be achieved by the usual translation of symbolic rules into neural networks, and problem descriptions can be obtained by rule extraction. A number of such translation and extraction algorithms have been proposed, e.g. [17, 18, 19, 20, 21, 22, 23].

Some of the challenges for neural-symbolic integration today still stem from the questions posed by McCarthy in 1988, but with the focus on the effective integration of expressive reasoning and robust learning. As such, we cannot afford to loose on learning capability as we add reasoning capability to neural models. This means that we cannot depart from the idea that neural networks are composed of simple processing units organised in a massively parallel fashion (and thus allow for some *clever* neurons to perform complex symbol processing). We also would like our models to be biologically plausible, not as a principle but in a pragmatic way. There have been recent advances in brain imaging, which offer us data we can make use of to get insight into new forms of representation. There are also computational challenges associated with more practical aspects of the application of neural-symbolic systems in areas such as engineering, robotics, semantic web, etc. [24, 25]. Challenges such as the effective, massively parallel computation of logical models, the efficient extraction of comprehensible rules, and striking the right balance between computational tractability and expressiveness.

11.2 Neural-Symbolic Learning Systems

For neural-symbolic integration to be effective in complex applications, we need to investigate how to represent, reason, and learn expressive logics in neural networks. We also need to find effective ways of expressing the knowledge encoded in a trained network in a comprehensible symbolic form. There are at least two lines of action in this direction. The first is to take standard neural networks and try and find out which logics they can represent. The other is to take well established logics and concepts (e.g. recursion) and try and encode those in a neural network architecture. Both require a principled approach, so that whenever we show that a particular logic can be represented by a particular neural network, we need to show that the network and the logic are in fact equivalent (a way of doing this is to prove that the network computes the formal semantics of the logic). Similarly, if we develop a knowledge extraction algorithm, we need to make sure that it is sound (i.e. correct) in the sense that it produces rules that are encoded in the network, and that it is *quasi*-complete in the sense that it produces rules that increasingly approximate the exact behaviour of the network.

During the past twenty years, a number of models for neural-symbolic integration have been proposed. Broadly speaking, researchers have made contributions to three main areas, providing either: (*i*) a logical characterisation of a connectionist system; (*ii*) a connectionist implementation of a logic; or (*iii*) a hybrid system bringing together advantages from connectionist systems and symbolic artificial intelligence [20]. Major contributions include [23, 26, 27, 14] on the knowledge representation side, [19, 28] on learning with background knowledge, and [17, 18, 21, 22, 29] on knowledge extraction, among others. The reader is referred to [4] for a detailed presentation of neural-symbolic learning systems and applications.

Neural-symbolic learning systems contain six main phases: (1) *background knowledge insertion*; (2) *inductive learning from examples*; (3) *massively parallel deduction*; (4) *theory fine-tuning*; (5) *symbolic knowledge extraction*; and (6) *feedback* (see Figure 11.1). In phase (1), symbolic knowledge is translated into the initial architecture of a neural network with the use of a *translation algorithm*. In phase (2), the neural network is trained with examples by a neural learning algorithm, which revises the theory given in phase (1) as *background knowledge*. In phase (3), the network can be used as a massively parallel system to compute the logical consequences of the theory encoded in it. In phase (4), information obtained from the computation carried out in phase (3) may be used to help fine-tuning the network to better represent the problem domain. This mechanism can be used, for example, to resolve inconsistencies between the background knowledge and the training examples. In phase (5), the result of training is explained by the extraction of revised symbolic knowledge. As with the insertion of rules, the *extraction algorithm* must be provably correct, so that each rule extracted is guaranteed to be a rule of the network. Finally, in phase (6), the knowledge extracted may be analysed by an expert who decides if it should feed the system once again, closing the learning cycle.

Our neural network model consists of feedforward and partially recurrent networks, as opposed to the symmetric networks investigated, e.g., in [30]. It uses a localist rather than a distributed representation[1], and it works with backpropagation, the neural learning algorithm most successfully used in industry-strength applications [32].

A typical application of neural-symbolic systems is in safety-critical domains (e.g. power plant fault diagnosis), where the neural network can be used to detect a fault quickly, triggering safety procedures, while the knowledge extracted can be used to explain the reasons for the fault later on. If mistaken, this information can be used to fine-tune the learning system.

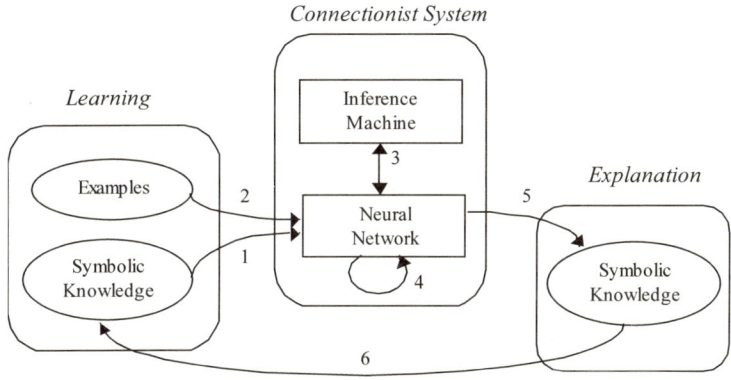

Fig. 11.1. Neural-Symbolic Learning Systems

In this paper, we focus on knowledge representation (phase (1) above) since appropriate knowledge representation precedes effective reasoning and learning. First, let us see how the *translation algorithm* works in the case of logic programming. We consider theories T_1 that are sets of Horn clauses, each clause of the form $A_1, \ldots, A_n \to A_0$, where the A_i are ground atoms. By convention, A_0 is referred to as the *head* atom of the clause and $A_1, \ldots, A_n$ are called *body* atoms. Intuitively, this rule states that "if all of the body atoms $A_1, \ldots, A_n$ are true, then the head atom A_0 must also be true". If the body is

---

[1] We depart from distributed representations for two main reasons: localist representations can be associated with highly effective learning algorithms such as backpropagation, and in our view localist networks are at an appropriate level of abstraction for symbolic knowledge representation. As advocated in [31], we believe one should be able to achieve the goals of distributed representations by properly changing the levels of abstraction of localist networks, while some of the desirable properties of localist models cannot be exhibited by fully distributed ones.

empty, then the head is called a *fact* and is simply written A_0. A set of such clauses is usually called a (propositional) *Horn theory* or *logic program* [33]. For ease of reference, we normally prefix the clauses (sometimes also called rules) by a label r_i, giving $r_i : A_1, ..., A_n \to A_0$. We also consider theories T_2 that are sets of general logic programs, which are finite sets of general clauses, a general clause being a rule of the form $L_1, ..., L_n \to A_0$, where A_0 is an atom and L_i ($1 \leq i \leq n$) is a literal (i.e. an atom or its negation).

Let us now define the type of neural network that we use more precisely. An artificial neural network is a directed graph. A unit in this graph is characterised, at time t, by its input vector $I_i(t)$, its input potential $U_i(t)$, its activation state $A_i(t)$, and its output $O_i(t)$. The units (neurons) of the network are interconnected via a set of directed and weighted connections. If there is a connection from unit i to unit j then $W_{ji} \in \mathbb{R}$ denotes the *weight* associated with such a connection. The input potential of neuron i ($U_i(t)$) is generally obtained by applying the *propagation rule* $U_i(t) = \sum_j((\mathbf{W}_{ij} \cdot x_i(t)) - \theta_i)$, where $\mathbf{W}_{ij}$ denotes the weight vector $(W_{i1}, W_{i2}, ..., W_{in})$ to neuron i, and $-\theta_i$ (θ_i denoting an extra weight with input always fixed at 1) is known as the *threshold* of neuron i. The *activation state* of neuron i ($A_i(t)$) is a bounded real or integer number according to an *activation function* (h_i) s. t. $A_i(t) = h_i(U_i(t))$. We say that neuron i is *active* at time t (or equivalently that the literal we associate with neuron i is *true*) if $A_i(t) > \theta_i$. Finally, the neuron's output is generally given by the identity function s. t. $O_i(t) = A_i(t)$.

A multilayer feedforward network computes a function $\varphi : \mathbb{R}^r \to \mathbb{R}^s$, where r and s are the number of neurons occurring, respectively, in the input and output layers of the network. In the case of *single hidden layer networks*, the computation of φ occurs as follows: at time t_1, the input vector is presented to the input layer. At time t_2, the input vector is propagated to the hidden layer, and the neurons in the hidden layer update their input potential and activation state. At time t_3, the hidden layer activation state is propagated to the output layer, and the neurons in the output layer update their input potential and activation state, producing the associated output vector.

Most neural models have a *learning rule*, responsible for changing the weights of the network progressively so that it learns to approximate φ given a number of *training examples* (input vectors and their respective target output vectors). In the case of *backpropagation* [32], an *error* is calculated as the difference between the network's actual output vector and the target vector for each input vector in the set of examples. This error $\mathbf{E}$ is then propagated back through the network, and used to calculate the variation of the weights $\triangle \mathbf{W}$. This calculation is such that the weights vary according to the *gradient* of the error, i.e. $\triangle \mathbf{W} = -\eta \nabla \mathbf{E}$, where $0 < \eta < 1$ is called the *learning rate*. The process is repeated a number of times in an attempt to minimise the error, and thus approximate the network's actual output to the target output for each example. Notice that h_i is required to be a continuous and typically a monotonically increasing function such as $tanh(x)$ since the derivative of h_i has to be calculated in the computation of $\nabla \mathbf{E}$.

Bringing together logic programming and neural networks, let $\mathcal{P}$ be a general logic program, and let $\mathcal{N}$ be a single hidden layer neural network. Each clause (r_l) of $\mathcal{P}$ can be mapped from the input layer to the output layer of $\mathcal{N}$ through one neuron (N_l) in the single hidden layer of $\mathcal{N}$. Intuitively, the *translation algorithm* from $\mathcal{P}$ to $\mathcal{N}$ implements the following conditions: (**C1**) The input potential of a hidden neuron (N_l) can only exceed N_l's threshold (θ_l), activating N_l, when all the positive literals in the body of r_l are assigned the truth-value *true* while all the negative literals in the body of r_l are assigned *false*; and (**C2**) The input potential of an output neuron (A) can only exceed A's threshold (θ_A), activating A, when at least one hidden neuron N_l that is connected to A is activated. The following example illustrates this.

Example 1. Consider the logic program $\mathcal{P} = \{B, C, \neg D \rightarrow A; E, F \rightarrow A; B\}^2$. The *translation algorithm* derives the network $\mathcal{N}$ of Figure 11.2, setting weights (W) and thresholds (θ) in such a way that conditions (**C1**) and (**C2**) above are satisfied. Note that, if $\mathcal{N}$ ought to be fully-connected, any other link (not shown in Figure 11.2) should receive weight zero initially.

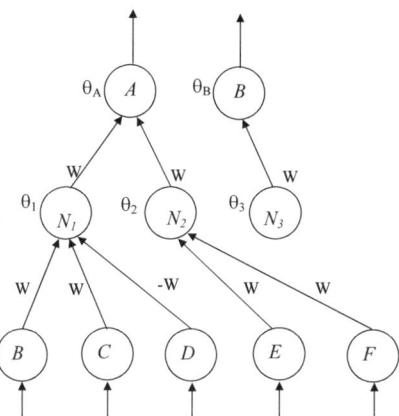

Fig. 11.2. Neural Network for Logic Programming

In the above example, each input and output neuron of $\mathcal{N}$ is associated with an atom of $\mathcal{P}$. As a result, each input and output vector of $\mathcal{N}$ can be associated with an interpretation for $\mathcal{P}$. Notice that each hidden neuron N_l corresponds to a rule r_l of $\mathcal{P}$. Notice how negative weights are used to represent negative literals. This structure has been first proposed in [27] for networks with discontinuous activation functions. In order for $\mathcal{N}$ to compute the stable model semantics of $\mathcal{P}$ [34], the output of neuron B should feed the input of

---
[2] To avoid ambiguity, we use semi-colons at the end of clauses.

neuron B such that $\mathcal{N}$ is used to iterate $\mathcal{T}_P$, the fixed-point operator of $\mathcal{P}$. $\mathcal{N}$ will eventually converge to a stable state which is identical to a stable model of $\mathcal{P}$.

Details about the translation algorithm, the soundness proof in the case of continuous activation functions, and extensions to other types of logic programs can be found in [4], together with algorithms to deal with inconsistencies, rule extraction, and experimental results in DNA sequence analysis, power systems fault diagnosis, and the evolution of requirements in software engineering.

11.3 Connectionist Non-Classical Reasoning

We now turn to the question of expressiveness. We believe that for neural computation to achieve its promise, connectionist models must be able to cater for non-classical reasoning. We believe that the neural-symbolic community cannot ignore the achievements and impact that non-classical logics have had in computer science [35]. While non-monotonic reasoning has dominated research in AI in the eighties and nineties, temporal logic has had large an impact in both academia and industry, and modal logics have become a *lingua franca* for the specification and analysis of knowledge and communication in multi-agent and distributed systems [36]. In this section, we consider modal and temporal reasoning as key representatives of non-classical reasoning.

The basic idea behind more expressive, connectionist non-classical reasoning is simple. Instead of having a single network, if we now consider a set of networks like the one in Figure 11.2, and we label them, say, as ω_1, ω_2, etc., then we can talk about a concept L holding at ω_1 and the same concept L holding at ω_2 separately. In this way, we can see ω_1 as a possible world and ω_2 as another, and this allows us to represent modalities such as necessity and possibility, and time (as detailed below) and also argumentation [37], epistemic states [38], and intuitionistic reasoning [6] (which we do not discuss in this paper). It is important noting, however, that this avenue of research is of interest in connection with McCarthy's conjecture on the propositional fixation of neural networks because there is a well established translation between propositional modal logic and the two-variable fragment of first-order logic[3] [39] indicating that relatively simple neural-symbolic systems may go beyond propositional logic.

---

[3] In [39], Vardi states that "(propositional) modal logic, in spite of its apparent propositional syntax, is essentially a first-order logic, since the necessity and possibility modalities quantify over the set of possible worlds... the states in a Kripke structure correspond to domain elements in a relational structure, and modalities are nothing but a limited form of quantifiers". In the same paper, Vardi then proves that propositional modal logics correspond to fragments of first-order logic.

11.3.1 Connectionist Modal Reasoning

Modal logic deals with the analysis of concepts such as *necessity* (represented by $\Box L$, read "box L", and meaning that L is *necessarily true*), and *possibility* (represented by $\Diamond L$, read "diamond L", and meaning that L is *possibly true*). A key aspect of modal logic is the use of *possible worlds* and a binary (accessibility) relation $\mathcal{R}(\omega_i, \omega_j)$ between possible worlds ω_i and ω_j. In possible world semantics, a proposition is necessary in a world if it is true in all worlds which are possible in relation to that world, whereas it is possible in a world if it is true in at least one world which is possible in relation to that same world.

Connectionist Modal Logic (CML) uses ensembles of neural networks (instead of single networks) to represent the language of modal logic programming [40]. The theories T_3 are now sets of modal clauses each of the form $\omega_i : ML_1, ..., ML_n \rightarrow MA$, where ω_i is a label representing a world in which the associated clause holds, and $M \in \{\Box, \Diamond\}$, together with a finite set of relations $\mathcal{R}(\omega_i, \omega_j)$ between worlds ω_i and ω_j in T_3. Such theories are implemented in a network ensemble, each network representing a possible world, with the use of ensembles and labels allowing for the representation of the accessibility relations.

In CML, each network in the ensemble is a simple, single-hidden layer network like the network of Figure 11.2 to which standard neural learning algorithms can be applied. Learning, in this setting, can be seen as learning the concepts that hold in each possible world independently, with the accessibility relation providing the information on how the networks should interact. For example, take three networks all related to each other. If neuron $\Diamond a$ is activated in one of these networks then neuron a must be activated in at least one of the networks. If neuron $\Box a$ is activated in one network then neuron a must be activated in all the networks. This implements in a connectionist setting the possible world semantics mentioned above. It is achieved by defining the connections and the weights of the network ensemble, following a translation algorithm for modal clauses which extends that for Horn and general clauses. Details of the translation algorithm along with a soundness proof can be found in [5].

As an example, consider Figure 11.3. It shows an ensemble of three neural networks labelled N_1, N_2, N_3, which might *communicate* in different ways. We look at N_1, N_2 and N_3 as *possible worlds*. Input and output neurons may now represent $\Box L$, $\Diamond L$ or L, where L is a literal. $\Box A$ will be *true* in a world ω_i if A is *true* in all worlds ω_j to which ω_i is related. Similarly, $\Diamond A$ will be *true* in a world ω_i if A is *true* in some world ω_j to which ω_i is related. As a result, if neuron $\Box A$ is activated at network N_1, denoted by $\omega_1 : \Box A$, and world ω_1 is related to worlds ω_2 and ω_3 then neuron A must be activated in networks N_2 and N_3. Similarly, if neuron $\Diamond A$ is activated in N_1 then a neuron A must be activated in an arbitrary network that is related to N_1.

It is also possible to make use of CML to compute that $\Box A$ holds at a possible world, say ω_i, whenever A holds at *all* possible worlds related to ω_i, by connecting the output neurons of the related networks to a hidden neuron in ω_i which connects to an output neuron labelled as $\Box A$. Dually for $\Diamond A$, whenever A holds at *some* possible world related to ω_i, we connect the output neuron representing A to a hidden neuron in ω_i which connects to an output neuron labelled as $\Diamond A$. Due to the simplicity of each network in the ensemble, when it comes to learning, we can still use backpropagation on each network to learn the local knowledge in each possible world.

Example 2. Let $\mathcal{P} = \{\omega_1 : r \to \Box q;\ \omega_1 : \Diamond s \to r;\ \omega_2 : s;\ \omega_3 : q \to \Diamond p;\ \mathcal{R}(\omega_1;\omega_2),\ \mathcal{R}(\omega_1,\omega_3)\}$. The network ensemble $\mathcal{N}$ in Figure 11.3 is equivalent to $\mathcal{P}$. Take network N_1 (representing ω_1). To implement the semantics of $\Diamond$, output neurons of the form $\Diamond \alpha$ should be connected to output neurons α in an arbitrary network N_i (representing ω_i) to which N_1 is related. For example, taking $i = 2$, $\Diamond s$ in N_1 is connected to s in N_2. To implement the semantics of $\Box$, output neurons $\Box \alpha$ should be connected to output neurons α in every network N_i to which N_1 is related. For example, $\Box q$ in N_1 is connected to q in both N_2 and N_3. Dually, taking N_2, output neurons α need to be connected to output neurons $\Diamond \alpha$ and $\Box \alpha$ in every network N_j related to N_2. For example, s in N_2 is connected to $\Diamond s$ in N_1 via the hidden neuron denoted by $\vee$ in Figure 11.3, while q in N_2 is connected to $\Box q$ in N_1 via the hidden neuron denoted by $\wedge$. Similarly, q in N_3 is connected to $\Box q$ in N_1 via $\wedge$. The translation terminates when all output neurons have been considered. The translation algorithm defines the weights and thresholds of the network in such a way that it can be shown to compute a fixed-point semantics of the modal logic program associated with it (for any extended modal program $\mathcal{P}$ there exists an ensemble of neural networks $\mathcal{N}$ with a single hidden layer and semi-linear neurons, such that $\mathcal{N}$ computes the modal fixed-point operator $MT_\mathcal{P}$ of $\mathcal{P}$). Finally, as we link the neurons in the output layer to the corresponding neurons in the input layer of each network N_i, the ensemble can be used to compute the modal program in parallel. In this example, we connect output neurons $\Diamond s$ and r to input neurons $\Diamond s$ and r, respectively, in N_1, and output neuron q to input neuron q in N_3. The ensemble converges to a stable state containing $\{\Diamond s, r, \Box q\}$ in ω_1, $\{s, q\}$ in ω_2, and $\{q, \Diamond s\}$ in ω_3.

11.3.2 Connectionist Temporal Reasoning

An extension of CML considers temporal and epistemic knowledge [7]. Generally speaking, the idea is to allow, instead of a single ensemble, a number n of ensembles, each representing the knowledge held by a number of agents at a given time point t. Figure 11.4 illustrates how this dynamic feature can be combined with the symbolic features of the knowledge represented in each network, allowing not only for the analysis of the current state (possible world

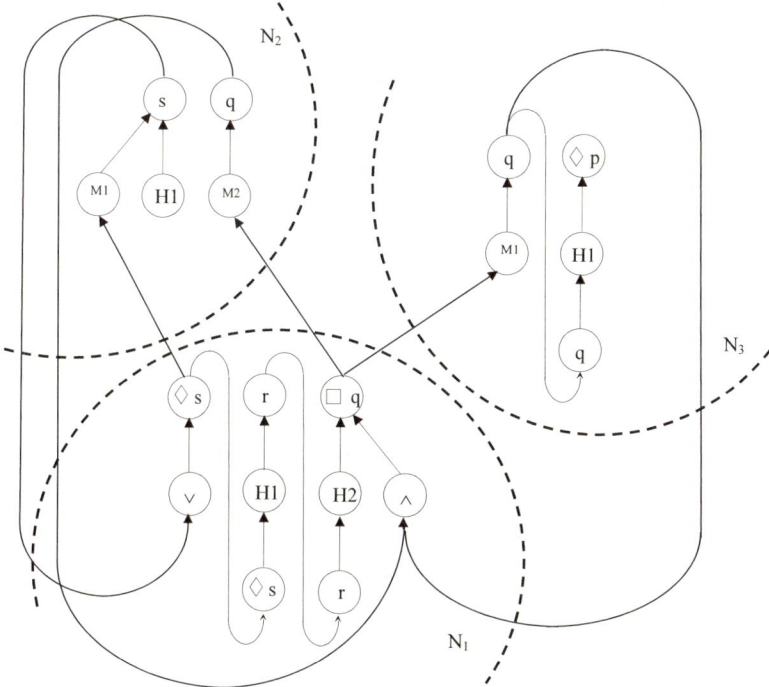

Fig. 11.3. Neural Network Ensemble for Modal Reasoning

or time point), but also for the analysis of how knowledge changes through time. As before, in this paper we discuss the idea in the context of an example (below).

Example 3. One of the typical axioms of temporal logics of knowledge is $K_i \bigcirc \alpha \rightarrow \bigcirc K_i \alpha$, where $\bigcirc$ denotes the *next time* temporal operator. This means that what an agent i knows today (K_i) about tomorrow ($\bigcirc \alpha$), she still knows tomorrow ($\bigcirc K_i \alpha$). In other words, this axiom states that an agent would not forget what she knew. This can be represented in a network ensemble with the use of a network that represents the agent's knowledge today, a network that represents the agent's knowledge tomorrow (despite partially), and the appropriate connections between networks. Clearly, an output neuron $K \bigcirc \alpha$ of a network that represents agent i at time t needs to be connected to an output neuron $K\alpha$ of a network that represents agent i at time $t+1$ in such a way that, whenever $K \bigcirc \alpha$ is activated, $K\alpha$ is also activated.

Of course there are important issues to do with (1) the optimisation of the model of Figure 11.4 in practice, (2) the fact that the number of networks may be bounded, and (3) the trade-off between space and time computational complexity. The fact, however, that this model is sufficient to deal with such a variety of reasoning tasks is encouraging to us.

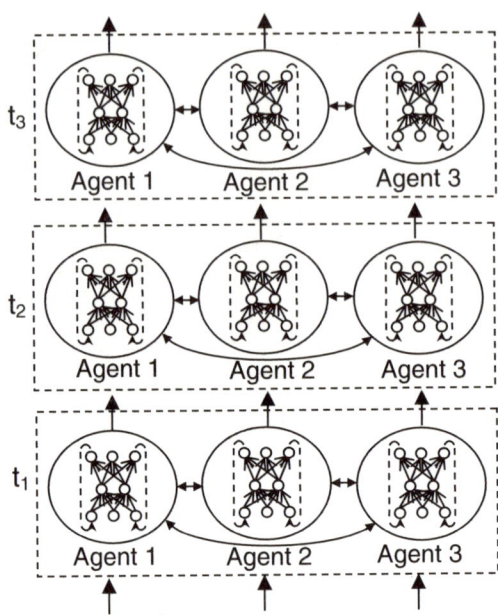

Fig. 11.4. Neural Network Ensemble for Temporal Reasoning

The definition of the number of ensembles s that are necessary to solve a given problem clearly depends on the problem domain, and on the number of time points that are relevant for reasoning about the problem. For example, in the case of the well-known *muddy children* puzzle of distributed knowledge representation [36, 38], we know that it suffices to have s equal to the number of children that are muddy. The definition of s in a different domain might not be as straightforward, possibly requiring a fine-tuning process similar to that performed at learning, but with a varying network architecture. These and other considerations, including more extensive evaluations of the model w.r.t. learning, are still required.

11.4 Fibring Neural Networks

In Connectionist Modal Logic (CML), one may need to create copies of certain concepts. As a result, CML does not deal with infinite domains, since this would require infinitely many copies. An alternative is to map the instances of a variable onto the real numbers, and then use the real numbers as inputs to a neural network as a way of representing the instances. This has been done in [41]. However, the question of how to construct a neural network to compute and learn a given first-order program remained unanswered, since no translation algorithm has been given in [41]. A follow-up paper providing such an algorithm for first-order covered programs has appeared recently [14] – see also the corresponding chapter by Bader et al. in this volume.

In [13], we have also followed the idea of representing variables as real numbers, and proposed a translation algorithm from first-order acyclic programs to neural network ensembles. The algorithm makes use of *fibring* of neural networks [8], which we discuss in this section. Briefly, the idea is to use a neural network to iterate a global counter n. For each clause C_i in the logic program, this counter is combined (fibred) with another neural network, which determines whether C_i outputs an atom of level n for a given interpretation I. This allows us to translate programs with an infinite number of ground instances into a finite neural network structure (e.g. $\neg even(x) \rightarrow even(s(x))$ for $x \in \mathbb{N}, s(x) = x+1$), and to prove that indeed the network approximates the fixed-point semantics of the program. The translation is made possible because fibring allows one to implement a key feature of symbolic computation in neural networks, namely, *recursion*, as we describe below.

The idea of fibring neural networks is simple. Fibred networks may be composed not only of interconnected neurons but also of other networks, forming a recursive architecture. A fibring function then defines how this architecture behaves by defining how the networks in the ensemble relate to each other. Typically, the fibring function will allow the activation of neurons in one network (A) to influence the change of weights in another network (B) (e.g. by multiplying the activation state of a neuron in A by the weights of neurons in B). Intuitively, this may be seen as training network B at the same time that one runs network A. Interestingly, albeit being a combination of simple and standard neural networks, fibred networks can approximate any polynomial function in an unbounded domain, thus being more expressive than standard feedforward networks (which are universal approximators of functions in compact, i.e. closed and bounded domains only) [8]. For example, fibred networks compute $f(x) = x^2$ exactly for $x \in \mathbb{R}$.

Figure 11.5 exemplifies how a network (B) can be fibred onto a network (A). Of course, the idea of fibring is not only to organise networks as a number of subnetworks (A, B, etc.). In Figure 11.5, for example, the output neuron of A is expected to be a neural network (B) in its own right. The input, weights, and output of B may depend on the activation state of A's output neuron, according to the fibring function φ. As mentioned above, one such function may be simply to *multiply* the weights of B by the activation state of A's output neuron.

Fibred networks can be trained by examples in the same way that standard networks are (for example, with the use of *backpropagation* [32]). Networks A and B mentioned above, e.g., could have been trained separately before having been fibred. Notice also that, in addition to using different fibring functions, networks can be fibred in a number of different ways as far as their architectures are concerned. Network B, e.g., could have been fibred onto a hidden neuron of network A.

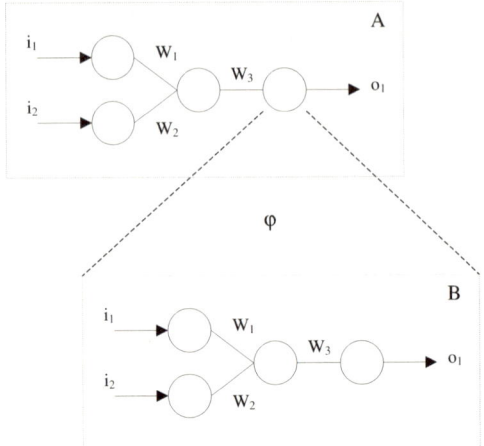

Fig. 11.5. Fibring Neural Networks

Fibring contributes to the solution of the problem of elaboration tolerance by offering a principled and modular way of combining networks. Network A could have been trained, e.g., with a robot's visual system, while network B would have been trained with its planning system, and fibring would serve to perform the composition of the two systems (along the lines of Gabbay's methodology for fibring logical systems [42]). Fibring can be very powerful. It offers the extra expressiveness required by complex applications at low computational cost (that of computing fibring function φ). Of course, we would like to keep φ as simple as possible so that it can be implemented itself by simple neurons in a fully connectionist model. Interesting work remains to be done in this area, particularly in what regards the question of how one should go about fibring networks in real applications.

11.5 Concluding Remarks

CML is an example of how neural networks can contribute to logic, and fibring is an example of how logic can bring insight into neural computation. CML offers a parallel model of computation to modal logic that, at the same time, can be integrated with an efficient learning algorithm. Fibring is a clear example of how concepts from symbolic computation may help in the development of neural network models (this is not necessarily conflicting with the ambition of biological plausibility, e.g. fibring functions can be understood as a model of *presynaptic weights*, which play an important role in biological neural networks).

Connectionist non-classical reasoning and network fibring bring us to our overall cognitive model, which we may call *fibred network ensembles*. In this model, a network ensemble A (representing, e.g., a temporal theory) may be combined with another network ensemble B (representing, e.g., an intuitionistic theory). In the same spirit, meta-level concepts (in A) may be combined and brought into the object-level (B), without necessarily blurring the distinction between the two levels. One may reason in the meta-level and use that information in the object-level, a typical example being (meta-level) reasoning about actions in (object-level) databases containing inconsistencies [43].

Another example is *reasoning with variables*, where one would represent variables at the meta-level network A, and instantiate them as needed at the object-level network B. In the propositional case, when we look at a neural network, we can see the literals and their relationship with other literals explicitly represented as neurons and their connections with other neurons in the network. In the same way, we would like to be able to look at a *first-order* neural network and see the variables and their relationship with other variables represented explicitly in the network. A translation algorithm having this property may be called a *constructive* algorithm. This would allow us to reason about variables and to learn relations on variables possibly as efficiently as in the propositional case. At closer inspection, the real number representation of variables used in [13] is not completely satisfactory. Although it allows one to translate first-order programs into neural networks, its translation algorithm is not constructive, and this may complicate the learning task.

Together with the algorithms for learning from examples and background knowledge [4] and for rule extraction from trained neural networks [18] (which I have neglected in this paper), neural-symbolic integration addresses the challenges put forward by McCarthy in [9]. On the other hand, there are new challenges now, which arise directly from our goal of integrating reasoning and learning in a principled way, as put forward by Valiant in [3].

In my view, the key challenges ahead include how to obtain a constructive translation for relational learning within simple neural networks, and how to have a sound and quasi-complete extraction method that is efficient for large-scale networks. These challenges emerge from the goal of sound, expressive reasoning in an uncertain environment and robust, fault-tolerant learning. There are also more practical, computational challenges associated with the effective computation of logical models, efficient rule extraction in practical applications, and striking the right balance between computational tractability and expressiveness. These challenges guide our research and inform the material in our forthcoming book pon *connectionist non-classical reasoning* [44].

Of course, the question of how we humans integrate reasoning and learning remains unanswered. But I argue that the prospects of answering it are better as we investigate the connectionist processes of the brain together with the logical processes of symbolic computation, and not as two isolated paradigms. Of course, we will need to be precise as we develop the work, and test the *fibred network ensembles* model in real cognitive tasks. And we will need to keep an eye on the developments in *inductive logic programming* (ILP) [45], since in some respects the goals of ILP are the same as ours when we study the first-order case. We hope ours is a more robust and efficient setting, but experimental comparisons are needed here.

In summary, by paying attention to the developments on either side of the division between the symbolic and the sub-symbolic paradigms, we are getting closer to a unifying theory, or at least promoting a faster and principled development of AI. This paper described a family of connectionist non-classical reasoning systems and hinted at how they may be combined at different levels of abstraction by fibring. I hope it serves as a stepping stone towards such a theory to reconcile the symbolic and connectionist approaches. Human beings are quite extraordinary at performing practical reasoning as they go about their daily business. *There are cases where the human computer, slow as it is, is faster than artificial intelligence systems. Why are we faster? Is it the way we perceive knowledge as opposed to the way we represent it? Do we know immediately which rules to select and apply? We must look for the correct representation in the sense that it mirrors the way we perceive and apply the rules* [46]. Ultimately, neural-symbolic integration is about asking and trying to answer these questions, and the about associated provision of neural-symbolic systems with integrated expressive reasoning and robust learning capabilities.

Acknowledgements

I am grateful to Luis Lamb, Dov Gabbay, and Oliver Ray for many useful discussions.

References

1. d'Avila Garcez, A.S.: Fewer epistemological challenges for connectionism. In Cooper, S.B., Lowe, B., Torenvliet, L., eds.: Proceedings of Computability in Europe, CiE 2005. Volume LNCS 3526., Amsterdam, The Netherlands, Springer-Verlag (2005) 139–149
2. d'Avila Garcez, A.S., Gabbay, D.M., Ray, O., Woods, J.: Abductive reasoning in neural-symbolic systems. TOPOI: An International Review of Philosophy (2007) doi:10.1007/s11245-006-9005-5.
3. Valiant, L.G.: Three problems in computer science. Journal of the ACM **50** (2003) 96–99
4. d'Avila Garcez, A.S., Broda, K., Gabbay, D.M.: Neural-Symbolic Learning Systems: Foundations and Applications. Perspectives in Neural Computing. Springer-Verlag (2002)
5. d'Avila Garcez, A.S., Lamb, L.C., Gabbay, D.M.: Connectionist modal logic: Representing modalities in neural networks. Theoretical Computer Science **371** (2007) 34–53
6. d'Avila Garcez, A.S., Lamb, L.C., Gabbay, D.M.: Connectionist computations of intuitionistic reasoning. Theoretical Computer Science **358** (2006) 34–55

7. d'Avila Garcez, A.S., Lamb, L.C.: A connectionist computational model for epistemic and temporal reasoning. Neural Computation **18** (2006) 1711–1738
8. d'Avila Garcez, A.S., Gabbay, D.M.: Fibring neural networks. In: Proceedings of 19th National Conference on Artificial Intelligence AAAI'04, San Jose, California, USA, AAAI Press (2004) 342–347
9. McCarthy, J.: Epistemological challenges for connectionism. Behaviour and Brain Sciences **11** (1988) 44
10. Smolensky, P.: On the proper treatment of connectionism. Behavioral and Brain Sciences **11** (1988) 1–74
11. Gardenfors, P.: Conceptual Spaces: The Geometry of Thought. MIT Press (2000)
12. Haykin, S.: Neural Networks: A Comprehensive Foundation. Prentice Hall (1999)
13. Bader, S., d'Avila Garcez, A.S., Hitzler, P.: Computing first order logic programs by fibring artificial neural networks. In: Proceedings of International FLAIRS Conference, Florida, USA, AAAI Press (2005) 314–319
14. Bader, S., Hitzler, P., Hölldobler, S., Witzel, A.: A fully connectionist model generator for covered first-order logic programs. In: M. Veloso, ed.: Proceedings of International Joint Conference on Artificial Intelligence IJCAI07, Hyderabad, India (2007) 666–671
15. Hölldobler, S.: Automated inferencing and connectionist models. Postdoctoral Thesis, Intellektik, Informatik, TH Darmstadt (1993)
16. Shastri, L., Ajjanagadde, V.: A connectionist representation of rules, variables and dynamic binding. Technical report, Dept of Computer and Information Science, University of Pennsylvania (1990)
17. Bologna, G.: Is it worth generating rules from neural network ensembles? Journal of Applied Logic **2** (2004) 325–348
18. d'Avila Garcez, A.S., Broda, K., Gabbay, D.M.: Symbolic knowledge extraction from trained neural networks: A sound approach. Artificial Intelligence **125** (2001) 155–207
19. d'Avila Garcez, A.S., Zaverucha, G.: The connectionist inductive learning and logic programming system. Applied Intelligence **11** (1999) 59–77
20. Hitzler, P., Hölldobler, S., Seda, A.K.: Logic programs and connectionist networks. Journal of Applied Logic **2** (2004) 245–272 Special Issue on Neural-Symbolic Systems.
21. Jacobsson, H.: Rule extraction from recurrent neural networks: A taxonomy and review. Neural Computation **17** (2005) 1223–1263
22. Setiono, R.: Extracting rules from neural networks by pruning and hidden-unit splitting. Neural Computation **9** (1997) 205–225
23. Sun, R.: Robust reasoning: integrating rule-based and similarity-based reasoning. Artificial Intelligence **75** (1995) 241–296
24. d'Avila Garcez, A.S., Elman, J., Hitzler, P., eds.: Proceedings of IJCAI International Workshop on Neural-Symbolic Learning and Reasoning NeSy05, Edinburgh, Scotland (2005)
25. d'Avila Garcez, A.S., Hitzler, P., Tamburrini, G., eds.: Proceedings of ECAI International Workshop on Neural-Symbolic Learning and Reasoning NeSy06, Trento, Italy (2006)
26. Shastri, L.: Advances in SHRUTI: a neurally motivated model of relational knowledge representation and rapid inference using temporal synchrony. Applied Intelligence **11** (1999) 79–108

27. Hölldobler, S., Kalinke, Y.: Toward a new massively parallel computational model for logic programming. In: Proceedings of the Workshop on Combining Symbolic and Connectionist Processing, ECAI 94. (1994) 68–77
28. Towell, G.G., Shavlik, J.W.: Knowledge-based artificial neural networks. Artificial Intelligence **70** (1994) 119–165
29. Thrun, S.B.: Extracting provably correct rules from artificial neural networks. Technical report, Institut fur Informatik, Universitat Bonn (1994)
30. Smolensky, P., Legendre, G.: The Harmonic Mind: From Neural Computation to Optimality-Theoretic Grammar. MIT Press, Cambridge, Mass (2006)
31. Page, M.: Connectionist modelling in psychology: A localist manifesto. Behavioral and Brain Sciences **23** (2000) 443–467
32. Rumelhart, D.E., Hinton, G.E., Williams, R.J.: Learning internal representations by error propagation. In Rumelhart, D.E., McClelland, J.L., eds.: Parallel Distributed Processing. Volume 1. MIT Press (1986) 318–362
33. Lloyd, J.W.: Foundations of Logic Programming. Springer-Verlag (1987)
34. Gelfond, M., Lifschitz, V.: Classical negation in logic programs and disjunctive databases. New Generation Computing **9** (1991) 365–385
35. d'Avila Garcez, A.S., Lamb, L.C.: Neural-symbolic systems and the case for nonclassical reasoning. In Artemov, S., Barringer, H., d'Avila Garcez, A.S., Lamb, L.C., Woods, J., eds.: We will show them! Essays in Honour of Dov Gabbay. International Federation for Computational Logic, College Publications, London (2005) 469–488
36. Fagin, R., Halpern, J., Moses, Y., Vardi, M.: Reasoning about Knowledge. MIT Press (1995)
37. d'Avila Garcez, A.S., Gabbay, D.M., Lamb, L.C.: Value-based argumentation frameworks as neural-symbolic learning systems. Journal of Logic and Computation **15** (2005) 1041–1058
38. d'Avila Garcez, A.S., Lamb, L.C., Broda, K., Gabbay, D.M.: Applying connectionist modal logics to distributed knowledge representation problems. International Journal on Artificial Intelligence Tools **13** (2004) 115–139
39. Vardi, M.: Why is modal logic so robustly decidable? In: Descriptive Complexity and Finite Models. DIMACS Series in Discrete Mathematics and Theoretical Computer Science, American Mathematical Society (1996) 149–184
40. Orgun, M.A., Ma, W.: An overview of temporal and modal logic programming. In: Proc. Inl. Conf. Temporal Logic. LNAI 827, Springer (1994) 445–479
41. Hölldobler, S., Kalinke, Y., Storr, H.P.: Approximating the semantics of logic programs by recurrent neural networks. Applied Intelligence **11** (1999) 45–58
42. Gabbay, D.M.: Fibring Logics. Oxford Univesity Press (1999)
43. Gabbay, D.M., Hunter, A.: Making inconsistency respectable: Part 2 - metalevel handling of inconsistency. In: Symbolic and Quantitative Approaches to Reasoning and Uncertainty ECSQARU'93. Volume LNCS 747., Springer-Verlag (1993) 129–136
44. d'Avila Garcez, A.S., Lamb, L., Gabbay, D.M.: Connectionist Non-Classical Reasoning. Cognitive Technologies. Springer-Verlag (2007) to appear.
45. Muggleton, S., Raedt, L.: Inductive logic programming: theory and methods. Journal of Logic Programming **19** (1994) 629–679
46. Gabbay, D.M.: Elementary Logics: a Procedural Perspective. Prentice Hall, London (1998)

12

Connectionist Representation of Multi-Valued Logic Programs

Ekaterina Komendantskaya, Máire Lane and Anthony Karel Seda

Department of Mathematics, University College Cork, Cork, Ireland
komendantskaya@gmail.com, maireln@bcri.ucc.ie, a.seda@ucc.ie[†]

Summary. Hölldobler and Kalinke showed how, given a propositional logic program P, a 3-layer feedforward artificial neural network may be constructed, using only binary threshold units, which can compute the familiar immediate-consequence operator T_P associated with P. In this chapter, essentially these results are established for a class of logic programs which can handle many-valued logics, constraints and uncertainty; these programs therefore represent a considerable extension of conventional propositional programs. The work of the chapter basically falls into two parts. In the first of these, the programs considered extend the syntax of conventional logic programs by allowing elements of quite general algebraic structures to be present in clause bodies. Such programs include many-valued logic programs, and semiring-based constraint logic programs. In the second part, the programs considered are bilattice-based annotated logic programs in which body literals are annotated by elements drawn from bilattices. These programs are well-suited to handling uncertainty. Appropriate semantic operators are defined for the programs considered in both parts of the chapter, and it is shown that one may construct artificial neural networks for computing these operators. In fact, in both cases only binary threshold units are used, but it simplifies the treatment conceptually to arrange them in so-called multiplication and addition units in the case of the programs of the first part.

12.1 Introduction

In their seminal paper [1], Hölldobler and Kalinke showed how, given a propositional logic program P, one may construct a 3-layer feedforward artificial neural network (ANN), having only binary threshold units, which can compute the familiar immediate-consequence operator T_P associated with P. This result has been taken as the starting point of a line of research which forms one component of the general problem of integrating the logic-based and connectionist or neural-network-based approaches to

---

[†] Author for correspondence: A.K. Seda.

computation. Of course, the objective of integrating these two computing paradigms is to combine the advantages to be gained from connectionism with those to be gained from symbolic computation. The papers [3, 4, 5, 6, 7, 8, 9, 10, 11, 12, 1, 2, 13, 14, 15, 16, 17, 18, 19] represent a small sample of the literature on this topic. In particular, the papers just cited contain extensions and refinements of the basic work of Hölldobler et al. in a number of directions including inductive logic programming, modal-logic programming, and distributed knowledge representation. In the large, the problem of integrating or combining these two approaches to computation has many aspects and challenges, some of which are discussed in [3, 12].

Our objective and theme in this chapter is to present generalizations of the basic results of [1] to classes of extended logic programs P which can handle many-valued logics, constraints and uncertainty; in fact, the programs we discuss are annotated logic programs. Thus, we are concerned primarily with the computation by ANN of the semantic operators determined by such extended logic programs. Therefore, we begin with a brief discussion of these programs, their properties and powers of knowledge representation.

As is well-known, there exists a great variety of many-valued logics of interest in computation, see [20] for a survey. For example, Lukasiewicz and Kleene introduced several three-valued logics [20]. Then infinite-valued Lukasiewicz logic appeared, and there are various other many-valued logics of interest, such as fuzzy logic and intuitionistic logic. In another direction, Belnap studied bilattice-based logics for reasoning with uncertainties, and his work was further developed in [21, 22]. Most of these logics have been adapted to logic programming: for example, see [23] for logic programs interpreted by arbitrary sets, [24] for applications of Kleene's logic to logic programming, [25] for semiring-based constraint logic programming, [26, 27] for fuzzy logic programming, and [28, 29, 30] for (bi)lattice-based logic programming. See also [31] for a very general algebraic analysis of different many-valued logic programs. However, in the main, there have been three approaches to many-valued logic programming, as follows.

First, *annotation-free logic programs* were introduced by Fitting in [24] and further developed in [25, 28, 29, 32]. They are formally the same as two-valued logic programs in that the clauses of an annotation-free logic program are exactly the same as those for a two-valued logic program. But, whilst each atomic ground formula of a two-valued logic program is given an interpretation in {true, false}, an atomic annotated formula of an annotation-free logic program receives its interpretation in an arbitrary set carrying idempotent, commutative and associative operations which model logical connectives, as we illustrate in more detail in Section 12.3. Next, *implication-based logic programs* were introduced by Van Emden in [33] and were designed in order to obtain a simple and effective proof procedure. Van Emden considered the particular case of the set $[0, 1]$ of truth values, but this has since been extended, see for example [34]. Much of this work carries over without difficulty to an arbitrary set with idempotent, commutative and associative operations. Van

Emden used the conventional syntax for logic programs, except for having clauses of the form
$$A \leftarrow\boxed{f}\!\!- B_1,\ldots,B_n,$$
where f is a *factor* or *threshold* taken from the interval $[0,1]$ of real numbers. The atoms $A, B_1, \ldots, B_n$ receive their interpretation from the interval $[0,1]$, and are such that the value of the head A of a clause has to be greater than or equal to $f \times \min(|B_1|,\ldots,|B_n|)$. Finally, *annotated (or signed) logic programs* require each atom in a given clause to be annotated (or signed) by a truth value. Most commonly, an annotated clause ([35]) has the following form
$$A : \mu \leftarrow B_1 : \mu_1, \ldots, B_n : \mu_n,$$
where each μ_i is an *annotation term*, which means that it is either an annotation constant, or an annotation variable, or a function over annotation constants and/or variables. For ease of implementation, this approach has been very popular and many variations of annotated and signed logic programs have appeared, see [36, 22, 35, 15, 23, 37, 38, 39, 40], for example. We consider logic programs of this type in Section 12.4 of the chapter.

The work of this chapter falls into two main sections: Section 12.3, concerned with annotation-free logic programs, and Section 12.4, concerned with annotated logic programs. However, there is considerable interaction between these two sections, as we show, and the final result, Theorem 4, strongly connects them. However, our approach in Section 12.3 is more algebraic than usual and we develop an abstract, general semantic operator $\mathfrak{T}_{P,\mathfrak{C}}$ defined over certain algebraic structures $\mathfrak{C}$. This theory not only gives a unified view of conventional many-valued logic programming, and of simple logic-based models of uncertainty in logic programming and databases, see [32], but it also includes semiring-based constraint logic programming as considered in [25]; one simply chooses $\mathfrak{C}$ suitably. In defining this operator, we are inspired by Fitting's paper [41] and we work over the set P^{**}, see Section 12.3.1. We are therefore faced with the problem of handling countable (possibly infinite) products $\bigodot_{i\in\mathbb{N}} c_i$, where $\odot$ is a binary operation defined on $\mathfrak{C}$, and with determining the value of $\bigodot_{i\in\mathbb{N}} c_i$ finitely, in some sense. Related problems also arise in constructing the neural networks we present to compute $\mathfrak{T}_{P,\mathfrak{C}}$ when P is propositional. We solve these problems by introducing the notion of finitely-determined binary operations $\odot$, see Definition 1 and Theorem 1, and it turns out that this notion is well-suited to these purposes and to building ANN, somewhat in the style of [1], to compute $\mathfrak{T}_{P,\mathfrak{C}}$, see Theorem 2. In fact, to construct the networks in this case, we introduce the notion of addition and multiplication units which are 2-layer ANN composed of binary threshold units, and this approach produces conceptually quite simple networks.

In Section 12.4, we focus on bilattice-based annotated logic programs and introduce a semantic operator $\mathcal{T}_P$ for them. Again, we build ANN in the style of [1], but this time to simulate $\mathcal{T}_P$. The operator $\mathcal{T}_P$ is simpler than $\mathfrak{T}_{P,\mathfrak{C}}$

to define in so much as it employs P directly, rather than P^{**}, but its simulation by ANN is more difficult than is the case for $\mathfrak{T}_{P,\mathfrak{e}}$. Indeed, the ANN used in Section 12.4 do not require additional layers, but instead employ two learning functions in order to perform computations of T_P. In this sense, one has a tradeoff: more complicated units and simpler connections versus simpler units and more complicated connections. This reduction of architecture and the use of learning functions builds a bridge between the essentially deductive neural networks of [1] and the learning neural networks implemented in Neurocomputing [42]. It also raises questions on the comparative time and space complexity of the architectures of the ANN used to compute $\mathfrak{T}_{P,\mathfrak{e}}$ and T_P, see Section 12.6 summarizing our conclusions, but we do not pursue this issue here in detail.

Finally, we show in Theorem 4 how the computations of $\mathfrak{T}_{P,\mathfrak{e}}$ by the ANN of Section 12.3 can be simulated by computations of T_P and the corresponding neural networks of Section 12.4. This result unites the two main sections of the chapter and brings it to a conclusion.

For reasons of lack of space, we do not consider here the extension of our results to the case of first-order logic programs P. However, the papers by the present authors listed at the end of the chapter contain some results in this direction.

Acknowlegement The authors thank the Boole Centre for Research in Informatics at University College Cork for its substantial financial support of the research presented here.

12.2 Neural Networks

We begin by briefly summarizing what we need relating to artificial neural networks or just neural networks for short; our terminology and notation are standard, and our treatment closely follows that of [10], but see also [9, 43].

A *neural network* or *connectionist network* is a weighted digraph. A typical *unit* (or node) k in this digraph is shown in Figure 12.1. We let $w_{kj} \in \mathbb{R}$ denote the weight of the connection from unit j to unit k (w_{kj} may be 0). Then the unit k is characterized, at time t, by the following data: its *inputs* $i_{kj}(t) = w_{kj}v_j(t)$ (the input received by k from j at time t) for $j = 1, \ldots, n_k$, its *threshold* $\theta_k \in \mathbb{R}$, its *potential* $p_k(t) = \left(\sum_{j=1}^{n_k} w_{kj}v_j(t)\right) - \theta_k \in \mathbb{R}$, and its *value* $v_k(t)$. The units are updated synchronously, time becomes $t + \Delta t$, and the output value for k, $v_k(t+\Delta t)$, is calculated from $p_k(t)$ by means of a given *output function* ψ, that is, $v_k(t+\Delta t) = \psi(p_k(t))$. The only output function ψ we use here is the Heaviside function H. Thus, $v_k(t+\Delta t) = H(p_k(t))$, where H is defined by $H(x)$ is equal to 1 if $x \geq 0$ and is equal to 0 otherwise. Units of this type are called *binary threshold units*.

As far as the architecture of neural networks is concerned, we will only consider networks where the units can be organized in layers. A *layer* is a

12 Connectionist Representation of Multi-Valued Logic Programs   287

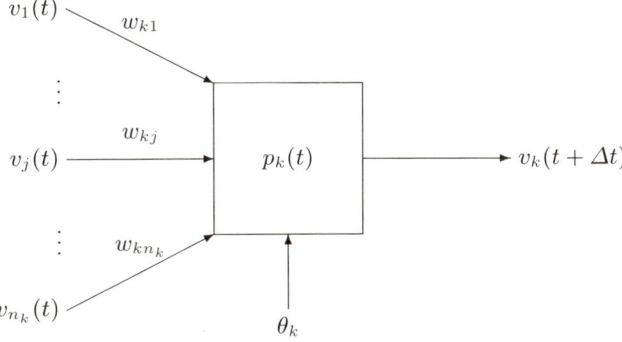

Fig. 12.1. Unit k in a connectionist network.

vector of units. An *n-layer feedforward network* $\mathcal{F}$ consists of the *input* layer, $n-2$ *hidden* layers, and the *output* layer, where $n \geq 2$. Here, we will mainly be concerned with $n = 3$. Each unit occurring in the i-th layer is connected to each unit occurring in the $(i+1)$-st layer, $1 \leq i < n$. A connectionist network $\mathcal{F}$ is called a *multilayer feedforward network* if it is an n-layer feedforward network for some n. Let r and s be the number of units occurring in the input and output layers, respectively, of a multilayer feedforward network $\mathcal{F}$. Then $\mathcal{F}$ computes a function $f_\mathcal{F} : \mathbb{R}^r \to \mathbb{R}^s$ as follows. The input vector (the argument of $f_\mathcal{F}$) is presented to the input layer at time t_0 and propagated through the hidden layers to the output layer. At each time point, all units update their potential and value. At time $t_0 + (n-1)\Delta t$, the output vector (the image under $f_\mathcal{F}$ of the input layer) is read off the output layer. Finally, a neural network is called *recurrent* or is *made recurrent* if the number of units in the input layer is equal to the number of units in the output layer and each unit in the output layer is connected with weight 1 to the corresponding unit in the input layer. A recurrent network can thus perform iterated computations because the output values can be returned to the input layer via the connections just described; it can thus perform computation of the iterates $T^k(I)$, $k \in \mathbb{N}$, for example, where I is an interpretation and T is a (semantic) operator.

12.3 Annotation-Free Logic Programs and the Operator $\mathfrak{T}_{P,\mathfrak{C}}$

12.3.1 Finitely-Determined Operations

Let $\mathfrak{C}$ denote a set endowed with binary operations $+$ and $\times$, and a unary operation $\neg$ satisfying $\neg(\neg c) = c$ for all $c \in \mathfrak{C}$[3]. In [44], the notion of *finitely-determined disjunctions and conjunctions* ($\vee$ and $\wedge$) was introduced for sets $\mathfrak{C}$ of truth values. We begin by giving this definition for a general binary operation $\odot$ on $\mathfrak{C}$. Note that we assume that $\odot$ has meaningfully been extended to include products $\bigodot_{i \in M} c_i$ of countably infinite families M of elements c_i of $\mathfrak{C}$. Indeed, the way in which we carry out this extension is the main point of the next definition and the discussion following it.

Definition 1. *Suppose that $\mathfrak{C}$ is a set equipped with a binary operation $\odot$. We say that $\odot$ is* finitely determined *or that* products (relative to $\odot$) are finitely determined in $\mathfrak{C}$ *if, for each $c \in \mathfrak{C}$, there exists a countable (possibly infinite) collection $\{(R_c^n, E_c^n) \mid n \in \mathcal{J}\}$ of pairs of sets $R_c^n \subseteq \mathfrak{C}$ and $E_c^n \subseteq \mathfrak{C}$, where each R_c^n is finite, such that a countable (possibly infinite) product $\bigodot_{i \in M} c_i$ in $\mathfrak{C}$ is equal to c if and only if for some $n \in \mathcal{J}$ we have*
(i) $R_c^n \subseteq \{c_i \mid i \in M\}$, and
(ii) for all $i \in M$, $c_i \notin E_c^n$, that is, $\{c_i \mid i \in M\} \subseteq (E_c^n)^{co}$, where $(E_c^n)^{co}$ denotes the complement of the set E_c^n.

We call the elements of E_c^n excluded values *for c, we call the elements of $\mathcal{A}_c^n = (E_c^n)^{co}$* allowable values *for c, and in particular we call the elements of R_c^n* required values *for c; note that, for each $n \in \mathcal{J}$, we have $R_c^n \subseteq \mathcal{A}_c^n$, so that each required value is also an allowable value (but not conversely). More generally, given $c \in \mathfrak{C}$, we call $s \in \mathfrak{C}$ an* excluded value for c *if no product $\bigodot_{i \in M} c_i$ with $\bigodot_{i \in M} c_i = c$ contains s, that is, in any product $\bigodot_{i \in M} c_i$ whose value is equal to c, we have $c_i = s$ for no $i \in M$. We let E_c denote the set of all excluded values for c, and let $\mathcal{A}_c$ denote the complement $(E_c)^{co}$ of E_c and call it the set of all* allowable values *for c. Note finally that when confusion might otherwise result, we will superscript each of the sets introduced above with the operation in question. Thus, for example, $\mathcal{A}_c^\odot$ denotes the allowable set for c relative to the operation $\odot$.*

This definition was originally motivated by the results of [45], and the following example shows the thinking behind it.

Example 1. Consider Belnap's well-known four-valued logic with set $\mathfrak{C} = \mathcal{FOUR} = \{t, u, b, f\}$ of truth values and connectives as defined in Table 12.1, where t denotes *true*, u denotes *undefined* or *none* (neither true nor false), b denotes *both* (true and false), and f denotes *false*.

Taking $\odot$ to be disjunction $\vee$, the sets E and R are as follows.
(a) For t, n takes values 1 and 2, $E_t^\vee = \emptyset$, $R_t^{\vee,1} = \{t\}$, and $R_t^{\vee,2} = \{u, b\}$.

---

[3] When $\mathfrak{C}$ has no natural negation $\neg$, we simply take $\neg$ to be the identity.

12 Connectionist Representation of Multi-Valued Logic Programs   289

Table 12.1. Truth table for the logic $\mathcal{FOUR}$

p	q	$\neg p$	$p \wedge q$	$p \vee q$
t	t	f	t	t
t	u	f	u	t
t	b	f	b	t
t	f	f	f	t
u	t	u	u	t
u	u	u	u	u
u	b	u	f	t
u	f	u	f	u
b	t	b	b	t
b	u	b	f	t
b	b	b	b	b
b	f	b	f	b
f	t	t	f	t
f	u	t	f	u
f	b	t	f	b
f	f	t	f	f

(b) For u, we have $n = 1$, $E_u^\vee = \{t, b\}$ and $R_u^\vee = \{u\}$.
(c) For b, we have $n = 1$, $E_b^\vee = \{t, u\}$ and $R_b^\vee = \{b\}$.
(d) For f, we have $n = 1$, $E_f^\vee = \{t, u, b\}$ and $R_f^\vee = \{f\}$.

Thus, a countable disjunction $\bigvee_{i \in M} c_i$ takes value t if and only if either (i) at least one of the c_i is equal to t or (ii) at least one of the c_i takes value b and at least one takes value u; no truth value is excluded. As another example, $\bigvee_{i \in M} c_i$ takes value u if and only if at least one of the c_i is u, none are equal to t and none are equal to b.

Now taking $\odot$ to be conjunction $\wedge$, the sets E and R are as follows.
(a) For t, we have $n = 1$, $E_t^\wedge = \{u, b, f\}$ and $R_t^\wedge = \{t\}$.
(b) For u, we have $n = 1$, $E_u^\wedge = \{b, f\}$ and $R_u^\wedge = \{u\}$.
(c) For b, we have $n = 1$, $E_b^\wedge = \{u, f\}$ and $R_b^\wedge = \{b\}$.
(d) For f, n takes values 1 and 2, $E_f^\wedge = \emptyset$, $R_f^{\wedge,1} = \{f\}$, and $R_f^{\wedge,2} = \{u, b\}$.

Notice finally that, if we restrict the connectives in $\mathcal{FOUR}$ to $\{t, u, f\}$, we obtain Kleene's well-known strong three-valued logic.

It turns out that the connectives in all the logics commonly encountered in logic programming, and indeed in many other logics, satisfy Definition 1, and it will be convenient to state next the main facts we need concerning arbitrary finitely-determined operations, see [44] for all proofs.

Theorem 1. *Suppose that $\odot$ is a binary operation defined on a set $\mathfrak{C}$. Then the following statements hold.*

1. *If $\odot$ is finitely determined, then it is idempotent, commutative and associative.*

2. Suppose that $\odot$ is finitely determined and that $\mathfrak{C}$ contains finitely many elements $\{c_1, \ldots, c_n\}$. Then, for any collection $\{s_i \mid i \in M\}$, where each of the $s_i \in \mathfrak{C}$ and M is a denumerable set, the sequence $s_1, s_1 \odot s_2, s_1 \odot s_2 \odot s_3, \ldots$ is eventually constant with value s, say. Therefore, setting $\bigodot_{i \in M} s_i = s$ gives each countably infinite product in $\mathfrak{C}$ a well-defined meaning which extends the usual meaning of finite products.
3. Suppose $\odot$ is finitely determined and that $\bigodot_{i \in M} s_i = c$, where M is a countable set. Then the sequence $s_1, s_1 \odot s_2, s_1 \odot s_2 \odot s_3, \ldots$ is eventually constant with value c.
4. Suppose that $\mathfrak{C}$ is a countable set and $\odot$ is idempotent, commutative and associative. Suppose further that, for any set $\{s_i \mid i \in M\}$ of elements of $\mathfrak{C}$ where M is countable, the sequence $s_1, s_1 \odot s_2, s_1 \odot s_2 \odot s_3, \ldots$ is eventually constant. Then all products in $\mathfrak{C}$ are (well-defined and are) finitely determined.
5. Suppose that $\mathfrak{C}$ is finite. Then $\odot$ is finitely determined if and only if it is idempotent, associative and commutative.

For a finitely-determined binary operation $\odot$ on $\mathfrak{C}$, we define the partial order $\leq_\odot$ on $\mathfrak{C}$ by $s \leq_\odot t$ if and only if $s \odot t = t$. (So that $s \leq_+ t$ if and only if $s + t = t$, and $s \leq_\times t$ if and only if $s \times t = t$, for finitely-determined operations $+$ and $\times$.) Note (i) that the orderings $\leq_+$ and $\leq_\times$ are dual to each other if and only if the absorption law holds for $+$ and $\times$, in which case $(\mathfrak{C}, \leq_+, \leq_\times)$ is a lattice, and (ii) that finitely-determined operations $+$ and $\times$ need not satisfy the distributive laws. Notice also that because $+$ and $\times$ are finitely determined, $\sum_{c \in \mathfrak{C}} c \in \mathfrak{C}$ is the top element of $\mathfrak{C}$ relative to $\leq_+$, and $\prod_{c \in \mathfrak{C}} c \in \mathfrak{C}$ is the top element of $\mathfrak{C}$ relative to $\leq_\times$. Note, however, that it does not follow that we have bottom elements for these orderings. We further suppose that two elements $\bar{c}$ and $\underline{c}$ are distinguished in $\mathfrak{C}$, and we will make use of these elements later on. (In some, but not all, situations when $\mathfrak{C}$ is a logic, $\bar{c}$ is taken to be *true*, and $\underline{c}$ is taken to be *false*.)

Example 2. In $\mathcal{FOUR}$, we have $t \leq_\wedge u \leq_\wedge f$, and $t \leq_\wedge b \leq_\wedge f$. Also, $f \leq_\vee u \leq_\vee t$, and $f \leq_\vee b \leq_\vee t$. In this case, we take $\bar{c} = t$, and $\underline{c} = f$.

Although we will not need to suppose here that $\mathfrak{C}$ is a complete partial order in the orders just defined, that assumption is often made and then a least element with respect to $\leq_+$ must be present in $\mathfrak{C}$ (we add this element to $\mathfrak{C}$ if necessary). In particular, to calculate the least fixed point of $\mathfrak{T}_{P,\mathfrak{C}}$, it is common practice to iterate on the least element. However, if we require the least fixed point to coincide with any useful semantics, it will usually be necessary to choose the default value $\underline{c} \in \mathfrak{C}$ to be the least element in the ordering $\leq_+$.

Furthermore, the allowable and excluded sets for $s \in \mathfrak{C}$ can easily be characterized in terms of these partial orders: $s \in \mathcal{A}_t^\odot$ if and only if $s \leq_\odot t$, see [44, Proposition 3.10]. Because of this fact, the following result plays an important role in the construction of the network to compute $\mathfrak{T}_{P,\mathfrak{C}}$, as we see later.

Proposition 1. *Suppose that $\odot$ is a finitely-determined binary operation on $\mathfrak{C}$ and that M is a countable set. Then a product $\bigodot_{i \in M} t_i$ evaluates to the element $s \in \mathfrak{C}$, where s is the least element in the ordering $\leq_\odot$ such that $t_i \in \mathcal{A}_s^\odot$ for all $i \in M$.*

Proof. Assume that $\bigodot_{i \in M} t_i = s$. Then each t_i is an allowable value for s and so we certainly have $t_i \in \mathcal{A}_s^\odot$ for all $i \in M$.

Now assume that $\{t_i \mid i \in M\} \subseteq \mathcal{A}_t^\odot$; we want to show that $s = \bigodot_{i \in M} t_i \in \mathcal{A}_t^\odot$, for then it will follow that $s \leq_\odot t$, as required. By Theorem 1, we may suppose that M is finite, $M = \{1, \ldots, n\}$, say. Since $t_i \leq_\odot t$ for $i = 1, \ldots, n$, we have $t_i \odot t = t$ for $i = 1, \ldots, n$. Therefore, by Statement 1 of Theorem 1, we obtain $t = \bigodot_{i \in M}(t_i \odot t) = (\bigodot_{i \in M} t_i) \odot t = s \odot t$. It follows that $s \leq_\odot t$, and the proof is complete.

In what follows throughout this section, $\mathfrak{C}$ will denote a set endowed with binary operations $+$ and $\times$; furthermore, $+$ at least will be assumed to be finitely determined and $\times$ will be assumed to be associative for simplicity. In § 12.3.3 and § 12.3.4, we will also need to assume that $\times$ is finitely determined. In fact, it transpires that our main definition (but not all our results) can be made simply in the context of the set $\mathfrak{C}$ with sufficient completeness properties, namely, that arbitrary countable sums can be defined.

Of particular interest to us are the following three cases.
(1) $\mathfrak{C}$ is a set of truth values, $+$ is disjunction $\vee$ and $\times$ is conjunction $\wedge$.
(2) $\mathfrak{C}$ is a c-semiring (constraint-based semiring) as considered in [25]. Thus, $\mathfrak{C}$ is a semiring, where the top element in the order $\leq_\times$ is the identity element $\mathbf{0}$ for $+$, and the top element in the order $\leq_+$ is the identity element $\mathbf{1}$ for $\times$. In addition, $+$ is idempotent, $\times$ is commutative, and $\mathbf{1}$ annihilates $\mathfrak{C}$ relative to $+$, that is, $\mathbf{1} + c = c + \mathbf{1} = \mathbf{1}$ for all elements $c \in \mathfrak{C}$.
(3) $\mathfrak{C}$ is the set L_m of truth values considered in [32], $+$ is max and $\times$ is min. These will be discussed briefly in Example 3.

12.3.2 The operator $\mathfrak{T}_{P,\mathfrak{C}}$

We next turn to giving the definition of the operator $\mathfrak{T}_{P,\mathfrak{C}}$.

Let $\mathcal{L}$ be a first-order language, see [46] for notation and undefined terms relating to conventional logic programming, and suppose that $\mathfrak{C}$ is given. By a $\mathfrak{C}$-*normal logic program* P or a *normal logic program* P *defined over* $\mathfrak{C}$, we mean a finite set of clauses or rules of the type $A \leftarrow L_1, \ldots, L_n$ (n may be 0, by the usual abuse of notation), where A is an atom in $\mathcal{L}$ and the L_j, for $1 \leq j \leq n$, are either literals in $\mathcal{L}$ or are elements of $\mathfrak{C}$. By a $\mathfrak{C}$-*interpretation* or just *interpretation* I for P, we mean a mapping $I : B_P \to \mathfrak{C}$, where B_P denotes the Herbrand base for P. We immediately extend I to $\neg \cdot B_P$ by $I(\neg A) = \neg I(A)$, for all $A \in B_P$, and to $B_P \cup \neg \cdot B_P \cup \mathfrak{C}$ by setting $I(c) = c$ for all $c \in \mathfrak{C}$. Finally, we let $I_{P,\mathfrak{C}}$ or simply I_P denote the set of all $\mathfrak{C}$-interpretations for P ordered by $\sqsubseteq_+$, that is, by the pointwise ordering relative to $\leq_+$. Notice that the value $I(L_1,\ldots,L_n)$ of I on any clause body is uniquely determined[4] by $I(L_1,\ldots,L_n) = I(L_1) \times \ldots \times I(L_n)$.

To define the *semantic operator* $\mathfrak{T}_{P,\mathfrak{C}}$, we essentially follow [41], allowing for our extra generality, in first defining the sets P^* and P^{**} associated with P. To define P^*, we first put in P^* all ground instances of clauses of P whose bodies are non-empty. Second, if a clause $A \leftarrow$ with empty body occurs in P, add $A \leftarrow \bar{c}$ to P^*. Finally, if the ground atom A is not yet the head of any member of P^*, add $A \leftarrow \underline{c}$ to P^*. To define P^{**}, we note that there may be many, even denumerably many, elements $A \leftarrow C_1, A \leftarrow C_2, \ldots$ of P^* having the same head A. We replace them with $A \leftarrow C_1 + C_2 + \ldots$, where $C_1 + C_2 + \ldots$ is to be thought of as a formal sum. Doing this for each A gives us the set P^{**}. Now, each ground atom A is the *head* of exactly one element $A \leftarrow C_1 + C_2 + \ldots$ of P^{**}, and it is common practice to work with P^{**} in place of P. Indeed, $A \leftarrow C_1 + C_2 + \ldots$ may be written $A \leftarrow \sum_i C_i$ and referred to as a (or as the) *pseudo-clause* with *head* A and *body* $\sum_i C_i$.

Definition 2. *(See [41]) Let P be a $\mathfrak{C}$-normal logic program. We define $\mathfrak{T}_{P,\mathfrak{C}}$: $I_{P,\mathfrak{C}} \to I_{P,\mathfrak{C}}$ as follows. For any $I \in I_{P,\mathfrak{C}}$ and $A \in B_P$, we set*

$$\mathfrak{T}_{P,\mathfrak{C}}(I)(A) = I(\sum_i C_i) = \sum_i I(C_i),$$

where $A \leftarrow \sum_i C_i$ is the unique pseudo-clause in P^{**} whose head is A. Note that when $\mathfrak{C}$ is understood, we may denote $\mathfrak{T}_{P,\mathfrak{C}}$ simply by $\mathfrak{T}_P$.

We note that $I(\sum_i C_i) = \sum_i I(C_i)$ is well-defined in $\mathfrak{C}$ by Theorem 1. Indeed, $\sum_i I(C_i)$ may be a denumerable sum in $\mathfrak{C}$, and it is this observation which motivates the introduction of the notion of finite determinedness.

Example 3. Some special cases of $\mathfrak{T}_{P,\mathfrak{C}}$. As mentioned in the introduction, the operator $\mathfrak{T}_{P,\mathfrak{C}}$ includes a number of important cases simply by choosing $\mathfrak{C}$ suitably, and we briefly consider this point next.
(1) The standard semantics of logic programming. Choosing $\mathfrak{C}$ to be classical two-valued logic, Kleene's strong three-valued logic, and $\mathcal{FOUR}$, one recovers respectively the usual single-step operator T_P, Fitting's three-valued operator Φ_P, and Fitting's four-valued operator Ψ_P, see [41]. Hence, one recovers the associated semantics as the least fixed points of $\mathfrak{T}_{P,\mathfrak{C}}$.

Furthermore, in [47], Wendt studied the fixpoint completion, fix(P), of a normal logic program P introduced by Dung and Kanchanasut in [48]. The fixpoint completion is a normal logic program in which all body literals are negated, and is obtained by a complete unfolding of the recursion through positive literals in the clauses of a program. In fact, Wendt obtained interesting connections between various semantic operators by means of fix(P). Specifically, he showed that for any normal logic program P, we have (i) $GL_P(I) = T_{\text{fix}(P)}(I)$ for any two-valued interpretation I, and (ii)

---

[4] This is meaningful even if $\times$ is not associative provided bracketing is introduced and respected.

$\overline{\Psi}_P(I) = \Phi_{\text{fix}(P)}(I)$ for any three-valued interpretation I, where GL_P is the well-known operator of Gelfond and Lifschitz used in defining the stable-model semantics, and $\overline{\Psi}_P$ is the operator used in [49] to characterize the well-founded semantics of P. These connections have the immediate corollary that GL_P and $\overline{\Psi}_P$ can be seen as special cases of $\mathfrak{T}_{P,\mathfrak{C}}$, and hence that the well-founded and stable-model semantics can be viewed as special cases of the fixed points of $\mathfrak{T}_{P,\mathfrak{C}}$.

(2) Constraint logic programs. A *semiring-based constraint logic program* P, see [25], consists of a finite set of clauses each of which is of the form

$$A \leftarrow L_1, L_2, \ldots, L_k, \tag{12.1}$$

where A is an atom and the L_i are literals or is of the form

$$A \leftarrow a, \tag{12.2}$$

where A is an atom and a is any semiring value. Those clauses with a semiring value in the body constitute the constraints and are also known as "facts". The distinguished values $\bar{c}$ and $\underline{c}$ in a c-semiring are $\mathbf{1}$ and $\mathbf{0}$ respectively. Thus, when constructing P^* for a semiring-based constraint logic program P, unit clauses $A \leftarrow$ are replaced by $A \leftarrow \mathbf{1}$ and for any atom A not the head of a clause, we add the clause $A \leftarrow \mathbf{0}$ to P^*.

In this context, an interpretation I is a mapping $I : B_P \to S$, where $S = \mathfrak{C}$ is the underlying c-semiring, and we denote by $I_{P,S}$ the set of all such interpretations. Finally, associated with each semiring-based constraint logic program P is a consequence operator $T_{P,S} : I_{P,S} \to I_{P,S}$ defined in [25] essentially as follows.

Definition 3. *Given an interpretation I and a ground atom A, we define $T_{P,S}(I)$ by*

$$T_{P,S}(I)(A) = \sum_i I(C_i),$$

where $A \leftarrow \sum_i C_i$ is the unique pseudo-clause whose head is A, and $I(C_i)$ is defined as follows. We set $I(C_i) = a$ when $A \leftarrow \sum_i C_i$ is the fact $A \leftarrow a$, and otherwise when $A \leftarrow \sum_i C_i$ is not a fact of the form $A \leftarrow a$, we set $I(C_i) = \prod_{j=1}^{n_i} I(L_j^i)$, where $C_i = L_1^i, \ldots, L_{n_i}^i$, say.

It is easy to see that if P is a semiring-based constraint logic program, then the semantic operator $\mathfrak{T}_{P,\mathfrak{C}}$ coincides with $T_{P,S}$ when we take $\mathfrak{C}$ to be the c-semiring S underlying P, as already observed.

Another example of interest in this context concerns uncertainty in rule-based systems, such as those considered in [32], but we omit the details.

12.3.3 Towards the Computation of $\mathfrak{T}_{P,\mathfrak{C}}$ by ANN

As noted in the introduction, an algorithm is presented in [1] for constructing an ANN which computes T_P exactly for any given propositional logic program

P. Indeed, it is further shown in [1] that 2-layer binary threshold feedforward networks cannot do this. In the algorithms discussed in [1], single units hold the truth values of atoms and clauses in P. A unit outputs 1 if the corresponding atom/clause is true with respect to the interpretation presented to the network, and outputs 0 if the corresponding atom/clause is false. However, when dealing with many-valued logics or with sets $\mathfrak{C}$ with more than two elements, single units can compute neither products nor sums of elements.

In this subsection and the next, we discuss the computation by ANN of the semantic operator $\mathfrak{T}_{P,\mathfrak{C}}$ determined by a propositional normal logic program P defined over some set $\mathfrak{C}$. This is done by simulating the clauses and the connections between the body literals in the clauses of P. This requires that we represent elements of $\mathfrak{C}$ by units, or combinations of them, in ANN and then simulate the operations of $+$ and $\times$ in $\mathfrak{C}$ by computations in ANN. To keep matters simple, and since we are not employing learning algorithms here, we shall focus our attention on binary threshold units only.

In order to define $\mathfrak{T}_{P,\mathfrak{C}}$, it is not even necessary for multiplication to be commutative because only finite products occur in the bodies of elements of P and P^{**}. Nevertheless, it will be necessary here in computing $\mathfrak{T}_{P,\mathfrak{C}}$ by ANN to impose the condition that multiplication is in fact finitely determined, since we need to make use of Proposition 1 relative to both addition and multiplication.

We shall focus on finite sets $\mathfrak{C}$ with n elements listed in some fixed order, $\mathfrak{C} = \{c_1, c_2, \ldots, c_n\}$ or $\mathfrak{C} = \{t_1, t_2, \ldots, t_n\}$, say. In order to simulate the operations in $\mathfrak{C}$ by means of ANN, we need to represent the elements of $\mathfrak{C}$ in a form amenable to their manipulation by ANN. To do this, we represent elements of $\mathfrak{C}$ by vectors of n units[5], where the first unit represents c_1, the second unit represents c_2, and so on. Hence, a vector of units with the first unit activated, or containing 1, represents c_1, a vector with the second unit activated, or containing 1, represents c_2, etc. Indeed, it will sometimes be convenient to denote such vectors by binary strings of length n, and to refer to the unit in the i-th position of a string as the i-th unit or the c_i-unit or the unit c_i; as is common, we represent these vectors geometrically by strings of not-necessarily adjacent rectangles. Note that we do not allow more than one unit to be activated at any given time in any of the vectors representing elements of $\mathfrak{C}$, and hence all but one of the units in such vectors contain 0. Furthermore, when the input is consistent with this, we shall see from the constructions we make that the output of any network we employ is consistent with it also.

Example 4. Suppose that $\mathfrak{C} = \mathcal{FOUR} = \{t, u, b, f\}$, listed as shown. Then t is represented by 1000, u by 0100, b by 0010, and f by 0001.

In general, the operations in $\mathfrak{C}$ are not linearly separable, and therefore we need two layers to compute addition and two to compute multiplication. As usual, we take the standard threshold for binary threshold units to be

---

[5] It is convenient sometimes to view them as column vectors.

0.5. This ensures that the Heaviside function outputs 1 if the input is strictly greater than 0, rather than greater than or equal to 0.

Definition 4. *A multiplication (×) unit or a conjunction (∧) unit $\mathcal{MU}$ for a given set $\mathfrak{C}$ is a 2-layer ANN in which each layer is a vector of n binary threshold units $c_1, c_2, \ldots, c_n$ corresponding to the n elements of $\mathfrak{C}$. The units in the input layer have thresholds $l - 0.5$, where l is the number of elements being multiplied or conjoined, and all output units have threshold 0.5. We connect input unit c_i to the output unit c_i with weight 1 and to any unit c_j in the output layer, where $c_i <_\times c_j$, with weight -1.*

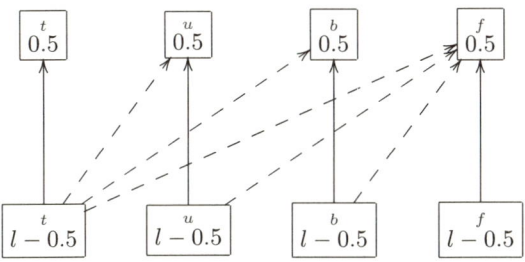

Fig. 12.2. A conjunction unit for $\mathcal{FOUR}$. The full arrows represent connections with weight 1, and the broken arrows represent connections with weight -1.

An input layer representing a product of l elements of $\mathfrak{C}$ is connected to a multiplication unit $\mathcal{MU}$ in the following way. For each element c of the product, where c is represented by the n units $c_1, c_2, \ldots, c_n$, the unit c_j is connected, with weight 1, to the c_j-unit in the input layer of $\mathcal{MU}$ and is also connected, with weight 1, to any unit c_k in the input layer of $\mathcal{MU}$ for which $c_j <_\times c_k$. For a negated element $d = \neg c$ in the product, we connect, with weight 1, c_j to the unit representing $\neg c_j$ in the input layer of $\mathcal{MU}$ and also, with weight 1, to any unit c_k in the input layer of $\mathcal{MU}$ for which $\neg c_j <_\times c_k$.

Proposition 2. *A multiplication or conjunction unit $\mathcal{MU}$ computes the value of a product or conjunction of l elements of $\mathfrak{C}$ when it is connected to an input layer as just described.*

Proof. The proof ultimately depends on Proposition 1 and indeed $\mathcal{MU}$, in effect, counts the number of elements of $\mathfrak{C}$ which are in the allowable set $\mathcal{A}_s^\times$ for each $s \in \mathfrak{C}$. Suppose given a product $s_1 \times s_2 \times \cdots \times s_l$, of l elements of $\mathfrak{C}$, whose value is equal to $c_i \in \mathfrak{C}$, so that each of the elements in this product is an allowable value for c_i. By the manner of connecting the units representing the s_k to the input layer of $\mathcal{MU}$, the c_j-unit in the input layer of $\mathcal{MU}$ will be activated if and only if $c_i \leq_\times c_j$. By construction of $\mathcal{MU}$, any c_j-unit in the output layer of $\mathcal{MU}$ for which the c_j-unit in the input layer is activated will receive an input of 1 ($= 1 \times 1$). However, any c_j-unit in the output layer of $\mathcal{MU}$ for which $c_i <_\times c_j$ will also receive negative input whereas the unit c_i itself will receive no such input. Therefore, the unit c_i is the only unit activated in the output layer of $\mathcal{MU}$, as required.

Example 5. Consider again $\mathfrak{C} = \mathcal{FOUR} = \{t, u, b, f\}$, and input the two elements u and b to a multiplication unit $\mathcal{MU}$, where $l = 2$. It is readily checked that the potentials of the units t, u and b in the input layer of $\mathcal{MU}$ are respectively -1.5, -0.5 and -0.5, that their outputs are all equal to 0, and that the outputs of the units t, u and b in the output layer of $\mathcal{MU}$ are also all equal to 0. On the other hand, the f-unit in the input layer of $\mathcal{MU}$ has potential $1 \times 0 + 1 \times 0 + 1 \times 1 + 1 \times 0 + 1 \times 0 + 1 \times 1 + 1 \times 0 + 1 \times 0 - 1.5 = 0.5$, and therefore the output of this unit is $H(0.5) = 1$. Furthermore, the input to the f-unit in the output layer of $\mathcal{MU}$ is $-1 \times 0 - 1 \times 0 - 1 \times 0 + 1 \times 1 = 1$. Hence, the output of this unit is $H(1 - 0.5) = 1$, and so $\mathcal{MU}$ outputs 0001 or f and this indeed is the value of $u \wedge b$, as required.

Note 1. (1) If we take $l = 1$, that is, if we consider a product of just one element c of $\mathfrak{C}$, then a multiplication unit outputs c whenever c is input to that unit. The same comment applies to addition units also, and these observations will be used in the network we construct in Theorem 2 to handle clauses whose body contains just one element.

(2) Suppose that $\mathcal{MU}$ is a multiplication unit with $l \geq 2$ and that $c \in \mathfrak{C}$. By permanently connected a vector representing c to $\mathcal{MU}$, we obtain a multiplication unit $\mathcal{MU}(c)$ which multiplies any input to it by c. We shall refer to $\mathcal{MU}$ as a *multiplication unit with one factor fixed at c*. One can similarly construct multiplication units with any number of factors fixed at elements of $\mathfrak{C}$. Such units will be needed in Step 2.2 of the translation algorithm used in Theorem 2.

The ideas behind multiplication units work, with minor changes, for addition or disjunction, and we consider this point next.

Definition 5. *An* addition (+) unit *or a* disjunction ($\vee$) unit $\mathcal{AU}$ *for a given set $\mathfrak{C}$ is a 2-layer ANN in which each layer is a vector of n binary threshold units $c_1, c_2, \ldots, c_n$ corresponding to the n elements of $\mathfrak{C}$. The units in the input layer have threshold $k - 0.5$, where k is the number of elements to be added or disjoined, and all output units have threshold 0.5. We connect input unit c_i to the output unit c_i with weight 1 and to any unit c_j in the output layer, where $c_i <_+ c_j$, with weight -1.*

The manner of connecting an input layer to an addition unit, and the calculation of a sum or disjunction of elements proceeds exactly as for multiplication, and again in effect makes use of Proposition 1.

Proposition 3. *An addition or disjunction unit $\mathcal{AU}$ computes the value of a sum or disjunction of k elements of $\mathfrak{C}$ when it is connected to an input layer as just described.*

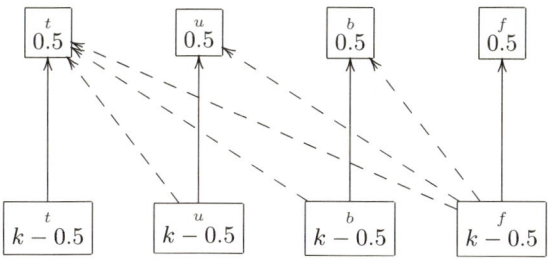

Fig. 12.3. A disjunction unit for $\mathcal{FOUR}$. The full arrows represent connections with weight 1, and the broken arrows represent connections with weight -1.

Proof. The proof follows that of Proposition 2 with the necessary minor changes.

Therefore, both operations in $\mathfrak{C}$ can be simulated by ANN. However, there are occasions when we desire the addition and multiplication units to be independent of the number of elements being added or multiplied: for instance, if we wish to add more clauses after the network has been constructed. To handle this situation, "constant threshold" units for an arbitrary finitely-determined $\mathfrak{C}$ were introduced and studied in [50], but they will not be considered here.

When clauses of the form $A \leftarrow c$, where $c \in \mathfrak{C}$, are present in the programs we consider, we need to compute c in the middle layer of the networks of Theorem 2, and this leads to the introduction of $\mathfrak{C}$-element units.

Definition 6. *A $\mathfrak{C}$-element unit for $c \in \mathfrak{C}$ is an ANN whose layers and connections are the same as those in a multiplication unit except for the thresholds. The thresholds for all units in the input layer are taken to be equal to 1 apart from the unit representing the element c, which has threshold taken to be -0.5. All thresholds in the output layer are taken to be 0.5.*

Note 2. Given any input $a_1, a_2, \ldots, a_n$ to a $\mathfrak{C}$-element unit for c, where all the a_i are either 0 or 1 and only one of them is equal to 1, the unit outputs c. These units will also be needed in Step 2.2 of the translation algorithm of Theorem 2.

Example 6. When $A \leftarrow u$ is present in a program defined over $\mathcal{FOUR}$, the following unit is placed in the middle layer of the networks required to compute $\mathfrak{T}_{P,\mathfrak{C}}$, see Figure 12.4. Only u will be activated at any time and the unit will always output 0100, or u, as desired.

12.3.4 Networks for the Computation of $\mathfrak{T}_{P,\mathfrak{C}}$

Suppose that P is a propositional logic program defined over $\mathfrak{C}$. Then B_P is finite with m elements $\{A_1, A_2, \ldots, A_m\}$, say. However, we need to fix

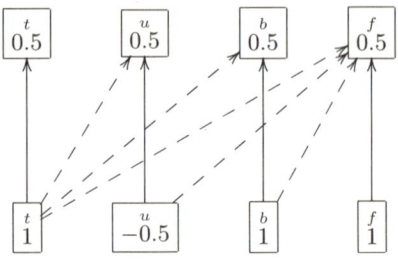

Fig. 12.4. u-unit for $\mathcal{FOUR}$.

an order on the elements of B_P, and henceforth will write B_P as the list $(A_1, A_2, \ldots, A_m)$. For finite sets $\mathfrak{C}$ of cardinality n, we can view interpretations $I \in I_{P,\mathfrak{C}}$ as vectors of length m in which each entry is itself a vector or binary string of length n holding the value $I(A)$ of a particular atom A. Thus, the j-th entry corresponds to A_j in the sense that the string in question will have 1 in its i-th place if $I(A_j) = c_i$, and 0 otherwise. It follows that the input layer of the network we are about to construct contains $m \times n$ units: the first n units correspond to the values of the first atom A_1, the second n correspond to the values of the second atom A_2, and so on.

Example 7. Take $\mathfrak{C}$ to be $\mathcal{FOUR} = \{t, u, b, f\}$, again ordered as listed, and suppose that $B_P = (A_1, A_2, A_3, A_4, A_5)$ has five elements. Then the following vector $(0100, 1000, 0001, 0100, 0010)$ represents a four-valued interpretation I such that $I(A_1) = u$, $I(A_2) = t$, $I(A_3) = f$, $I(A_4) = u$, and $I(A_5) = b$.

We are now in a position to present the following theorem.

Theorem 2. *Suppose that both operations in $\mathfrak{C}$ are finitely determined and that P is a propositional logic program defined over $\mathfrak{C}$. Then we can construct a 3-layer feedforward ANN $\mathcal{F}$ which contains multiplication units in its middle layer and addition units in its output layer such that $\mathcal{F}$ computes $\mathfrak{T}_{P,\mathfrak{C}}$.*

Proof. Because of the preponderance of symbols "c" in the proof, it will ease notation slightly to take $\mathfrak{C}$ as the set $\{t_1, t_2, \ldots, t_n\}$ instead of $\{c_1, c_2, \ldots, c_n\}$.

With the notation already established, let K be the number of clauses in P^*.

1. Set the first and third layers to be vectors of length m. Each entry in the vector in the first layer is itself a vector of n binary threshold units representing an element of B_P, and each entry in the third layer is an addition unit corresponding to an element of B_P. Each unit in each vector corresponds to the (truth) value of an atom $A_j \in B_P$, as described above.
2. For the second layer:
   2.1 Set the second layer to be a vector of length K, with each entry determining the value of the corresponding clause body C by means of a multiplication unit with input layer $C^{t_1}, C^{t_2}, \ldots, C^{t_n}$.

2.2 For each clause in P^*, connect each atom B in its body, C, with weight 1, as follows. Connect the unit B^{t_j} in the first layer of the network to the unit C^{t_j} in the input layer of the multiplication unit corresponding to C, and to any unit C^s in the input layer of that same multiplication unit for which $t_j <_\times s$, where $s \in \mathfrak{C}$. For a negated literal $L = \neg B$ in the body C of the given clause, connect, with weight 1, the unit B^{t_j} in the first layer of the network to the unit $C^{\neg t_j}$ in the input layer of the multiplication unit corresponding to C, and, also with weight 1, to any unit C^s in the input layer of that same multiplication unit for which $\neg t_j <_\times s$, where $s \in \mathfrak{C}$. Note that the connections specified here are precisely those given earlier for connecting a layer to a multiplication unit. Furthermore:

(i) If the body C of the given clause contains elements c of $\mathfrak{C}$ as well as atoms B or literals L, then the corresponding multiplication unit is a multiplication unit with one factor fixed at c for each such c, see Note 1.

(ii) If the given clause is of the form $A \leftarrow c$ for some $c \in \mathfrak{C}$, and hence contains no atoms nor literals, the the corresponding "multiplication unit" is a $\mathfrak{C}$-element unit for c, see Note 2. This unit is connected as a multiplication unit to any vector in the first layer of the network representing an element of B_P.

3. For the clause $A \leftarrow C$ in P^*, connect the unit C^{t_j} in the output layer of the second layer of the network, with weight 1, to the unit A^{t_j} in the input layer of the addition unit corresponding to A in the third layer of the network, and, also with weight 1, to any unit A^s in the input layer of that same addition unit for which $t_j <_+ s$, where $s \in \mathfrak{C}$. Note that the connections specified here are precisely those given earlier for connecting a layer to an addition unit.

Suppose that the interpretation I is presented to the input layer, and that $\mathfrak{T}_{P,\mathfrak{C}}(I)(A) = s$. Then $I(\sum_{i=1}^{h} C_i) = s$, where $A \leftarrow \sum_{i=1}^{h} C_i$ is the unique pseudo-clause in P^{**} with head A. Thus, $I(C_i) \in \mathcal{A}_s^+$ for all $1 \leq i \leq h$, as in Definition 1. In particular, there exists j with $R_s^{+,j} \subseteq \{I(C_i) \mid 1 \leq i \leq h\}$. If $R_s^{+,j} = \{r_1, r_2, \ldots, r_g\}$, then there are clauses $A \leftarrow C_{i_1}, A \leftarrow C_{i_2}, \ldots, A \leftarrow C_{i_g}$ in P^* such that $I(C_{i_u}) = r_u$, for $1 \leq u \leq g$. For each of these $C_{i_u} = L_1, L_2, \ldots, L_l$, say, we have $I(L_i) \in \mathcal{A}_{r_u}^\times$. Therefore, there is a j' such that $R_{r_u}^{\times, j'} = \{q_1, q_2, \ldots, q_v\} \subseteq \mathcal{A}_{r_u}^\times$ and $I(L_{\lambda_1}) = q_1, I(L_{\lambda_2}) = q_2, \ldots, I(L_{\lambda_v}) = q_v$, and all other $I(L_\mu) \in \mathcal{A}_{r_u}^\times$. Thus, each $C_{i_u}^{r_u}$ in the input layer of the corresponding C_{i_u} multiplication unit in the second layer is activated since it has a threshold of $l - 0.5$ and receives input from the l literals. The same applies to any of the $C_{i_u}^{r_w}$, but they will also get input -1 from $C_{i_u}^{r_u}$ in the output layer of the multiplication unit. Equally, $C_{i_u}^{r_u}$ in the output layer will receive no negative input since no $C_{i_u}^{r}$ with $r <_\times r_u$ can be activated in the input layer, and thus only $C_{i_u}^{r_u}$ will be activated in the output layer of the multiplication unit. Next, A^s in the input layer of the A addition unit in the third layer receives positive input from the h clauses of which A is the head. All A^t with $s <_+ t$ will also be activated in the input layer of the A addition unit, but not in the output layer of this unit, since A^t in the output layer will receive an input of -1 from A^s in the input layer. However, A^s will be activated, as required, because it receives an input from A^s in the input layer and no negative input because none of the A^t with $t <_+ s$ can be activated.

Conversely, if A^s is activated in the output layer of its addition unit in the third layer, then no A^t with $t <_+ s$ can be activated, by construction of an addition unit. Hence, the smallest t relative to $\leq_+$ such that all the activated $C_i^{t_j}$ have $t_j \in \mathcal{A}_t^+$ must be equal to s. Thus, the sum $\sum_{i=1}^h I(C_i) = s$, and accordingly there must exist a j with $R_s^{+,j} = \{r_1, r_2, \ldots, r_v\} \subseteq \{I(C_i) \mid 1 \leq i \leq h\}$. For any clause with C^r activated, where $r \in \mathcal{A}_s^+$, r is the smallest value relative to $\leq_+$ which is activated in the C multiplication unit in the second layer. There cannot be any C^t activated with $t <_\times r$ otherwise the body $L_1, L_2, \ldots, L_l$ would evaluate to t under the present interpretation I, and not to r as asserted. Thus, for this clause, we have that $\{I(L_i) \mid 1 \leq i \leq l\} \subseteq \mathcal{A}_r^\times$ and r is the smallest value relative to $\leq_\times$ for which this is the case. Therefore, $I(L_1) \times \ldots \times I(L_l) = r$, as desired.

Thus, $\mathfrak{T}_{P,\mathfrak{C}}(I)(A) = s$ if and only if A^s is activated in the third layer, and the proof is complete.

Notice that, as a corollary of this result, for each propositional logic program P, one can construct networks as given in Theorem 2, which compute, respectively, not only the classical consequence operator T_P, but also the three-valued operator Φ_P of Fitting, and the four-valued operator Ψ_P of Fitting. Indeed, as shown in [16], one can construct conventional 3-layer feedforward networks to compute Φ_P and Ψ_P containing only binary threshold units and not using multiplication and addition units. [6]

Example 8. Take $\mathfrak{C}$ as $\mathcal{FOUR}$ again and consider the program P whose clauses are $A \leftarrow A, b$; $A \leftarrow D, \neg E$; $A \leftarrow u$; $D \leftarrow$.

Here, $B_P = \{A, D, E\}$, and P^* contains the five clauses: $A \leftarrow A, b$; $A \leftarrow D, \neg E$; $A \leftarrow u$; $D \leftarrow t$; $E \leftarrow f$, which we list as 1 to 5 as shown. Also, P^{**} contains the three clauses: $A \leftarrow (A, b) + (D, \neg E) + u$; $D \leftarrow t$; $E \leftarrow f$. Thus, for any interpretation $I : B_P \to \mathcal{FOUR}$, we have $\mathfrak{T}_{P,\mathfrak{C}}(I)(A) = (I(A) \wedge b) \vee (I(D) \wedge \neg I(E)) \vee u$, $\mathfrak{T}_{P,\mathfrak{C}}(I)(D) = I(t) = t$, and $\mathfrak{T}_{P,\mathfrak{C}}(I)(E) = I(f) = f$.

For the network $\mathcal{F}$ produced by applying Theorem 2, we have $m = 3$, $n = 4$ and $K = 5$. Thus, the first layer contains three vectors (each of length 4) representing A, D and E, and the third layer contains three addition units corresponding to A, D, E with $k = 3, 1$ and 1 respectively. The middle layer contains five multiplication units corresponding to the five elements of P^*. The first of these has $l = 2$, has a factor fixed at b and is connected to the unit A in the first layer. The second multiplication unit has $l = 2$ and is

---

[6] See the thesis of Yvonne Kalinke: "Ein massiv paralleles Berechnungsmodell für normale logische Programme", Department of Computer Science, Dresden University of Technology, 1994, where these results are stated. We thank S. Hölldobler for drawing this reference to our attention.

connected to units D and E in the first layer. The other three units in the second layer are $\mathfrak{C}$-element units for u, t and f respectively, and each may be connected to E, say, in the first layer. On giving an interpretation I to $\mathcal{F}$ as input, it is readily checked that $\mathcal{F}$ computes $\mathfrak{T}_{P,\mathfrak{C}}(I)$.

12.4 Annoted Logic Programs and the Operator $\mathcal{T}_P$

12.4.1 Lattice and Bilattice-Based Annotated Logic Programs

We now turn to the second class of extended logic programs we wish to consider: annotated (bi)lattice-based logic programs. For a detailed exposition of lattice- and bilattice-based annotated logic programs, see [35, 15]. The so-called generalized annotated logic programs (GAPs) were introduced in [35] and were shown to generalize annotation-free, and implication-based logic programs as well as most of the various annotated logic programs introduced to date. In [15], GAPs were extended to bilattice-based annotated logic programs (BAPs), which allowed us to introduce a continuous semantic operator in place of the non-continuous semantic operator of [35]. We will therefore take BAPs as the most general approach to annotated logic programming, and use the semantic operator introduced for them, but we will concentrate, in the main, only on a one-lattice fragment of BAPs in order to bring uniformity into the discussion of the current section and the previous section which analysed sets of truth values with only one ordering defined on them. This permits us to claim that the results described in the previous section are extendable to BAPs with only variables allowed in annotations.

The notion of a bilattice was introduced in the 1980s as a generalization of the famous lattice $\mathcal{FOUR}$ of Belnap (see Example 1) as a suitable structure for interpreting different languages and programs when working with uncertainty and incomplete or inconsistent databases, see [28, 29, 21, 22, 35] for further details and further motivation.

Definition 7. *[21] A bilattice $\mathbf{B}$ is a sextuple $(\mathbf{B}, \vee, \wedge, \oplus, \otimes, \neg)$ such that $(\mathbf{B}, \vee, \wedge)$ and $(\mathbf{B}, \oplus, \otimes)$ are both complete lattices, and $\neg : \mathbf{B} \to \mathbf{B}$ is a mapping satisfying the following three properties: $(\neg)^2 = Id_\mathbf{B}$, $\neg$ is a dual lattice homomorphism from $(\mathbf{B}, \vee, \wedge)$ to $(\mathbf{B}, \wedge, \vee)$, and $\neg$ is a lattice homomorphism from $(\mathbf{B}, \oplus, \otimes)$ to itself.*

The lattice $(\mathbf{B}, \vee, \wedge)$ is traditionally thought of as generalizing the Boolean lattice {false, true}, and is used for describing measures of truth and falsity. The lattice $(\mathbf{B}, \oplus, \otimes)$ is thought of as measuring the amount of information (or knowledge) between none and both (as in $\mathcal{FOUR}$).

Let $(L_1, \leq_1)$ and $(L_2, \leq_2)$ denote two lattices, let x_1, x_2 denote arbitrary elements of the lattice L_1, and let y_1, y_2 denote arbitrary elements of the lattice L_2. Let $\cap_1, \cup_1$ denote the meet respectively join defined in the lattice L_1, and let $\cap_2, \cup_2$ denote the meet respectively join defined in the lattice L_2.

Now form the set of points $L_1 \times L_2$. We define the usual orderings $\leq_t$ (the truth ordering) and $\leq_k$ (the knowledge ordering) on $L_1 \times L_2$ as follows.
(1) $\langle x_1, y_1 \rangle \leq_t \langle x_2, y_2 \rangle$ if and only if $x_1 \leq_1 x_2$ and $y_2 \leq_2 y_1$.
(2) $\langle x_1, y_1 \rangle \leq_k \langle x_2, y_2 \rangle$ if and only if $x_1 \leq_1 x_2$ and $y_1 \leq_2 y_2$.

We use here the fact that each distributive bilattice can be regarded as a product of two lattices, see [29]. For convenience of presentation, we will treat each bilattice we work with as isomorphic to some subset of $\mathbf{B} = L_1 \times L_2 = ([0,1], \leq) \times ([0,1], \leq)$, where $[0,1]$ is the unit interval of real numbers with the linear ordering defined on it. Elements of such a bilattice are pairs. In particular, $(1,0)$ and $(0,1)$ are the analogues of true and false and are maximal respectively minimal in the truth ordering, whilst $(1,1)$ (or both) and $(0,0)$ (or none) are respectively maximal and minimal elements in the knowledge ordering.

We define an annotated bilattice-based language $\mathcal{L}$ to consist of individual variables, constants, functions and predicate symbols together with annotation terms which can consist of variables, constants and/or functions over a bilattice. Bilattice-based languages allow, in general, six connectives and four quantifiers, as follows: $\oplus, \otimes, \vee, \wedge, \neg, \sim, \Sigma, \Pi, \exists, \forall$. But in this chapter we restrict our attention to only one-lattice based BAPs and will work only with $\oplus, \otimes, \Sigma$, the latter being the existential quantifier with respect to the knowledge ordering. Returning to algebraic characterizations of many-valued logics, we make the remark that $\oplus$ and $\otimes$ correspond to the operations $+$ and $\times$ of Section 12.3.1, and that Σ corresponds to infinite summation $(+)$.

An *annotated formula* is defined inductively as follows: if R is an n-ary predicate symbol, $t_1, \ldots, t_n$ are terms, and μ is an annotation term, then $R(t_1, \ldots, t_n) : \mu$ is an *annotated formula* (called an *annotated atom*). Annotated atoms can be combined to form complex formulae using the connectives and quantifiers.

A *bilattice-based annotated logic program (BAP)* P consists of a finite set of *annotated program clauses* of the form

$$A : \mu \leftarrow L_1 : \mu_1, \ldots, L_n : \mu_n,$$

where $A : \mu$ denotes an annotated atom called the *head* of the clause, and $L_1 : \mu_1, \ldots, L_n : \mu_n$ denotes $L_1 : \mu_1 \otimes \ldots \otimes L_n : \mu_n$ and is called the *body* of the clause; each $L_i : \mu_i$ is an annotated literal called an *annotated body literal* of the clause. Individual and annotation variables in the body are thought of as being existentially quantified using Σ. In [15], we showed how the remaining connectives $\oplus, \vee, \wedge$ can be introduced into BAPs, but we will not address this issue here.

Each annotated atom $A : \mu$ is interpreted in two steps as follows: the first-order atomic formula A is interpreted in $\mathbf{B}$ (we may write $\mathcal{I}_\mathbf{B}(A) \to \mathbf{B}$ to indicate this process) using a domain of interpretation and a variable assignment, see [28, 29, 35, 15] for further details. Then we define the interpretation

$I_\mathbf{B}$ as follows: if $\mathcal{I}_\mathbf{B}(A) \geq \mu$, we put $I_\mathbf{B}(A : \mu) = 1$, and $I_\mathbf{B}(A : \mu) = 0$ otherwise.

Let $I_\mathbf{B}$ be an interpretation for $\mathcal{L}$ and let F be a closed annotated formula of $\mathcal{L}$. Then $I_\mathbf{B}$ is a *model* for F if $I_\mathbf{B}(F) = 1$. We say that $I_\mathbf{B}$ is a model for a set S of annotated formulae if $I_\mathbf{B}$ is a model for each annotated formula of S. We say that F *is a logical consequence of S* if, for every interpretation $I_\mathbf{B}$ of $\mathcal{L}$, $I_\mathbf{B}$ is a model for S implies $I_\mathbf{B}$ is a model for F.

Let B_P and U_P denote the annotation Herbrand base and the annotation Herbrand universe for a program P respectively; they are essentially B_P and U_P as defined in [46], but with annotation terms allowed in U_P and attached to ground formulae in B_P. In common with conventional logic programming, each Herbrand interpretation HI for P can be identified with the subset $\{R(t_1,\ldots,t_k) : \alpha \in B_P \mid R(t_1,\ldots,t_k) : \alpha$ receives the value 1 with respect to $I_\mathbf{B}\}$ of B_P, where $R(t_1,\ldots,t_k) : \alpha$ denotes a typical element of B_P. This set constitutes an *annotation Herbrand model for P*. Finally, we let $\text{HI}_{P,\mathbf{B}}$ denote the set of all annotation Herbrand interpretations for P.

It was observed in [22, 14, 15, 37, 38], that the non-linear ordering of (bi)lattices influences both model-theoretic properties and proof procedures for (bi)lattice-based logics, and this distinguishes them from classical and even fuzzy logic. In particular, both the semantic operator and SLD-resolution for BAPs must reflect the non-linear ordering of bilattices, see [13, 14, 15].

In [15], we introduced a semantic operator $\mathcal{T}_P$ for BAPs, proved its continuity and showed that it computes the least Herbrand model for a given BAP as its least fixed point.

Definition 8. *We define the mapping $\mathcal{T}_P : \text{HI}_{P,\mathbf{B}} \to \text{HI}_{P,\mathbf{B}}$ as follows: $\mathcal{T}_P(\text{HI})$ denotes the set of all $A : \mu \in B_P$ such that either*

1. *There is a strictly ground instance of a clause $A : \mu \leftarrow L_1 : \mu_1, \ldots, L_n : \mu_n$ such that $\{L_1 : \mu'_1, \ldots, L_n : \mu'_n\} \subseteq \text{HI}$ for some annotations $\mu'_1, \ldots, \mu'_n$, and one of the following conditions holds for each μ'_i:*
    a) $\mu'_i \geq_k \mu_i$,
    b) $\mu'_i \geq_k \otimes_{j \in J_i} \mu_j$, *where J_i is the finite set of those indices $i, j \in \{1, \ldots n\}$ such that $L_j = L_i$*

or

2. *there are annotated strictly ground atoms $A : \mu^*_1, \ldots, A : \mu^*_k \in \text{HI}$ such that $\mu \leq_k \mu^*_1 \oplus \ldots \oplus \mu^*_k$.*[7]

Item 1a is the analogue of the conventional T_P operator, see [46] for example, and of the generalized semantic operator $\mathfrak{T}_{P,\mathfrak{C}}$. Items 1b and 2 reflect properties of the non-linear ordering defined on the set $\text{HI}_{P,\mathbf{B}}$ of all interpretations, as further illustrated in the next example. Note that the absence of conditions 1b and 2 in the formulation of $\mathfrak{T}_{P,\mathfrak{C}}$ given in Section 12.3 is

---

[7] Note that whenever $F : \mu \in HI$ and $\mu' \leq_k \mu$, then $F : \mu' \in HI$. Also, for each formula F, $F : (0,0) \in HI$.

compensated for by the use of the ground completion P^{**} of a program P whenever $\mathfrak{T}_{P,\mathfrak{C}}$ is applied.

Example 9. Consider a bilattice-based annotated logic program P which can collect and process information about connectivity of some (probabilistic) graph G. Suppose we have received information from two different sources: one reports that there is an edge between nodes a and b, the other, that there is no such. This is represented by the two unit clauses edge$(a,b) : (1,0) \leftarrow$, edge$(a,b) : (0,1) \leftarrow$. It is reasonable to conclude that the information is contradictory, that is, to conclude that edge$(a,b) : (1,1)$, and this fact is captured by item 2. If, on the other hand, the program contains some clause of the form disconnected$(G) : (1,1) \leftarrow$ connected$(a,c) : (1,0)$, connected$(a,c) : (0,1)$, we may regard the clause disconnected$(G) : (1,1) \leftarrow$ connected$(a,c) : (0,0)$ as equal to the initial clause. This fact is captured by item 1b.

Let B, A and C denote, respectively, edge(a,b), connected(a,c) and disconnected(G). Consider the logic program: $B : (0,1) \leftarrow$, $B : (1,0) \leftarrow$, $A : (0,0) \leftarrow B : (1,1)$, $C : (1,1) \leftarrow A : (1,0)$, $A : (0,1)$. The least fixed point of $\mathcal{T}_P$ is $\mathcal{T}_P \uparrow 3 = \{B : (0,1), B : (1,0), B : (1,1), A : (0,0), C : (1,1)\}$. However, the item 1.$a$ (corresponding to the classical semantic operator) would allow us to compute only $\mathcal{T}_P \uparrow 1 = \{B : (0,1), B : (1,0)\}$, that is, to compute only explicit consequences of the program, which then leads to a contradiction in the two-valued case.

As was shown in [15], the BAPs introduced here generalize different sorts of implication-based and annotation-free logic programs, such as those of [28, 29]. The fragment of BAPs described here is computationally equivalent to GAPs - annotated logic programs based on one-lattice structures. This gives a very close connection between the structures used to interpret GAPs respectively BAPs and captured by the algebraic analysis of the operations $+$ and $\times$ on the set $\mathfrak{C}$, as described in Section 12.3.1. Namely, both GAPs and the fragment of BAPs described here, both taken only with variable annotations, yield the general semantic characterizations of Sections 12.3.1 and 12.3.2, as follows.

Definition 9. *Let P be an annotation-free logic program with clauses of the form $A \leftarrow B_1, \ldots, B_n$. We construct P', an annotated logic program derived from P, as follows. For each clause $A \leftarrow B_1, \ldots, B_n$ in P, add a clause $A : \mu \leftarrow B_1 : \mu_1, \ldots, B_n : \mu_n$ to P', where each μ_i is an annotation variable.*

The following proposition is an adaptation of the relevant proposition from [35, 15].

Proposition 4. *Let A be a ground first-order atomic formula and let α be an annotation constant. Then $I(A) = \alpha$ in the least fixed point of $\mathfrak{T}_{P,\mathfrak{C}}$ if and only if $A : \alpha$ is in the least fixed point of $\mathcal{T}_P$.*

We now turn to the description of the neural networks computing $\mathcal{T}_P$.

12.4.2 Neural Networks for (Bi)Lattice-Based Annotated Logic Programs

We extend the approach of [1] described in Section 12.3 to learning neural networks which can compute logical consequences of BAPs. This will allow us to introduce hypothetical and uncertain reasoning based on BAPs into the framework of neural-symbolic computation. Bilattice-based logic programs can work with conflicting sources of information and inconsistent databases. Therefore, neural networks corresponding to these logic programs should reflect this facility as well, and this is why we introduce some forms of learning into neural networks. These forms of learning can be seen as corresponding to unsupervised Hebbian learning which is widely implemented in neurocomputing. The general idea behind Hebbian learning is that positively correlated activities of two neurons strengthen the weight of the connection between them and uncorrelated or negatively correlated activities weaken the weight of the connection between them (the latter form is known as anti-Hebbian learning).

The general conventional definition of Hebbian learning is given as follows, see [51] for further details. Let k and j denote two units and w_{kj} denote the weight of a connection from j to k. We denote the value of j at time t by $v_j(t)$ and the potential of k at time t by $p_k(t)$. Then the rate of change in the weight between j and k is expressed in the form

$$\Delta w_{kj}(t) = F(v_j(t), p_k(t)),$$

where F is some function. As a special case of this formula, it is common to write

$$\Delta w_{kj}(t) = \eta(v_j(t))(p_k(t)),$$

where η is a constant that determines the *rate of learning* and is positive in the case of Hebbian learning and negative in the case of anti-Hebbian learning. Finally, we update by $w_{kj}(t+1) = w_{kj}(t) + \Delta w_{kj}(t)$.

In this section, we will compare the two learning functions we introduce with this conventional definition of Hebbian learning. First, we prove a theorem establishing a relationship between learning neural networks and BAPs with no function symbols occurring in either individual or annotation terms. (Since the annotation Herbrand base for these programs is finite, they can equivalently be seen as propositional bilattice-based logic programs with no functions allowed in the annotations.)

Theorem 3. *For each function-free BAP P, there exists a 3-layer feedforward learning neural network which computes T_P.*

Proof. Let m and n be the number of strictly ground annotated atoms from the annotation Herbrand base B_P and the number of clauses occurring in P respectively. Without loss of generality, we may assume that the annotated atoms are ordered. The network associated with P can now be constructed by the following translation algorithm.

1. The input and output layers are vectors of binary threshold units of length k, $1 \leq k \leq m$, where the i-th unit in the input and output layers represents the i-th strictly ground annotated atom. The threshold of each unit occurring in the input or output layer is set to 0.5.
2. For each clause of the form $A : (\alpha, \beta) \leftarrow B_1 : (\alpha_1, \beta_1), \ldots, B_m : (\alpha_m, \beta_m)$, $m \geq 0$, in P, do the following.
   2.1 Add a binary threshold unit c to the hidden layer.
   2.2 Connect c to the unit representing $A : (\alpha, \beta)$ in the output layer with weight 1. We will call connections of this type 1-*connections*.
   2.3 For each atom $B_j : (\alpha_j, \beta_j)$ in the input layer, connect the unit representing $B_j : (\alpha_j, \beta_j)$ to c and set the weight to 1. (We will call these connections 1-connections also.)
   2.4 Set the threshold θ_c of c to $l - 0.5$, where l is the number of atoms in $B_1 : (\alpha_1, \beta_1), \ldots, B_m : (\alpha_m, \beta_m)$.
   2.5 If some input unit representing $B : (\alpha, \beta)$ is connected to a hidden unit c, connect each of the input units representing annotated atoms $B : (\alpha_i, \beta_i), \ldots, B : (\alpha_j, \beta_j)$ to c. These connections will be called $\otimes$-*connections*. The weights of these connections will depend on a learning function. If the function is inactive, set the weight of each $\otimes$-*connection* to 0.
3. If there are units representing atoms of the form $B : (\alpha_i, \beta_i), \ldots, B : (\alpha_j, \beta_j)$ in the input and output layers, correlate them as follows. For each $B : (\alpha_i, \beta_i)$, connect the unit representing $B : (\alpha_i, \beta_i)$ in the input layer to each of the units representing $B : (\alpha_i, \beta_i), \ldots, B : (\alpha_j, \beta_j)$ in the output layer. These connections will be called $\oplus$-*connections*. If an $\oplus$-connection is set between two atoms with different annotations, we consider them as being connected via hidden units with thresholds 0. If an $\oplus$-connection is set between input and output units representing the same annotated atom $B : (\alpha, \beta)$, we set the threshold of the hidden unit connecting them to -0.5, and we will call them $\oplus$-hidden units, so as to distinguish the hidden units of this type. The weights of all these $\oplus$-connections will depend on a learning function. If the function is inactive, set the weight of each $\oplus$-*connection* to 0.
4. Set all the weights which are not covered by these rules to 0. For each annotated atom $A : (\alpha, \beta)$, connect the unit representing $A : (\alpha, \beta)$ in the output layer to the unit representing it in the input layer with weight 1.

Allow two learning functions to be embedded into the $\otimes$-connections and the $\oplus$-connections. We let v_i denote the value of the input unit representing $B : (\alpha_i, \beta_i)$ and let p_c denote the potential of the unit c.

Let a unit representing $B : (\alpha_i, \beta_i)$ in the input layer be denoted by i. If i is connected to a hidden unit c via an $\otimes$-connection, then a learning function ϕ_1 is associated to this connection as defined next. We let $\phi_1 = \Delta w_{ci}(t-1) = v_i(t-1)(-p_c(t-1)+0.5)$ become active and change the weight of the $\otimes$-connection from i to c at time t if i became activated at time $(t-1)$;

units representing atoms $B : (\alpha_j, \beta_j), \ldots, B : (\alpha_k, \beta_k)$ in the input layer are connected to c via 1-connections, and $(\alpha_i, \beta_i) \geq_k (\alpha_j, \beta_j) \otimes \ldots \otimes (\alpha_k, \beta_k)$.

Function ϕ_2 is embedded only into connections of type $\oplus$, namely, into $\oplus$-connections between hidden and output layers. Let o be an output unit representing an annotated atom $B : (\alpha_i, \beta_i)$. Apply $\phi_2 = \Delta w_{oc}(t-2) = v_c(t-2)(p_o(t-2) + 1.5)$ to change w_{oc} at time t if (i) ϕ_2 is embedded into an $\oplus$-connection from the $\oplus$-hidden unit c to o, and there are output units representing annotated atoms $B : (\alpha_j, \beta_j), \ldots, B : (\alpha_k, \beta_k)$ which are connected to the unit o via $\oplus$-connections, and (ii) these output units became activated at time $t-2$ and $(\alpha_i, \beta_i) \leq_k (\alpha_j, \beta_j) \oplus \ldots \oplus (\alpha_k, \beta_k)$.

Each annotation Herbrand interpretation HI for P can be represented by a binary vector $(v_1, \ldots, v_m)$. Such an interpretation is given as input to the network by externally activating corresponding units of the input layer at time t_0. It remains to show that $A : (\alpha, \beta) \in \mathcal{T}_P \uparrow n$ for some n if and only if the unit representing $A : (\alpha, \beta)$ becomes active at time $t + 2$, for some t. The proof that this is so proceeds by routine induction.

Example 10. The following diagram displays the neural network which computes $\mathcal{T}_P \uparrow 3$ from Example 9. Without the functions ϕ_1, ϕ_2, the neural network will compute only $\mathcal{T}_P \uparrow 1 = \{B : (0,1), B : (1,0)\}$, and these are explicit logical consequences of the program. Indeed, it is the use of ϕ_1 and ϕ_2 that allows the neural network to compute $\mathcal{T}_P \uparrow 3$. Note that the arrows ———→ , − − ≻ , ·······≻ denote respectively 1-connections, $\otimes$-connections and $\oplus$-connections, and we have marked by ϕ_1, ϕ_2 the connections which are activated by the learning functions.[8]

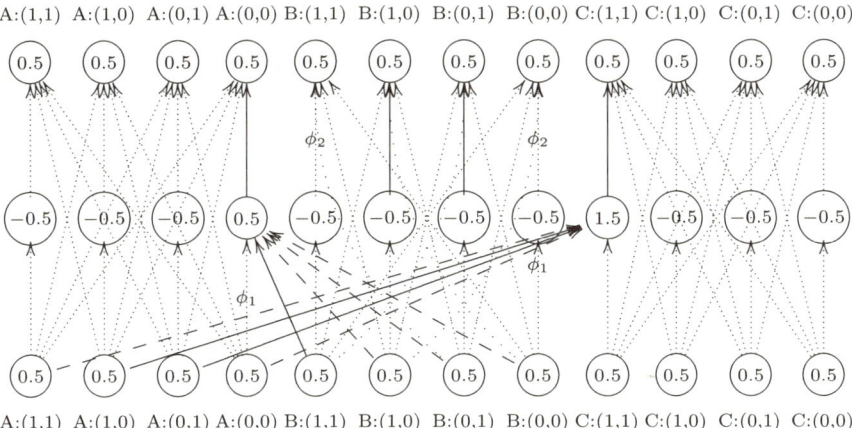

We can make several conclusions from the construction of Theorem 3.

---

[8] We do not draw here the connections which make this network recurrent.

- Neurons representing annotated atoms with identical first-order (or propositional) components are joined into multineurons in which units are correlated using $\oplus$- and $\otimes$-connections.
- The learning function ϕ_2 roughly corresponds to Hebbian learning, with the rate of learning $\eta_2 = 1$, the learning function ϕ_1 corresponds to anti-Hebbian learning with the rate of learning $\eta_1 = -1$, and we regard η_1 as negative because the factor p_c in the formula for ϕ_1 is multiplied by (-1).
- The main problem Hebbian learning causes is that the weights of connections with embedded learning functions tend to grow exponentially, which cannot fit the model of biological neurons. This is why traditionally functions are introduced to bound the growth. In the neural networks we have built, some of the weights may grow with iterations, but the growth will be very slow because we are using binary threshold units in the computation of each v_i.

12.5 Relationships Between the Neural Networks Simulating $\mathfrak{T}_{P,\mathfrak{C}}$ and $\mathcal{T}_P$

In this section, we will briefly compare the neural networks computing $\mathfrak{T}_{P,\mathfrak{C}}$ and $\mathcal{T}_P$, and establish a result which relates the computations performed by these two neural networks.

Both the ANNs constructed here can be seen as direct extensions of the work of [1] in that the fundamental processing unit in each case is the binary threshold unit, including those units employed in multiplication and adddition units. This property of the neural networks of Section 12.4 is a direct consequence of the fact that annotated logic programs eventually receive a two-valued meaning. The neural networks of Section 12.3 could be designed to process many-valued vectors, as they are built upon annotation-free logic programs which receive their meaning only via many-valued interpretations, but we use only binary vectors here.

The main difference between the two types of neural networks is that the ANNs of Section 12.3 require two additional layers in comparison with those of [1] and Section 12.4, although their connections are less complex. On the other hand, the ANNs of Section 12.4 have complex connections and compensate for "missing" layers by applying learning functions in the spirit of neurocomputing. The latter property can perhaps be seen as an optimization of the ANNs built in Section 12.3, although the ANNs of Section 12.3 can compute many-valued semantic operators without using learning functions.

We close the chapter with the following important result.

Theorem 4. *Let P denote an annotation-free logic program, and let P' denote the annotated logic program derived from P. Then there exist an ANN1 (as constructed in Section 12.3.4) simulating $\mathfrak{T}_{P,\mathfrak{C}}$ and an ANN2 (as constructed in Section 12.4.2) simulating $\mathcal{T}_{P'}$ such that the output vector of ANN1 at each time $t + 2$ is equal to the output vector of ANN2 at time t.*

Proof. We use Definition 9 and Proposition 4.

Without loss of generality, we assume that all the ground atoms and all the values are ordered. When constructing the networks ANN1 and ANN2, we assume also that all the input and output units of ANN1 and ANN2 are synchronized in that the order at which each m-th proposition with n-th value appears in ANN1 corresponds to the order in which each unit representing each $A_m : \mu_n$ appears in ANN2.

The rest of the proof makes use of the constructions of Theorems 2 and 3 and proceeds by routine induction on the number of iterations of $\mathfrak{T}_{P,\mathfrak{C}}$ and $\mathcal{T}_P$.

12.6 Conclusions and Further Work

We have given a very general algebraic characterization of many-valued annotation-free logic programs, and have shown how this analysis can be extended to other types of many-valued logic programs. We have also shown how two very general semantic operators, the generalized semantic operator $\mathfrak{T}_{P,\mathfrak{C}}$ for annotation-free logic programs and the enriched semantic operator $\mathcal{T}_P$ for annotated logic programs, can be defined, and we have established semantical and computational relationships between them.

Furthermore, we have proposed neural networks in the style of [1] for computing $\mathfrak{T}_{P,\mathfrak{C}}$ and $\mathcal{T}_P$. The neural networks we have given for computing $\mathfrak{T}_{P,\mathfrak{C}}$ require several additional layers in order to reflect the many-valued properties of $\mathfrak{T}_{P,\mathfrak{C}}$ that they simulate. On the other hand, the neural networks computing $\mathcal{T}_P$ have learning functions embedded in them which compensate for the use of additional layers. It would be interesting to carry out a detailed analysis of the complexity (time and space) of both of these neural networks and to compare them on complexity grounds.

Future work in the general direction of the chapter includes the following.

1. Further analysis of the properties of many-valued semantic operators, as given, for example, in [31], and its implications for the construction of the corresponding networks.
2. The neural networks computing the operators considered here could perhaps be optimized if transformed into non-binary neural networks. This might result, for example, in the removal of the annotations used in the representation of input and output units.
3. Another area of our future research is to investigate how learning can optimize and improve the representation of different neuro-symbolic systems and the computations performed by them.

References

1. Hölldobler, S., Kalinke, Y.: Towards a new massively parallel computational model for logic programming. In: Proceedings ECAI94 Workshop on Combining Symbolic and Connectionist Processing, ECCAI (1994) 68–77
2. Hölldobler, S., Kalinke, Y., Störr, H.P.: Approximating the semantics of logic programs by recurrent neural networks. Applied Intelligence **11** (1999) 45–58
3. Bader, S., Hitzler, P., Hölldobler, S.: The integration of connectionism and first-order knowledge representation and reasoning as a challenge for artificial intelligence. Information **9**(1) (2006) Invited paper.
4. Bornscheuer, S.E., Hölldobler, S., Kalinke, Y., Strohmaier, A.: Massively parallel reasoning. In: Automated Deduction – A Basis for Applications. Volume II. Kluwer Academic Publishers (1998) 291–321
5. d'Avila Garcez, A.S., Broda, K., Gabbay, D.M.: Symbolic knowledge extraction from trained neural networks: A sound approach. Artificial Intelligence **125** (2001) 155–207
6. d'Avila Garcez, A.S., Broda, K.B., Gabbay, D.M.: Neural-Symbolic Learning Systems — Foundations and Applications. Perspectives in Neural Computing. Springer, Berlin (2002)
7. d'Avila Garcez, A.S., Gabbay, D.: Fibring neural networks. In McGuinness, D., Ferguson, G., eds.: Proceedings of the 19th National Conference on Artificial Intelligence, 16th Conference on Innovative Applications of Artificial Intelligence, July 2004, San Jose, California, USA, AAAI Press/The MIT Press (2004) 342–347
8. d'Avila Garcez, A.S., Lamb, L.C., Gabbay, D.M.: A connectionist inductive learning system for modal logic programming. In: Proceedings of the IEEE International Conference on Neural Information Processing ICONIP'02, Singapore. (2002)
9. Fu, L.: Neural Networks in Computer Intelligence. McGraw-Hill, Inc. (1994)
10. Hitzler, P., Hölldobler, S., Seda, A.K.: Logic programs and connectionist networks. Journal of Applied Logic **2**(3) (2004) 245–272
11. Hitzler, P., Seda, A.K.: Continuity of semantic operators in logic programming and their approximation by artificial neural networks. In Günter, A., Krause, R., Neumann, B., eds.: Proceedings of the 26th German Conference on Artificial Intelligence, KI2003. Volume 2821 of Lecture Notes in Artificial Intelligence., Springer (2003) 105–119
12. Hölldobler, S.: Challenge problems for the integration of logic and connectionist systems. In Bry, F., Geske, U., Seipel, D., eds.: Proceedings 14. Workshop Logische Programmierung. Volume 90 of GMD Report., GMD (2000) 161–171
13. Komendantskaya, E., Seda, A.K.: Declarative and operational semantics for bilattice-based annotated logic programs. In Li, L., Ren, F., Hurley, T., Komendantsky, V., an Airchinnigh, M.M., Schellekens, M., Seda, A., Strong, G., Woods, D., eds.: Proceedings of the Fourth International Conference on Information and the Fourth Irish Conference on the Mathematical Foundations of Computer Science and Information Technology, NUI, Cork, Ireland. (2006) 229–232
14. Komendantskaya, E., Seda, A.K.: Logic programs with uncertainty: neural computations and automated reasoning. In: Proceedings of the International Conference Computability in Europe (CiE), Swansea, Wales (2006) 170–182

15. Komendantskaya, E., Seda, A.K., Komendantsky, V.: On approximation of the semantic operators determined by bilattice-based logic programs. In: Proc. 7th International Workshop on First-Order Theorem Proving (FTP'05), Koblenz, Germany (2005) 112–130
16. Lane, M., Seda, A.K.: Some aspects of the integration of connectionist and logic-based systems. Information **9**(4) (2006) 551–562
17. Seda, A.K.: Morphisms of ANN and the Computation of Least Fixed Points of Semantic Operators. In Mira, J., Álvarez, J.R., eds.: Proceedings of the Second International Work-Conference on the Interplay between Natural and Artificial Computation, IWINAC2007, Murcia, Spain, June 2007, Part 1. Volume 4527 of Lecture Notes in Computer Science, Springer (2007) 224–233
18. Seda, A.K.: On the integration of connectionist and logic-based systems. In Hurley, T., an Airchinnigh, M.M., Schellekens, M., Seda, A.K., Strong, G., eds.: Proceedings of MFCSIT'04, Trinity College Dublin, July, 2004. Volume 161 of Electronic Notes in Theoretical Computer Science (ENTCS)., Elsevier (2006) 109–130
19. Bader, S., Hitzler, P., Hölldobler, S., Witzel, A.: A fully connectionist model generator for covered first-order logic programs. In: M. Veloso, ed.: Proceedings of International Joint Conference on Artificial Intelligence IJCAI07, Hyderabad, India, AAAI Press (2007) 666–671
20. Resher, N.: Many-valued logic. Mac Graw Hill (1996)
21. Ginsberg, M.L.: Multivalued logics: A uniform approach to inference in artificial intelligence. Computational Intelligence **4**(3) (1992) 256–316
22. Kifer, M., Lozinskii, E.L.: RI: A logic for reasoning with inconsistency. In: Proceedings of the 4th IEEE Symposium on Logic in Computer Science (LICS), Asilomar, IEEE Computer Press (1989) 253–262
23. Lu, J.J.: Logic programming with signs and annotations. Journal of Logic and Computation **6**(6) (1996) 755–778
24. Fitting, M.: A Kripke-Kleene semantics for general logic programs. The Journal of Logic Programming **2** (1985) 295–312
25. Bistarelli, S., Montanari, U., Rossi, F.: Semiring-based constraint logic programming: Syntax and semantics. ACM Transactions on Programming Languages and Systems (TOPLAS) **23**(1) (2001) 1–29
26. Sessa, M.I.: Approximate reasoning by similarity-based SLD-resolution. Theoretical computer science **275** (2002) 389–426
27. Vojtás, P., Paulik, L.: Soundness and completeness of non-classical extended sld-resolution. In: Extensions of Logic Programming, 5th International Workshop ELP'96, Leipzig, Germany, March 28-30, 1996. Volume 1050 of Lecture notes in Computer Science., Springer (1996) 289–301
28. Fitting, M.: Bilattices and the semantics of logic programming. The Journal of Logic Programming **11** (1991) 91–116
29. Fitting, M.C.: Bilattices in logic programming. In Epstein, G., ed.: The twentieth International Symposium on Multiple-Valued Logic, IEEE (1990) 238–246
30. Fitting, M.C.: Kleene's logic, generalized. Journal of Logic and Computation **1** (1992) 797–810
31. Damásio, C.V., Pereira, L.M.: Sorted monotonic logic programs and their embeddings. In: Proceedings of the 10th International Conference on Information Processing and Management of Uncertainty in Knowledge-Based Systems (IPMU-04). (2004) 807–814

32. Stamate, D.: Quantitative datalog semantics for databases with uncertain information. In Pierro, A.D., Wiklicky, H., eds.: Proceedings of the 4th Workshop on Quantitative Aspects of Programming Languages (QAPL 2006). Electronic Notes in Theoretical Computer Science, Vienna, Austria, Elsevier (2006) To appear.
33. van Emden, M.: Quantitative deduction and its fixpoint theory. Journal of Logic Programming **3** (1986) 37–53
34. Lakshmanan, L.V.S., Sadri, F.: On a theory of probabilistic deductive databases. Theory and Practice of Logic Programming **1**(1) (2001) 5–42
35. Kifer, M., Subrahmanian, V.S.: Theory of generalized annotated logic programming and its applications. Journal of logic programming **12** (1991) 335–367
36. Calmet, J., Lu, J.J., Rodriguez, M., Schü, J.: Signed-formula logic programming: Operational semantics and applications. In: In Proceedings of the Ninth International Symposium on Foundations of Intelligent Systems. Volume 1079 of Lecture Notes in Artificial Intelligence., Berlin, Springer (1996) 202–211
37. Lu, J.J., Murray, N.V., Rosental, E.: A framework for automated reasoning in multiple-valued logics. Journal of Automated Reasoning **21**(1) (1998) 39–67
38. Lu, J.J., Murray, N.V., Rosental, E.: Deduction and search strategies for regular multiple-valued logics. Journal of Multiple-valued logic and soft computing **11** (2005) 375–406
39. Ng, R.: Reasoning with uncertainty in deductive databases and logic programs. International Journal of Uncertainty, Fuzziness and Knowledge-based Systems **2**(3) (1997) 261–316
40. Straccia, U.: Query answering in normal logic programs under uncertainty. In: 8th European Conference on Symbolic and Quantitative Approaches to Reasoning with Uncertainty (ECSQARU-05). Lecture Notes in Computer Science, Barcelona, Spain, Springer Verlag (2005) 687–700
41. Fitting, M.: Fixpoint semantics for logic programming — A survey. Theoretical Computer Science **278**(1–2) (2002) 25–51
42. Hecht-Nielsen, R.: Neurocomputing. Addison-Wesley (1990)
43. Hertz, J., Krogh, A., Palmer, R.: Introduction to the Theory of Neural Computation. Addison-Wesley Publishing Company (1991)
44. Seda, A.K., Lane, M.: On the measurability of the semantic operators determined by logic programs. Information **8**(1) (2005) 33–52
45. Hitzler, P., Seda, A.K.: Characterizations of classes of programs by three-valued operators. In Gelfond, M., Leone, N., Pfeifer, G., eds.: Logic Programming and Non-monotonic Reasoning, Proceedings of the 5th International Conference on Logic Programming and Non-Monotonic Reasoning, LPNMR'99, El Paso, Texas, USA. Volume 1730 of Lecture Notes in Artificial Intelligence., Springer, Berlin (1999) 357–371
46. Lloyd, J.W.: Foundations of Logic Programming. Springer, Berlin (1987)
47. Wendt, M.: Unfolding the well-founded semantics. Journal of Electrical Engineering, Slovak Academy of Sciences **53**(12/s) (2002) 56–59
48. Dung, P.M., Kanchanasut, K.: A fixpoint approach to declarative semantics of logic programs. In Lusk, E.L., Overbeek, R.A., eds.: Logic Programming, Proceedings of the North American Conference 1989, NACLP'89, Cleveland, Ohio, MIT Press (1989) 604–625
49. Bonnier, S., Nilsson, U., Näslund, T.: A simple fixed point characterization of the three-valued stable model semantics. Information Processing Letters **40**(2) (1991) 73–78

50. Lane, M.: 𝔈-Normal Logic Programs and Semantic Operators: Their Simulation by Artificial Neural Networks. PhD thesis, Department of Mathematics, University College Cork, Cork, Ireland (2006)
51. Haykin, S.: Neural Networks. A Comprehensive Foundation. Macmillan College Publishing Company (1994)

Index

𝔈-element unit, 297
𝔈-interpretation, 291
𝔈-normal logic program, 291

abduction, 265
abstract topos, 235
active forgetting, 173
acyclic, 212
addition unit, 296
adjointness condition, 237
allowable values, 288
AMI, 166, 172
ANN, 286
annotated bilattice-based language, 302
annotated formula, 302
annotated Herbrand interpretation, 303
annotated logic program, 301
architectural bias, 96
attractor, 100
authorities, 50
autoregressive model, 138
average cluster coefficient, 58

back-propagation through structure, 73
background knowledge, 267
backpropagation, 210, 251, 269
Banach's contraction mapping theorem, 212
behavioural model, 167
belief-net, 187
Bellman equation, 187
bi-directional recurrent network, 128
bilattice, 301

bilattice-based annotated logic program (BAP), 302
binding problem, 185
bioinformatics, 124
BLAST, 34
box-counting dimension, 99, 112
brain function, 50
brain imaging, 267

Cantor set, 214
cascade correlation, 73
 contextual recursive, 74
 recursive, 73
categorical construction, 249
category theory, 235
co-product, 237
cognitive computation, 266
cognitive robotics, 161
collector node, 188
collision relation, 77
complete kernel, 14
compositionality, 233
computation of T_P by ANN, 305
computation of $\mathfrak{T}_{P,\mathfrak{E}}$ by ANN, 297
computational model, 266
config interaction, 162
connectionist modal logic, 273
constraint logic program, 293
constructive translation, 279
contraction, 98, 117
controller, 166, 167
convolution kernel, 8, 24
coproduct arrow, 237

316   Index

Core Method, 206, 209, 227
core network, 212
cortex, 50
cover, 99
covered logic program, 207

Dale's principle, 55
DBAG, 69, 80
decomposition kernel, 23
deduction, 265
definite memory machine, 106
distributed coding, 163
distributed representation, 269
double vector quantization, 145
DPAG, 69
dynamic binding problem, 185

echo state machine, 106
embedding dimension, 138
empirical error, 109
enabler node, 188
energy minimization, 206
epi-mono factorisation, 238
epic, 237
episodic fact, 190
equalizer, 238
equational description, 249
excluded values, 288
explaining away, 192
explanatory inference, 192
exponent, 236, 237

factorization, 236
fault diagnosis, 272
feed-forward network, 209, 270
feedback SOM, 113
FFC, 187
fibred network ensembles, 278
fibred neural network, 277
fibring, 277
fibring function, 277
Fine Blend, 221, 225
finite state automaton, 85, 120, 164
finitely determined operation, 288
first-order logic programs, 205, 277
Fisher kernel, 11, 27
fixed-point, 272
fractal, 99
fractal dimension, 100

fractal prediction machine, 105
frontier to root tree automaton, 71
Funahashi's theorem, 212, 213
functional architecture, 160
functional building block, 175
functional focal-cluster, 187
fundamental theorem of topoi, 236

GENE, 243, 250
general logic program, 271
generalization ability, 108
generalization bound, 110
graph kernel, 13
grasp, 170
    all finger precision, 170
    power grasp, 170
    two finger pinch, 170
    two finger precision, 170

Hausdorff dimension, 100, 112
Hebbian learning, 193
Herbrand base, 208
Herbrand interpretation, 208
Herbrand model, 208
Herbrand universe, 208
hierarchical state machine, 164
histogram intersection kernel, 27
hubs, 50
human reasoning, 279
hybrid system, 268
hyper-square, 219

IDG, 194
imitation grasping, 168
incomplete belief, 187
inconsistency, 268
inconsistent belief, 187
induction, 265
inductive learning, 268
inductive logic programming, 280
inference, 192
inferential dependency, 194
inferential dependency graph, 194
information fusion, 173
initial object, 236
injection, 237
integrate-and-fire neuron, 51, 54
interaction pattern, 162
interconnectivity, 171

iterative function system, 100

k-nearest neighbor, 34
Kalman filter, 110
KBANN, 206, 210
kernel, 7, 23
    kernel for strings, 8
knowledge based artificial neural network, 206
knowledge extraction, 268
knowledge ordering, 302
knowledge representation, 269
Kolmogoriv's theorem, 249
KSOM, 150

learning, 249
learning algorithm, 268
learning cycle, 269
learning function, 305
learning rule, 270
level mapping, 208, 217
limit, 236
link mining, 50
liquid state machine, 106
local linear model, 147, 149
local variable, 207
localist representation, 269
logic program, 207, 269
logical consequence, 268
long-term prediction, 145

manual intelligence, 161
Markov model, 103, 106
Markovian, 99, 107, 114
mathcalSET, 236
McCulloch-Pitts networks, 205
Mealy machine, 71
mean firing rate, 56, 59
merge SOM, 113, 122
meta-level, 279
mismatch kernel, 10, 28
modal reasoning, 272
model completeness, 77
molecular graph, 13
monic, 237
Moore machine, 71
motif, 125
multifractal analysis, 112
multiplicative neuron, 87, 295

n-gram kernel, 9
NEST simulator, 56
network ensemble, 266
network motif, 51, 58
neural computation, 266
neural network, 270, 286
neural prediction machine, 101
neural-symbolic cycle, 206, 211, 229
neural-symbolic integration, 268
neural-symbolic learning systems, 266, 267
NeuronRank, 52, 56
non-classical logic, 266
nonlinear regressive model, 138
normalized root mean square error, 139

open set condition, 100
optimal assignment kernel, 19
out-of-sample prediction, 139

P-kernel, 11
pagerank, 50
parameterised SOM, 143
persistence of knowledge, 193
petri net, 164
possible worlds, 273
pre-grasp posture, 170
Predicate Logic, 233
prediction error, 139
prediction suffix tree, 104
predictive inference, 192
priming, 192
probabilistic knowledge, 187
probability product kernel, 27
processing unit, 267
product, 237
product arrow, 237
propositional fixation, 206, 229
propositional logic, 209
protein 3D structure, 33
protein function, 33
protein homology, 9
pullback, 237

QSAR, 67
QSPR, 67
quantifier, 240

radial basis function network, 143

318   Index

random network, 55
random walk kernel, 15
rational kernel, 11
RBF network, 213, 216
reactive grasp, 170
real error, 109
real time recurrent learning, 73, 110
reasoning, 227, 269
receptive field, 114
recurrent network, 55, 101, 209, 269
recurrent SOM, 113, 148
recursive neural network, 72, 277
recursive SOM, 113, 117
reflexive reasoning, 184, 185
relational knowledge, 187, 267
relational learning, 279
relational structures, 185
required values, 288
requirements engineering, 272
residual, 139
reward fact, 190
rhythmic activity, 194
robust learning, 267
robustness, 226
role node, 187
rooted positional tree, 69
rule extraction, 143, 279

safety-critical systems, 269
self-organizing map, 113, 135, 136
semantic operator $\mathcal{T}_P$, 303
semantic operator $\mathcal{T}_{P,\mathcal{C}}$, 292
SGNG, 217, 225
SHRUTI, 183, 184, 187
SHRUTI-CM5, 196
sigmoidal network, 213
signal peptide, 127
single-step operator, 208, 227
small world network, 51, 55
Socrates' Mortality, 253
SOM for structured data, 113, 119
sparse network, 55
spectrum kernel, 9, 26, 28
stable state, 272
standard semantics of logic programming, 292
statechart, 165
Steamroller Problem, 253
structural risk minimization, 109

subcelullar localization, 34
subobject classifier, 236
subsequences kernel
   gap weighted, 9
   non-contiguous, 10
subtree kernel, 26
Supervised Growing Neural Gas, 217
support vector machine, 34
supported model, 208
symbolic logic, 266
symbolic representation, 163

taxon fact, 190
temporal Kohonen map, 113
temporal reasoning, 274
temporal synchrony, 187, 195
terminal object, 237
text categorization, 9
TF-IDF features, 9
threshold AR, 150
time series prediction, 135
topos, 235, 236, 248, 261
Topos Theory, 233
training, 221
transduction, 69
   adaptive, 70
   causal, 70
   contextual, 70
   I/O-isomorphic, 70, 78
   supersource, 70, 77
translation algorithm, 268
tree graph kernel, 19
tree kernel, 12
truth ordering, 302
truth-value object, 237

unit failure, 226
universal approximation, 72, 82, 108, 211, 249
utility node, 188

value fact, 190
Vapnik Chervonenkis dimension, 109
variable binding, 185, 194
variable memory length Markov model, 103
variable-free logic, 235
variables, 276
vector space, 249

vector-based network, 217
vector-quantized temporal associative memory, 140

weighted decomposition kernel, 28

WEKA, 59

XML, 166, 172